Finance Construction-12, *Tim Asikin, Steve Asikin, Indra Senihardja*

FINANCE

Construction-12

Corporate IFRS-GAAP (B/S-I/S) Engineering

Technologies No. 8,001-8,500 of 111,111 Laws

by

Tim Asikin, Steve Asikin,

Indra Senihardja

0

Corporate IFRS-GAAP (B/S-I/S),ISBN-13: **978-1721066308**, ISBN-10: **1721066306**

FINANCE CONSTRUCTION-12

Corporate IFRS-GAAP (B/S-I/S) Engineering
Technology No. 8,001-8,500 of 111,111 Laws
By Tim Asikin, Steve Asikin, Indra Senihardja

Creative Team:Steve Asikin
         & Asikin Business Quant, New York
An Amazon Paperback
First Published in United States 2018
By Amazon Createspace
This 1st edition published in 2018
AMZN, PO. Box 81226, Seattle
WA 98108-1226, United States

ISBN-13: **978-1721066308**
ISBN-10: **1721066306**

PREFACE

This one is the only Real Finance Engineering Book, written by Engineering & Business Experts with strong Accounting Maths. It comes from a very long world wide experience, analyzing more than 100,000 Corporate Balancesheet and Income Statement reports, in conservative IFRS (International Finance Reporting Standard) and GAAP (Generally Accepted Accounting Principles) of 9,000 days works as Financing Banker, plus Business CEO expert & fresh Engineering views.

This is a manual for healthy Corporate Governance assurance dan practical business management. It is best for Government Authorities, Accounting Regulators, Business Owners & Small Public Investors (all majority & minority share holders as well).

This is not just a tool for making debt instrument unrelated to daily Corporate Finance. This only good for sincere Directors & Executive, and truly bad for the lazy and /or incapable one.

The nicest thing about it, is that only requires simple Junior Highschool Algebra. We just need to be consistent with its less number treatment of constellations. Its saddest part is that many professors and doctors are no longer able to und erstand Junior Highschool Algebra, and hiding in Accounting jargons. Please go back to its Introduction for glossary, at anywhere inside the book, so then we can still understand what are those mean with simple Algebraic Rule there. We try to avoid some sentences multiplied, added, eliminated divided, rooted or powered by another sentences, that make many traditional finance book successful (delivering few simple things, seen as difficult and scary), but less useful.

2

Corporate IFRS-GAAP (B/S-I/S),ISBN-13: **978-1721066308**, ISBN-10: **1721066306**

TABLE OF CONTENTS

Preface, p.002

Table of Contents, p.003

INTRODUCTION: The Math Fin Law Terminologies, p.004

Chapter-01: Growth of **S**= SALES, in Optimum Math Fin Law,

Chapter-02: Manageable **V**= VARIABLE COST, Optimization,

Chapter-03: **M**= MARGIN of CONTRIBUTION, Optimization,

Chapter-04: Manageable **F**= FIXED COST, Optimization,

Chapter-05: **O**= OPERATIONAL SURPLUS, Optimization,

Chapter-06: Manageable **I**= INTEREST COST, Optimization,

Chapter-07: **B**= BEFORE TAX INCOME, Optimization,

Chapter-08: Manageable **T**= TAX, Planning Optimization,

Chapter-09: **A**= AFTER TAX INCOME, Optimization,

Chapter-10: Manageable **D**= DIVIDEND, Optimization.

Chapter-11: Manageable **Y**= YIELDING TAX, Optimization,

Chapter-12: **H**= HOME TAKEN DIVIDEND, Optimization,

Chapter-13: **R**= RETAINED EARNINGS, Optimization,

Chapter-14: **E**= EQUITY, or Capital Optimization,

Chapter-15: **U**= Utilized, or Starting Capital Optimization,

Chapter-16: **L**= LIABILITIES, or Debt Optimization,

Chapter-17: **W**= WEALTH, or Total Assets Optimization,

Chapter-18: **P**= PROCURED TRADE INVENTORY, Optimization,

Chapter-19: **X**= XPRESS or Current Debt Optimization,

***Chapter-20*: Q**= QUOTED LONGTERM, Debt Optimization, p.6

Chapter-21: **G**= GOODS or Trade Payable Optimization,

Chapter-22: **Z**= ZERO TRADE or Current Debt Optimization,

Chapter-23: **J**= JOB or Trade Account Receivable Optimization,

Chapter-24: **C**= CURRENT ASSETS, Optimization,

Chapter-25: **K**=KIND of CASH, Optimization,

Chapter-26: **N**=NON CURRENT ASSETS, Optimization,

Corporate IFRS-GAAP (B/S-I/S),ISBN-13: **978-1721066308**, ISBN-10: **1721066306**

INTRODUCTION:
The Math Fin Law Terminologies

Income Statement

S	V				
	M	F			
		O	I		
			B	T	
				A	D	Y
						H
					R

Balance Sheet

K	C	W	L	X	G
J					Z
P				Q	
	N		E	U	
				R	

Please carefully look that all of 26 characters (**A** to **Z**) used once, except the **R**, that in Accountancy must be appeared in both diagrams at same identical value, with glossary (of well known abbreviations) in the next page. Only 2 (two) of them are historical as $U=E'$ and S' means carried forward legal numbers from previous year.

The 13 (thirteen) common measures (**c, d, f, g, i, j, l, p, q, s, t, v,** and **y**), are put up front as desired criteria along the way, so then later analyst could be surprised.

In Balance Sheet, we found that **W= C+N= K+J+P+N= L+E= X+Q+U+R= G+Z+Q+U+R** as well, so then we put **W** in the middle of its T-Account balance.

In Income Statement, same applied as **S= V+F+I +T+Y+H+R= V+M= V+F+ O= V+F+I+B= V+F+I+T+A= V+F+I+B= V+F+I+T+D+R**.

Corporate IFRS-GAAP (B/S-I/S),ISBN-13: **978-1721066308**, ISBN-10: **1721066306**

A= After Tax Income;

B= Before Tax Income,

C= Current Assets, **c**= C/X= Current Ratio,

D= Dividend Paid, **d**= D/A=Dividend Portion or Payout,

E= Equity or Capital, **E'**= Equity of Past Year,

F= Fixed Cost, **f**= F/S=Fixed Portion,

G= Goods or Trade Payables,

       g= 360G/V= Goods Payable Days,

H= Home Taken Dividend,

I= Interest Expense, **i**= I/S= Interest Portion,

J= Jobs or Account Receivable,

       j= 360J/S= Jobs or Trade Account ReceivableDays,

K= Kind of Cash,

L= Liabilities or Debt, **l**= L/E= Leverage or Gearing Ratio,

M= Margin of Contribution,

N= Non Current Assets,

O= Operational Surplus,

P= Procured Inventories,

       p= 360P/V= Procured Inventory Days,

Q= Quoted Longterm Liabilities,

       q= [C-P]/X= Quick or Acid Test Ratio,

R= Retained Earnings,

S= Sales or Revenues, **S'**= Sales of Past Year,

       s= [S/S']-1= Sales Growth,

T= Tax Paid, **t**= T/B= Tax Rate,

U= Utilized or Starting Capital,

V= Variable Cost, **v**= V/S= Variable Portion,

W= Wealth or Total Assets,

X= Xpress or Current Debt,

Y= Yielding Tax of Dividend,

       y= Y/D= Yielding Tax Rate of Dividend,

Z= Zero Trade Current Debt.

Corporate IFRS-GAAP (B/S-I/S),ISBN-13: **978-1721066308**, ISBN-10: **1721066306**

CHAPTER-20:
Q= QUOTED Longterm Debt Optimization

Based on 12/31/15 Pfizer Annual Report
(http://www.nasdaq.com/symbol/pfe/financials?query=income-statement)

Continuing the Laws No.1-8,000
at Finance Construction 1-11

Law-8001:

If both (**U**= Utilized or Starting Capital= 64,720 millions USD), (**I**= L/E= Leverage or Gearing Ratio= 99.00%), (**q**= [C-P]/X= Quick or Acid Test Ratio= 1.01 times), (**p**= 360 P/V= Procured Inventory Days= 240 Days), (**c**= C/X= Current Ratio= 1.20 times), (**V**= Variable Cost= 9,746 millions USD), (**D**= Dividend Paid= 2,370 millions USD), (**M**= Margin of Contribution= 41,548 millions USD), (**t**= T/B= Tax Rate= 30.00%), (**$**= Sales or Revenues= 51,294 millions USD), (**i**= I/S= Interest Portion= 1.00%), and (**F**= Fixed Cost= 29,740 millions USD), then it's (**Q**= Quoted Longterm Liabilities planned), is:

$$Q = I\{U+[M-F-\$i][1-t]-D\}-Vp/\{360[c-q]\}$$
$$= 0.9900\{64,720+[41,548-29,750$$
$$-51,294*0.0100][1-0.3000]$$
$$-2,370\}-9,746*240/\{360$$
$$[1.2000-1.0100]\}$$
$$= 35,351 \text{ millions USD}$$

Corporate IFRS-GAAP (B/S-I/S),ISBN-13: **978-1721066308**, ISBN-10: **1721066306**

Law-8002:

If both (**U**= Utilized or Starting Capital= 64,720 millions USD), (**Q**= Quoted Longterm Liabilities= 35,351 millions USD), (**q**= [C-P]/X= Quick or Acid Test Ratio= 1.01 times), (**p**= 360 P/V= Procured Inventory Days= 240 Days), (**c**= C/X= Current Ratio= 1.20 times), (**V**= Variable Cost= 9,746 millions USD), (**M**= Margin of Contribution= 41,548 millions USD), (**t**= T/B= Tax Rate= 30.00%), (**$**= Sales or Revenues= 51,294 millions USD), (**i**= I/S= Interest Portion= 1.00%), and (**F**= Fixed Cost= 29,740 millions USD), then it's (**I** = Leverage or Gearing Ratio planned), is:

$$I = (\mathbf{Q} + \mathbf{Vp}/\{360[\mathbf{c} - \mathbf{q}]\})$$
$$/\{\mathbf{U} + [\mathbf{M} - \mathbf{F} - \mathbf{Si}][1 - \mathbf{t}] - \mathbf{D}\}$$
$$= (35,351 + 9,746*240/\{360[1.2000$$
$$-1.0100]\})/\{64,720 + [41,548$$
$$-29,750 \ 51,294*0.0100]$$
$$[1 - 0.3000] - 2,370\}$$
$$= 99.00\%$$

Corporate IFRS-GAAP (B/S-I/S),ISBN-13: **978-1721066308**, ISBN-10: **1721066306**

Law-8003:

If both (I= L/E= Leverage or Gearing Ratio= 99.00%), (Q= Quoted Longterm Liabilities= 35,351 millions USD), (q= [C-P]/X= Quick or Acid Test Ratio= 1.01 times), (p= 360 P/V= Procured Inventory Days= 240 Days), (c= C/X= Current Ratio= 1.20 times), (V= Variable Cost= 9,746 millions USD), (D= Dividend Paid= 2,370 millions USD), (M= Margin of Contribution= 41,548 millions USD), (t= T/B= Tax Rate= 30.00%), (i= I/S= Interest Portion= 1.00%), and (F= Fixed Cost= 29,740 millions USD), then it's (U=Utilized or Starting Capital planned), is:

$$U = D+(Q+Vp/\{360[c-q]\})/I$$
$$-[M-F-Si][1-t]$$
$$= 2,370+(35,351+9,746*240/\{360$$
$$[1.2000-1.0100]\})/0.9900$$
$$-[41,548-29,750-51,294$$
$$*0.0100][1-0.3000]$$
$$= 64,720 \text{ millions USD}$$

Corporate IFRS-GAAP (B/S-I/S),ISBN-13: **978-1721066308**, ISBN-10: **1721066306**

Law-8004:

If both (I= L/E= Leverage or Gearing Ratio= 99.00%), (Q= Quoted Longterm Liabilities= 35,351 millions USD), (q= [C-P]/X= Quick or Acid Test Ratio= 1.01 times), (p= 360 P/V= Procured Inventory Days= 240 Days), (c= C/X= Current Ratio= 1.20 times), (V= Variable Cost= 9,746 millions USD), (D= Dividend Paid= 2,370 millions USD), (U= Utilized or Starting Capital= 64,720 millions USD), (t= T/B= Tax Rate= 30.00%), (i= I/S= Interest Portion= 1.00%), and (F= Fixed Cost= 29,740 millions USD), then it's (Utilized or Starting Capital planned), is:

$$U= F+Si+[D-U+(Q+Vp/\{360[c-q]\})/I]$$
$$/[1-t]$$
$$= 29,750+51,294*0.0100+[2,370$$
$$-64,720+(35,351+9,746$$
$$*240/\{360[1.2000$$
$$-1.0100]\})/0.9900]$$
$$/[1-0.3000]$$
$$= 41,548 \text{ millions USD}$$

Corporate IFRS-GAAP (B/S-I/S),ISBN-13: **978-1721066308**, ISBN-10: **1721066306**

Law-8005:

If both (**I**= L/E= Leverage or Gearing Ratio= 99.00%), (**Q**= Quoted Longterm Liabilities= 35,351 millions USD), (**q**= [C-P]/X= Quick or Acid Test Ratio= 1.01 times), (**p**= 360 P/V= Procured Inventory Days= 240 Days), (**c**= C/X= Current Ratio= 1.20 times), (**V**= Variable Cost= 9,746 millions USD), (**D**= Dividend Paid= 2,370 millions USD), (**U**= Utilized or Starting Capital= 64,720 millions USD), (**t**= T/B= Tax Rate= 30.00%), (**i**= I/S= Interest Portion= 1.00%), and (**M**= Margin of Contribution= 41,548 millions USD), then it's (**F**=Fixed Cost planned), is:

$$F= M\text{-}Si\text{-}[D\text{-}U+(Q+Vp/\{360[c\text{-}q]\})/I\,]$$
$$/[1\text{-}t]$$
$$= 41,548\text{-}51,294*0.0100\text{-}[2,370$$
$$-64,720+(35,351+9,746$$
$$*240/\{360[1.2000\text{-}1.0100]\})$$
$$/0.9900]/[1\text{-}0.3000]$$
$$= \underline{29,750} \text{ millions USD}$$

Corporate IFRS-GAAP (B/S-I/S),ISBN-13: **978-1721066308**, ISBN-10: **1721066306**

Law-8006:

If both (**I**= L/E= Leverage or Gearing Ratio=
99.00%), (**Q**= Quoted Longterm Liabilities= 35,351
millions USD), (**q**= [C-P]/X= Quick or Acid Test
Ratio= 1.01 times), (**p**= 360 P/V= Procured Inventory
Days= 240 Days), (**c**= C/X= Current Ratio= 1.20
times), (**V**= Variable Cost= 9,746 millions USD), (**D**=
Dividend Paid= 2,370 millions USD), (**U**= Utilized or
Starting Capital= 64,720 millions USD), (**t**= T/B=
Tax Rate= 30.00%), (**i**= I/S= Interest Portion=
1.00%), (**F**= Fixed Cost= 29,740 millions USD), and
(**M**= Margin of Contribution= 41,548 millions USD),
then it's (**S**=Sales or Revenues planned), is:

$$S= \{M\text{-}F\text{-}[D\text{-}U+(Q+Vp/\{360[c\text{-}q]\})/I]$$
$$/[1\text{-}t]\}/i$$
$$= \{41,548\text{-}29,750\text{-}[2,370\text{-}64,720$$
$$+(35,351+9,746*240/\{360$$
$$[1.2000\text{-}1.0100]\})/0.9900]$$
$$/[1\text{-}0.3000]\}/0.0100$$
$$= 51,294 \text{ millions USD}$$

Corporate IFRS-GAAP (B/S-I/S),ISBN-13: **978-1721066308**, ISBN-10: **1721066306**

Law-8007:

If both (I= L/E= Leverage or Gearing Ratio= 99.00%), (Q= Quoted Longterm Liabilities= 35,351 millions USD), (q= [C-P]/X= Quick or Acid Test Ratio= 1.01 times), (p= 360 P/V= Procured Inventory Days= 240 Days), (c= C/X= Current Ratio= 1.20 times), (V= Variable Cost= 9,746 millions USD), (D= Dividend Paid= 2,370 millions USD), (U= Utilized or Starting Capital= 64,720 millions USD), (t= T/B= Tax Rate= 30.00%), (S= Sales or Revenues= 51,294 millions USD), (F= Fixed Cost= 29,740 millions USD), and (M= Margin of Contribution= 41,548 millions USD), then it's (i=Interest Portion planned), is:

$$S= \{M\text{-}F\text{-}[D\text{-}U+(Q+Vp/\{360[c\text{-}q]\})/I\,] /[1\text{-}t]\}/S$$
$$= \{41,548\text{-}29,750\text{-}[2,370\text{-}64,720 +(35,351+9,746*240/\{360 [1.2000\text{-}1.0100]\})/0.9900] /[1\text{-}0.3000]\}/51,294$$
$$= 1.00\%$$

Law-8008:

If both (**I**= L/E= Leverage or Gearing Ratio= 99.00%), (**Q**= Quoted Longterm Liabilities= 35,351 millions USD), (**q**= [C-P]/X= Quick or Acid Test Ratio= 1.01 times), (**p**= 360 P/V= Procured Inventory Days= 240 Days), (**c**= C/X= Current Ratio= 1.20 times), (**V**= Variable Cost= 9,746 millions USD), (**D**= Dividend Paid= 2,370 millions USD), (**U**= Utilized or Starting Capital= 64,720 millions USD), (**i**= I/S= Interest Portion= 1.00%), (**S**= Sales or Revenues= 51,294 millions USD), (**F**= Fixed Cost= 29,740 millions USD), and (**M**= Margin of Contribution= 41,548 millions USD), then it's (**t**=Tax Rate planned), is:

$$t= 1-[D-U+(Q+Vp/\{360[c-q]\})/I\,]/[M-F-Si]$$
$$= 1-[2,370-64,720+(35,351+9,746$$
$$*240/\{360[1.2000-1.0100]\})$$
$$/0.9900]/[41,548-29,750$$
$$-51,294*0.0100]$$
$$= 30.00\%$$

`

Corporate IFRS-GAAP (B/S-I/S),ISBN-13: **978-1721066308**, ISBN-10: **1721066306**

Law-8009:

If both (**I**= L/E= Leverage or Gearing Ratio= 99.00%), (**Q**= Quoted Longterm Liabilities= 35,351 millions USD), (**q**= [C-P]/X= Quick or Acid Test Ratio= 1.01 times), (**p**= 360 P/V= Procured Inventory Days= 240 Days), (**c**= C/X= Current Ratio= 1.20 times), (**V**= Variable Cost= 9,746 millions USD), (**t**= T/B= Tax Rate= 30.00%), (**U**= Utilized or Starting Capital= 64,720 millions USD), (**i**= I/S= Interest Portion= 1.00%), (**$**= Sales or Revenues= 51,294 millions USD), (**F**= Fixed Cost= 29,740 millions USD), and (**M**= Margin of Contribution= 41,548 millions USD), then it's (**D**=Dividend Paid planned), is:

$$D= U+[M\text{-}F\text{-}\$i][1\text{-}t]\}\text{-}(Q+Vp/\{360[c\text{-}q]\})/I$$
$$= 64,720+[41,548\text{-}29,750\text{-}51,294$$
$$*0.0100][1\text{-}0.3000]\}$$
$$\text{-}(35,351+9,746*240$$
$$/\{360[1.2000\text{-}1.0100]\})$$
$$/0.9900$$
$$= 2,370 \text{ millions USD}$$

Corporate IFRS-GAAP (B/S-I/S),ISBN-13: **978-1721066308**, ISBN-10: **1721066306**

Law-8010:

If both (**I**= L/E= Leverage or Gearing Ratio= 99.00%), (**Q**= Quoted Longterm Liabilities= 35,351 millions USD), (**q**= [C-P]/X= Quick or Acid Test Ratio= 1.01 times), (**p**= 360 P/V= Procured Inventory Days= 240 Days), (**c**= C/X= Current Ratio= 1.20 times), (**D**= Dividend Paid= 2,370 millions USD), (**t**= T/B= Tax Rate= 30.00%), (**U**= Utilized or Starting Capital= 64,720 millions USD), (**i**= I/S= Interest Portion= 1.00%), (**S**= Sales or Revenues= 51,294 millions USD), (**F**= Fixed Cost= 29,740 millions USD), and (**M**= Margin of Contribution= 41,548 millions USD), then it's (**V**=Variable Cost planned), is:

$$\mathbf{V} = 360[\mathbf{c}\text{-}\mathbf{q}](\mathbf{I}\ \{\mathbf{U}+[\mathbf{M}\text{-}\mathbf{F}\text{-}\mathbf{Si}][1\text{-}\mathbf{t}]\text{-}\mathbf{D}\}\text{-}\mathbf{Q})/\mathbf{p}$$

$$= 360[1.2000\text{-}1.0100](0.9900$$
$$\{64,720+[41,548\text{-}29,750$$
$$\text{-}51,294*0.0100][1\text{-}0.3000]$$
$$\text{-}2,370\}\text{-}35,351)/240$$

$$= \underline{9,746}\ \text{millions USD}$$

Corporate IFRS-GAAP (B/S-I/S),ISBN-13: **978-1721066308**, ISBN-10: **1721066306**

Law-8011:

If both (**I**= L/E= Leverage or Gearing Ratio= 99.00%), (**Q**= Quoted Longterm Liabilities= 35,351 millions USD), (**q**= [C-P]/X= Quick or Acid Test Ratio= 1.01 times), (**V**= Variable Cost= 9,746 millions USD), (**c**= C/X= Current Ratio= 1.20 times), (**D**= Dividend Paid= 2,370 millions USD), (**t**= T/B= Tax Rate= 30.00%), (**U**= Utilized or Starting Capital= 64,720 millions USD), (**i**= I/S= Interest Portion= 1.00%), (**S**= Sales or Revenues= 51,294 millions USD), (**F**= Fixed Cost= 29,740 millions USD), and (**M**= Margin of Contribution= 41,548 millions USD), then it's (**p**=Procured Inventories Days planned), is:

$$p= 360[c-q](I \{U+[M-F-Si][1-t]-D\}-Q)/V$$
$$= 360[1.2000-1.0100](0.9900$$
$$\{64,720+[41,548-29,750$$
$$-51,294*0.0100][1-0.3000]$$
$$-2,370\}- 35,351)/9,746$$

$$= \underline{240} \text{ days}$$

Corporate IFRS-GAAP (B/S-I/S),ISBN-13: **978-1721066308**, ISBN-10: **1721066306**

Law-8012:

If both (**I**= L/E= Leverage or Gearing Ratio= 99.00%), (**Q**= Quoted Longterm Liabilities= 35,351 millions USD), (**q**= [C-P]/X= Quick or Acid Test Ratio= 1.01 times), (**V**= Variable Cost= 9,746 millions USD), (**p**= 360 P/V= Procured Inventory Days= 240 Days), (**D**= Dividend Paid= 2,370 millions USD), (**t**= T/B= Tax Rate= 30.00%), (**U**= Utilized or Starting Capital= 64,720 millions USD), (**i**= I/S= Interest Portion= 1.00%), (**S**= Sales or Revenues= 51,294 millions USD), (**F**= Fixed Cost= 29,740 millions USD), and (**M**= Margin of Contribution= 41,548 millions USD), then it's (**c**= Current Ratio planned), is:

$$c= q+Vp/[360(I\,\{U+[M-F-Si][1-t]-D\}-Q)]$$
$$= 1.0100+9,746*240/[360(0.9900$$
$$\{64,720+[41,548-29,750$$
$$-51,294*0.0100][1-0.3000]$$
$$-2,370\}-35,351)]$$
$$= 1.20 \text{ times}$$

Corporate IFRS-GAAP (B/S-I/S),ISBN-13: **978-1721066308**, ISBN-10: **1721066306**

Law-8013:

If both (l= L/E= Leverage or Gearing Ratio= 99.00%), (Q= Quoted Longterm Liabilities= 35,351 millions USD), (c= C/X= Current Ratio= 1.20 times), (V= Variable Cost= 9,746 millions USD), (p= 360 P/V= Procured Inventory Days= 240 Days), (D= Dividend Paid= 2,370 millions USD), (t= T/B= Tax Rate= 30.00%), (U= Utilized or Starting Capital= 64,720 millions USD), (i= I/S= Interest Portion= 1.00%), (S= Sales or Revenues= 51,294 millions USD), (F= Fixed Cost= 29,740 millions USD), and (M= Margin of Contribution= 41,548 millions USD), then it's (q= Quick or Acid Test Ratio planned), is:

$$q= c\text{-}Vp/[360(l\ \{U+[M\text{-}F\text{-}Si][1\text{-}t]\text{-}D\}\text{-}Q)]$$
$$= 1.2000\text{-}9,746*240/[360(0.9900$$
$$\{64,720+[41,548\text{-}29,750$$
$$-51,294*0.0100][1\text{-}0.3000]$$
$$-2,370\}\text{-}35,351)]$$

$$= 1.01 \text{ times}$$

Corporate IFRS-GAAP (B/S-I/S),ISBN-13: **978-1721066308**, ISBN-10: **1721066306**

<u>Law-8014</u>:

If both (I= L/E= Leverage or Gearing Ratio= 99.00%), (q= [C-P]/X= Quick or Acid Test Ratio= 1.01 times), (c= C/X= Current Ratio= 1.20 times), (v= V/S= Variable Portion= 19.00%), (p= 360 P/V= Procured Inventory Days= 240 Days), (D= Dividend Paid= 2,370 millions USD), (t= T/B= Tax Rate= 30.00%), (U= Utilized or Starting Capital= 64,720 millions USD), (i= I/S= Interest Portion= 1.00%), (S= Sales or Revenues= 51,294 millions USD), (F= Fixed Cost= 29,740 millions USD), and (M= Margin of Contribution= 41,548 millions USD), then it's (Q= Quoted Longterm Liabilities planned), is:

$$Q = I\{U+[M-F-Si][1-t]-D\}-Svp/\{360[c-q]\}$$
$$= 0.9900\{64,720+[41,548-29,750$$
$$-51,294*0.0100][1-0.3000]$$
$$-2,370\}-51,294*0.1900$$
$$*240/\{360[1.2000-1.0100]\}$$
$$= \underline{35,351} \text{ millions USD}$$

Law-8015:

If both (**Q**= Quoted Longterm Liabilities= 35,351 millions USD), (**q**= [C-P]/X= Quick or Acid Test Ratio= 1.01 times), (**c**= C/X= Current Ratio= 1.20 times), (**v**= V/S= Variable Portion= 19.00%), (**p**= 360 P/V= Procured Inventory Days= 240 Days), (**D**= Dividend Paid= 2,370 millions USD), (**t**= T/B= Tax Rate= 30.00%), (**U**= Utilized or Starting Capital= 64,720 millions USD), (**i**= I/S= Interest Portion= 1.00%), (**S**= Sales or Revenues= 51,294 millions USD), (**F**= Fixed Cost= 29,740 millions USD), and (**M**= Margin of Contribution= 41,548 millions USD), then it's (**I** = Leverage or Gearing Ratio planned), is:

$$I = (Q+Svp/\{360[c-q]\})$$
$$/\{U+[M-F-Si][1-t]-D\}$$
$$= (35,351+51,294*0.1900*240/\{360$$
$$[1.2000-1.0100]\})/\{64,720$$
$$+[41,548-29,750-51,294$$
$$*0.0100][1-0.3000]-2,370\}$$
$$= 99.00\%$$

Corporate IFRS-GAAP (B/S-I/S),ISBN-13: **978-1721066308**, ISBN-10: **1721066306**

Law-8016:

If both (**Q**= Quoted Longterm Liabilities= 35,351 millions USD), (**q**= [C-P]/X= Quick or Acid Test Ratio= 1.01 times), (**c**= C/X= Current Ratio= 1.20 times), (**v**= V/S= Variable Portion= 19.00%), (**p**= 360 P/V= Procured Inventory Days= 240 Days), (**D**= Dividend Paid= 2,370 millions USD), (**t**= T/B= Tax Rate= 30.00%), (**l**= L/E= Leverage or Gearing Ratio= 99.00%), (**i**= I/S= Interest Portion= 1.00%), (**S**= Sales or Revenues= 51,294 millions USD), (**F**= Fixed Cost= 29,740 millions USD), and (**M**= Margin of Contribution= 41,548 millions USD), then it's (**U**= Utilized or Starting Capital planned), is:

$$U= D+(Q+Svp/\{360[c-q]\})/l$$
$$-[M-F-Si][1-t]$$
$$= 2,370 +(35,351+51,294*0.1900$$
$$*240/\{360[1.2000-1.0100]\})$$
$$/0.9900-[41,548-29,750$$
$$-51,294*0.0100][1-0.3000]$$
$$= 64,720 \text{ millions USD}$$

Corporate IFRS-GAAP (B/S-I/S),ISBN-13: **978-1721066308**, ISBN-10: **1721066306**

Law-8017:

If both (**Q**= Quoted Longterm Liabilities= 35,351 millions USD), (**q**= [C-P]/X= Quick or Acid Test Ratio= 1.01 times), (**c**= C/X= Current Ratio= 1.20 times), (**v**= V/S= Variable Portion= 19.00%), (**p**= 360 P/V= Procured Inventory Days= 240 Days), (**D**= Dividend Paid= 2,370 millions USD), (**t**= T/B= Tax Rate= 30.00%), (**I**= L/E= Leverage or Gearing Ratio= 99.00%), (**i**= I/S= Interest Portion= 1.00%), (**S**= Sales or Revenues= 51,294 millions USD), (**F**= Fixed Cost= 29,740 millions USD), and (**U**= Utilized or Starting Capital= 64,720 millions USD), then it's (**M**= Margin of Contribution planned), is:

$$M= F+Si+[D-U+(Q+Svp/\{360[c-q]\})/I]/[1-t]$$

$$= 29,750+51,294*0.0100+[2,370$$
$$-64,720+(35,351+51,294$$
$$*0.1900*240/\{360[1.2000$$
$$-1.0100]\})/0.9900]$$
$$/[1-0.3000]$$

$$= 41,548 \text{ millions USD}$$

Corporate IFRS-GAAP (B/S-I/S),ISBN-13: **978-1721066308**, ISBN-10: **1721066306**

Law-8018:

If both (**Q**= Quoted Longterm Liabilities= 35,351 millions USD), (**q**= [C-P]/X= Quick or Acid Test Ratio= 1.01 times), (**c**= C/X= Current Ratio= 1.20 times), (**v**= V/S= Variable Portion= 19.00%), (**p**= 360 P/V= Procured Inventory Days= 240 Days), (**D**= Dividend Paid= 2,370 millions USD), (**t**= T/B= Tax Rate= 30.00%), (**I**= L/E= Leverage or Gearing Ratio= 99.00%), (**i**= I/S= Interest Portion= 1.00%), (**S**= Sales or Revenues= 51,294 millions USD), (**M**= Margin of Contribution= 41,548 millions USD), and (**U**= Utilized or Starting Capital= 64,720 millions USD), then it's (**F**= Fixed Cost planned), is:

$$F= M\text{-}Si\text{-}[D\text{-}U+(Q+Svp/\{360[c\text{-}q]\})/I\,]$$
$$/[1\text{-}t]$$
$$= 41,548\text{-}51,294*0.0100\text{-}[2,370$$
$$-64,720+(35,351+51,294$$
$$*0.1900*240/\{360[1.2000$$
$$-1.0100]\})/0.9900]$$
$$/[1\text{-}0.3000]$$
$$= 29,750 \text{ millions USD}$$

Corporate IFRS-GAAP (B/S-I/S),ISBN-13: **978-1721066308**, ISBN-10: **1721066306**

<u>Law-8019</u>:

If both (Q= Quoted Longterm Liabilities= <u>35,351</u> millions USD), (q= [C-P]/X= Quick or Acid Test Ratio= <u>1.01</u> times), (c= C/X= Current Ratio= <u>1.20</u> times), (v= V/S= Variable Portion= <u>19.00%</u>), (p= 360 P/V= Procured Inventory Days= <u>240</u> Days), (D= Dividend Paid= <u>2,370</u> millions USD), (t= T/B= Tax Rate= <u>30.00%</u>), (I= L/E= Leverage or Gearing Ratio= <u>99.00%</u>), (i= I/S= Interest Portion= <u>1.00%</u>), (F= Fixed Cost= <u>29,740</u> millions USD), (M= Margin of Contribution= <u>41,548</u> millions USD), and (U= Utilized or Starting Capital= <u>64,720</u> millions USD), then it's (S= Sales Growth planned), is:

$$S= \{M-F-[D-U+(Q+Svp/\{360[c-q]\})/I]$$
$$/[1-t]\}/i$$
$$= \{41,548-29,750-[2,370-64,720$$
$$+(35,351+51,294*0.1900$$
$$*240/\{360[1.2000-1.0100]\})$$
$$/0.9900]/[1-0.3000]\}/0.0100$$
$$= \underline{51,294} \text{ millions USD}$$

Corporate IFRS-GAAP (B/S-I/S),ISBN-13: **978-1721066308**, ISBN-10: **1721066306**

Law-8020:

If both (**Q**= Quoted Longterm Liabilities= 35,351 millions USD), (**q**= [C-P]/X= Quick or Acid Test Ratio= 1.01 times), (**c**= C/X= Current Ratio= 1.20 times), (**v**= V/S= Variable Portion= 19.00%), (**p**= 360 P/V= Procured Inventory Days= 240 Days), (**D**= Dividend Paid= 2,370 millions USD), (**t**= T/B= Tax Rate= 30.00%), (**l**= L/E= Leverage or Gearing Ratio= 99.00%), (**\$**= Sales or Revenues= 51,294 millions USD), (**F**= Fixed Cost= 29,740 millions USD), (**M**= Margin of Contribution= 41,548 millions USD), and (**U**= Utilized or Starting Capital= 64,720 millions USD), then it's (**i**= Interest Portion planned), is:

$$i= \{M\text{-}F\text{-}[D\text{-}U+(Q+\$vp/\{360[c\text{-}q]\})/l\,]$$
$$/[1\text{-}t]\}/\$$$
$$= \{41,548\text{-}29,750\text{-}[2,370\text{-}64,720$$
$$+(35,351+51,294*0.1900$$
$$*240/\{360[1.2000$$
$$-1.0100]\})/0.9900]$$
$$/[1\text{-}0.3000]\}/51,294$$
$$= 1.00\%$$

Corporate IFRS-GAAP (B/S-I/S),ISBN-13: **978-1721066308**, ISBN-10: **1721066306**

<u>Law-8021</u>:

If both (**Q**= Quoted Longterm Liabilities= <u>35,351</u> millions USD), (**q**= [C-P]/X= Quick or Acid Test Ratio= <u>1.01</u> times), (**c**= C/X= Current Ratio= <u>1.20</u> times), (**v**= V/S= Variable Portion= <u>19.00%</u>), (**p**= 360 P/V= Procured Inventory Days= <u>240</u> Days), (**D**= Dividend Paid= <u>2,370</u> millions USD), (**i**= I/S= Interest Portion= <u>1.00%</u>), (**f**= L/E= Leverage or Gearing Ratio= <u>99.00%</u>), (**S**= Sales or Revenues= <u>51,294</u> millions USD), (**F**= Fixed Cost= <u>29,740</u> millions USD), (**M**= Margin of Contribution= <u>41,548</u> millions USD), and (**U**= Utilized or Starting Capital= <u>64,720</u> millions USD), then it's (**t**= Tax Rate planned), is:

$$t= 1-[\mathbf{D\text{-}U}+(\mathbf{Q+Svp}/\{360[\mathbf{c\text{-}q}]\})/\mathbf{f}]$$
$$/[\mathbf{M\text{-}F\text{-}Si}]$$
$$= 1-[2,370-64,720+(35,351+51,294$$
$$*0.1900*240/\{360[1.2000$$
$$-1.0100]\})/0.9900]/[41,548$$
$$-29,750-51,294*0.0100]$$

$$= \underline{30.00\%}$$

Corporate IFRS-GAAP (B/S-I/S),ISBN-13: **978-1721066308**, ISBN-10: **1721066306**

Law-8022:

If both (**Q**= Quoted Longterm Liabilities= 35,351 millions USD), (**q**= [C-P]/X= Quick or Acid Test Ratio= 1.01 times), (**c**= C/X= Current Ratio= 1.20 times), (**v**= V/S= Variable Portion= 19.00%), (**p**= 360 P/V= Procured Inventory Days= 240 Days), (**t**= T/B= Tax Rate= 30.00%), (**i**= I/S= Interest Portion= 1.00%), (**f**= L/E= Leverage or Gearing Ratio= 99.00%), (**$**= Sales or Revenues= 51,294 millions USD), (**F**= Fixed Cost= 29,740 millions USD), (**M**= Margin of Contribution= 41,548 millions USD), and (**U**= Utilized or Starting Capital= 64,720 millions USD), then it's (**D**= Dividend Paid planned), is:

$$D= U+[M-F-Si][1-t]$$
$$-(Q+Svp/\{360[c-q]\})/f$$
$$= 64,720+[41,548-29,750-51,294$$
$$*0.0100][1-0.3000]-(35,351$$
$$+51,294*0.1900*240/\{360$$
$$[1.2000-1.0100]\})/0.9900$$
$$= 2,370 \text{ millions USD}$$

Corporate IFRS-GAAP (B/S-I/S),ISBN-13: **978-1721066308**, ISBN-10: **1721066306**

Law-8023:

If both (**Q**= Quoted Longterm Liabilities= 35,351 millions USD), (**q**= [C-P]/X= Quick or Acid Test Ratio= 1.01 times), (**c**= C/X= Current Ratio= 1.20 times), (**D**= Dividend Paid= 2,370 millions USD), (**p**= 360 P/V= Procured Inventory Days= 240 Days), (**t**= T/B= Tax Rate= 30.00%), (**i**= I/S= Interest Portion= 1.00%), (**I**= L/E= Leverage or Gearing Ratio= 99.00%), (**S**= Sales or Revenues= 51,294 millions USD), (**F**= Fixed Cost= 29,740 millions USD), (**M**= Margin of Contribution= 41,548 millions USD), and (**U**= Utilized or Starting Capital= 64,720 millions USD), then it's (**v**= Variable Portion planned), is:

$$v= 360[c\text{-}q](I\ \{U+[M\text{-}F\text{-}Si][1\text{-}t]\text{-}D\}\text{-}Q)$$
$$/[Sp]$$

$$= 360[1.2000\text{-}1.0100](0.9900$$
$$\{64,720+[41,548\text{-}29,750$$
$$-51,294*0.0100][1\text{-}0.3000]$$
$$-2,370\}\text{-}35,351)$$
$$/[51,294*240]$$

$$= 19.00\%$$

Corporate IFRS-GAAP (B/S-I/S),ISBN-13: **978-1721066308**, ISBN-10: **1721066306**

Law-8024:

If both (**Q**= Quoted Longterm Liabilities= 35,351 millions USD), (**q**= [C-P]/X= Quick or Acid Test Ratio= 1.01 times), (**c**= C/X= Current Ratio= 1.20 times), (**D**= Dividend Paid= 2,370 millions USD), (**v**= V/S= Variable Portion= 19.00%), (**t**= T/B= Tax Rate= 30.00%), (**i**= I/S= Interest Portion= 1.00%), (**l**= L/E= Leverage or Gearing Ratio= 99.00%), (**S**= Sales or Revenues= 51,294 millions USD), (**F**= Fixed Cost= 29,740 millions USD), (**M**= Margin of Contribution= 41,548 millions USD), and (**U**= Utilized or Starting Capital= 64,720 millions USD), then it's (**p**= Procured Inventory Days planned), is:

$$\mathbf{p}= 360[\mathbf{c}\text{-}\mathbf{q}](\mathbf{l}\,\{\mathbf{U}+[\mathbf{M}\text{-}\mathbf{F}\text{-}\mathbf{Si}][1\text{-}\mathbf{t}]\text{-}\mathbf{D}\}\text{-}\mathbf{Q})$$
$$/[\mathbf{Sv}]$$

$$= 360[1.2000\text{-}1.0100](0.9900$$
$$\{64,720+[41,548\text{-}29,750$$
$$51,294*0.0100][1\text{-}0.3000]$$
$$\text{-}2,370\}\text{-}35,351)/[51,294$$
$$*0.1900]$$

$$= \underline{240}\ \text{days}$$

Corporate IFRS-GAAP (B/S-I/S),ISBN-13: **978-1721066308**, ISBN-10: **1721066306**

Law-8025:

If both (**Q**= Quoted Longterm Liabilities= 35,351 millions USD), (**q**= [C-P]/X= Quick or Acid Test Ratio= 1.01 times), (**p**= 360 P/V= Procured Inventory Days= 240 Days), (**D**= Dividend Paid= 2,370 millions USD), (**v**= V/S= Variable Portion= 19.00%), (**t**= T/B= Tax Rate= 30.00%), (**i**= I/S= Interest Portion= 1.00%), (**F**= L/E= Leverage or Gearing Ratio= 99.00%), (**S**= Sales or Revenues= 51,294 millions USD), (**F**= Fixed Cost= 29,740 millions USD), (**M**= Margin of Contribution= 41,548 millions USD), and (**U**= Utilized or Starting Capital= 64,720 millions USD), then it's (**c**= Current Ratio planned), is:

$$c= q+Svp/[360(I \{U+[M-F-Si][1-t]-D\}-Q)]$$

$$= 1.0100+51,294*0.1900*240$$
$$/[360(0.9900\{64,720$$
$$+[41,548-29,750-51,294$$
$$*0.0100][1-0.3000]$$
$$-2,370\}-35,351)]$$

$$= \underline{1.20} \text{ times}$$

Corporate IFRS-GAAP (B/S-I/S),ISBN-13: **978-1721066308**, ISBN-10: **1721066306**

Law-8026:

If both (**Q**= Quoted Longterm Liabilities= 35,351 millions USD), (**c**= C/X= Current Ratio= 1.20 times), (**p**= 360 P/V= Procured Inventory Days= 240 Days), (**D**= Dividend Paid= 2,370 millions USD), (**v**= V/S= Variable Portion= 19.00%), (**t**= T/B= Tax Rate= 30.00%), (**i**= I/S= Interest Portion= 1.00%), (**l**= L/E= Leverage or Gearing Ratio= 99.00%), (**S**= Sales or Revenues= 51,294 millions USD), (**F**= Fixed Cost= 29,740 millions USD), (**M**= Margin of Contribution= 41,548 millions USD), and (**U**= Utilized or Starting Capital= 64,720 millions USD), then it's (**q**= Quick or Acid Test Ratio planned), is:

$$q= c\text{-}Svp/[360(l\{U+[M\text{-}F\text{-}Si][1\text{-}t]\text{-}D\}\text{-}Q)]$$
$$= 1.2000\text{-}51,294*0.1900*240/[360$$
$$(0.9900\{64,720+[41,548$$
$$-29,750\text{-}51,294*0.0100]$$
$$[1\text{-}0.3000]\text{-}2,370\}\text{-}35,351)]$$
$$= \underline{1.01} \text{ times}$$

Corporate IFRS-GAAP (B/S-I/S),ISBN-13: **978-1721066308**, ISBN-10: **1721066306**

Law-8027:

    If both (q= [C-P]/X= Quick or Acid Test Ratio= 1.01 times), (S'= Sales of Past Year= 48,851 millions USD), (s= [S/S']-1= Sales Growth= 5.00%), (c= C/X= Current Ratio= 1.20 times), (p= 360 P/V= Procured Inventory Days= 240 Days), (D= Dividend Paid= 2,370 millions USD), (v= V/S= Variable Portion= 19.00%), (t= T/B= Tax Rate= 30.00%), (i= I/S= Interest Portion= 1.00%), (l= L/E= Leverage or Gearing Ratio= 99.00%), (S= Sales or Revenues= 51,294 millions USD), (F= Fixed Cost= 29,740 millions USD), (M= Margin of Contribution= 41,548 millions USD), and (U= Utilized or Starting Capital= 64,720 millions USD), then it's (Q= Quoted Longterm Liabilities planned), is:

$$Q= l \{U+[M-F-Si][1-t]-D\}$$
$$-S'vp[1+s]/\{360[c-q]\}$$
$$= 0.9900\{64,720+[41,548-29,750$$
$$-51,294*0.0100][1-0.3000]$$
$$-2,370\}-48,851*0.1900*240$$
$$[1+0.0500]/\{360[1.2000$$
$$-1.0100]\}$$
$$= 35,351 \text{ millions USD}$$

Corporate IFRS-GAAP (B/S-I/S),ISBN-13: **978-1721066308**, ISBN-10: **1721066306**

Law-8028:

   If both (q= [C-P]/X= Quick or Acid Test Ratio= 1.01
times), (S'= Sales of Past Year= 48,851 millions
USD), (s= [S/S']-1= Sales Growth= 5.00%), (c=
C/X= Current Ratio= 1.20 times), (p= 360 P/V=
Procured Inventory Days= 240 Days), (D= Dividend
Paid= 2,370 millions USD), (v= V/S= Variable
Portion= 19.00%), (t= T/B= Tax Rate= 30.00%), (i=
I/S= Interest Portion= 1.00%), (Q= Quoted Longterm
Liabilities= 35,351 millions USD), (S= Sales or
Revenues= 51,294 millions USD), (F= Fixed Cost=
29,740 millions USD), (M= Margin of Contribution=
41,548 millions USD), and (U= Utilized or Starting
Capital= 64,720 millions USD), then it's (I=
Leverage or Gearing Ratio planned), is:

$$I = (Q+S'vp[1+s]/\{360[c-q]\})$$
$$/\{U+[M-F-Si][1-t]-D\}$$
$$- (35,351 + 48,851*0.1900*240$$
$$[1+0.0500]/\{360[1.2000$$
$$-1.0100]\})/\{64,720+[41,548$$
$$-29,750-51,294*0.0100]$$
$$[1-0.3000]-2,370\}$$
$$= 99.00\%$$

Corporate IFRS-GAAP (B/S-I/S),ISBN-13: **978-1721066308**, ISBN-10: **1721066306**

Law-8029:

If both (q= [C-P]/X= Quick or Acid Test Ratio= 1.01 times), (S'= Sales of Past Year= 48,851 millions USD), (s= [S/S']-1= Sales Growth= 5.00%), (c= C/X= Current Ratio= 1.20 times), (p= 360 P/V= Procured Inventory Days= 240 Days), (D= Dividend Paid= 2,370 millions USD), (v= V/S= Variable Portion= 19.00%), (t= T/B= Tax Rate= 30.00%), (i= I/S= Interest Portion= 1.00%), (Q= Quoted Longterm Liabilities= 35,351 millions USD), (S= Sales or Revenues= 51,294 millions USD), (F= Fixed Cost= 29,740 millions USD), (M= Margin of Contribution= 41,548 millions USD), and (I= L/E= Leverage or Gearing Ratio= 99.00%), then it's (U= Utilized or Starting Capital planned), is:

$$U= D+(Q+S'vp[1+s]/\{360[c-q]\})/I$$
$$-[M-F-Si][1-t]$$

$$= 2,370+(35,351+48,851*0.1900$$
$$*240[1+0.0500]/\{360[1.2000$$
$$-1.0100]\})/0.9900-[41,548$$
$$-29,750-51,294*0.0100]$$
$$[1-0.3000]$$

$$= 64,720 \text{ millions USD}$$

Corporate IFRS-GAAP (B/S-I/S),ISBN-13: **978-1721066308**, ISBN-10: **1721066306**

Law-8030:

If both (**q**= [C-P]/X= Quick or Acid Test Ratio= <u>1.01</u> times), (**S'**= Sales of Past Year= <u>48,851</u> millions USD), (**s**= [S/S']-1= Sales Growth= <u>5.00%</u>), (**c**= C/X= Current Ratio= <u>1.20</u> times), (**p**= 360 P/V= Procured Inventory Days= <u>240</u> Days), (**D**= Dividend Paid= <u>2,370</u> millions USD), (**v**= V/S= Variable Portion= <u>19.00%</u>), (**t**= T/B= Tax Rate= <u>30.00%</u>), (**i**= I/S= Interest Portion= <u>1.00%</u>), (**Q**= Quoted Longterm Liabilities= <u>35,351</u> millions USD), (**S**= Sales or Revenues= <u>51,294</u> millions USD), (**F**= Fixed Cost= <u>29,740</u> millions USD), (**U**= Utilized or Starting Capital= <u>64,720</u> millions USD), and (**I**= L/E= Leverage or Gearing Ratio= <u>99.00%</u>), then it's (**M**= Margin of Contribution planned), is :

$$M= F+Si+[D-U+(Q+S'vp[1+s]$$
$$/\{360[c-q]\})/I\ ]/[1-t]$$
$$- 29,750+51,294*0.0100+[2,370$$
$$-64,720+(35,351+48,851$$
$$*0.1900*240[1+0.0500]$$
$$/\{360[1.2000-1.0100]\})$$
$$/0.9900]/[1-0.3000]$$
$$= \underline{41,548} \text{ millions USD}$$

Corporate IFRS-GAAP (B/S-I/S),ISBN-13: **978-1721066308**, ISBN-10: **1721066306**

Law-8031:

If both (**q**= [C-P]/X= Quick or Acid Test Ratio= 1.01 times), (**S'**= Sales of Past Year= 48,851 millions USD), (**s**= [S/S']-1= Sales Growth= 5.00%), (**c**= C/X= Current Ratio= 1.20 times), (**p**= 360 P/V= Procured Inventory Days= 240 Days), (**D**= Dividend Paid= 2,370 millions USD), (**v**= V/S= Variable Portion= 19.00%), (**t**= T/B= Tax Rate= 30.00%), (**i**= I/S= Interest Portion= 1.00%), (**Q**= Quoted Longterm Liabilities= 35,351 millions USD), (**S**= Sales or Revenues= 51,294 millions USD), (**M**= Margin of Contribution= 41,548 millions USD), (**U**= Utilized or Starting Capital= 64,720 millions USD), and (**I**= L/E= Leverage or Gearing Ratio= 99.00%), then it's (**F**= Fixed Cost planned), is:

$$F= M\text{-}Si\text{-}[D\text{-}U+(Q+S'vp[1+s]$$
$$/\{360[c\text{-}q]\})/I\,]/[1\text{-}t]$$
$$= 41,548\text{-}51,294*0.0100\text{-}[2,370$$
$$-64,720+(35,351+48,851$$
$$*0.1900*240[1+0.0500]$$
$$/\{360[1.2000\text{-}1.0100]\})$$
$$/0.9900]/[1\text{-}0.3000]$$
$$= 29,750 \text{ millions USD}$$

<u>Law-8032</u>:

If both (**q**= [C-P]/X= Quick or Acid Test Ratio= <u>1.01</u> times), (**$'**= Sales of Past Year= <u>48,851</u> millions USD), (**s**= [S/S']-1= Sales Growth= <u>5.00%</u>), (**c**= C/X= Current Ratio= <u>1.20</u> times), (**p**= 360 P/V= Procured Inventory Days= <u>240</u> Days), (**D**= Dividend Paid= <u>2,370</u> millions USD), (**v**= V/S= Variable Portion= <u>19.00%</u>), (**t**= T/B= Tax Rate= <u>30.00%</u>), (**i**= I/S= Interest Portion= <u>1.00%</u>), (**Q**= Quoted Longterm Liabilities= <u>35,351</u> millions USD), (**F**= Fixed Cost= <u>29,740</u> millions USD), (**M**= Margin of Contribution= <u>41,548</u> millions USD), (**U**= Utilized or Starting Capital= <u>64,720</u> millions USD), and (**I**= L/E= Leverage or Gearing Ratio= <u>99.00%</u>), then it's (**$**= Sales or Revenues planned), is:

$$\$= \{M\text{-}F\text{-}[D\text{-}U+(Q+\$'vp[1+s]$$
$$/\{360[c\text{-}q]\})/I]/[1\text{-}t]\}/i$$
$$- \{41,548\ 29,750\text{-}[2,370\text{-}64,720$$
$$+(35,351+48,851*0.1900$$
$$*240[1+0.0500]/\{360$$
$$[1.2000\text{-}1.0100]\})/0.9900]$$
$$/[1\text{-}0.3000]\}/0.0100$$
$$= \underline{51,294}\ \text{millions USD}$$

<u>Law-8033</u>:

If both (**q**= [C-P]/X= Quick or Acid Test Ratio= <u>1.01</u> times), (**$'**= Sales of Past Year= <u>48,851</u> millions USD), (**s**= [S/S']-1= Sales Growth= <u>5.00%</u>), (**c**= C/X= Current Ratio= <u>1.20</u> times), (**p**= 360 P/V= Procured Inventory Days= <u>240</u> Days), (**D**= Dividend Paid= <u>2,370</u> millions USD), (**v**= V/S= Variable Portion= <u>19.00%</u>), (**t**= T/B= Tax Rate= <u>30.00%</u>), (**$**= Sales or Revenues= <u>51,294</u> millions USD), (**Q**= Quoted Longterm Liabilities= <u>35,351</u> millions USD), (**F**= Fixed Cost= <u>29,740</u> millions USD), (**M**= Margin of Contribution= <u>41,548</u> millions USD), (**U**= Utilized or Starting Capital= <u>64,720</u> millions USD), and (**$**= L/E= Leverage or Gearing Ratio= <u>99.00%</u>), then it's (**i**= Interest Portion planned), is:

$$i= \{M-F-[D-U+(Q+S'vp[1+s]$$
$$/\{360[c-q]\})/I\,]/[1-t]\}/S$$

$$= \{41,548-29,750-[2,370-64,720$$
$$+(35,351+48,851*0.1900$$
$$*240[1+0.0500]/\{360[1.2000$$
$$-1.0100]\})/0.9900]$$
$$/[1-0.3000]\}/51,294$$

$$= \underline{1.00\%}$$

Corporate IFRS-GAAP (B/S-I/S),ISBN-13: **978-1721066308**, ISBN-10: **1721066306**

<u>Law-8034</u>:

If both (q= [C-P]/X= Quick or Acid Test Ratio= <u>1.01</u> times), (S'= Sales of Past Year= <u>48,851</u> millions USD), (s= [S/S']-1= Sales Growth= <u>5.00%</u>), (c= C/X= Current Ratio= <u>1.20</u> times), (p= 360 P/V= Procured Inventory Days= <u>240</u> Days), (D= Dividend Paid= <u>2,370</u> millions USD), (v= V/S= Variable Portion= <u>19.00%</u>), (i= I/S= Interest Portion= <u>1.00%</u>), (S= Sales or Revenues= <u>51,294</u> millions USD), (Q= Quoted Longterm Liabilities= <u>35,351</u> millions USD), (F= Fixed Cost= <u>29,740</u> millions USD), (M= Margin of Contribution= <u>41,548</u> millions USD), (U= Utilized or Starting Capital= <u>64,720</u> millions USD), and (I= L/E= Leverage or Gearing Ratio= <u>99.00%</u>), then it's (t= Tax Rate planned), is:

$$t= 1-[D-U+(Q+S'vp[1+s]/\{360[c-q]\})/I]/[M-F-Si]$$

$$- 1-[2,370-64,720 + (35,351 + 48,851$$
$$*0.1900*240[1+0.0500]$$
$$/\{360[1.2000-1.0100]\})$$
$$/0.9900]/[41,548-29,750$$
$$-51,294*0.0100]$$

$$= \underline{30.00\%}$$

Corporate IFRS-GAAP (B/S-I/S),ISBN-13: **978-1721066308**, ISBN-10: **1721066306**

Law-8035:

If both (**q**= [C-P]/X= Quick or Acid Test Ratio= 1.01 times), (**$'**= Sales of Past Year= 48,851 millions USD), (**s**= [S/S']-1= Sales Growth= 5.00%), (**c**= C/X= Current Ratio= 1.20 times), (**p**= 360 P/V= Procured Inventory Days= 240 Days), (**t**= T/B= Tax Rate= 30.00%), (**v**= V/S= Variable Portion= 19.00%), (**i**= I/S= Interest Portion= 1.00%), (**$**= Sales or Revenues= 51,294 millions USD), (**Q**= Quoted Longterm Liabilities= 35,351 millions USD), (**F**= Fixed Cost= 29,740 millions USD), (**M**= Margin of Contribution= 41,548 millions USD), (**U**= Utilized or Starting Capital= 64,720 millions USD), and (**l**= L/E= Leverage or Gearing Ratio= 99.00%), then it's (**D**= Dividend Paid planned), is:

$$D= U+[M-F-Si][1-t]-(Q+S'vp[1+s]$$
$$/\{360[c-q]\})/l$$

$$= 64,720+[41,548-29,750-51,294$$
$$*0.0100][1-0.3000]-(35,351$$
$$+48,851*0.1900*240$$
$$[1+0.0500]/\{360[1.2000$$
$$-1.0100]\})/0.9900$$

$$= 2,370 \text{ millions USD}$$

Corporate IFRS-GAAP (B/S-I/S),ISBN-13: **978-1721066308**, ISBN-10: **1721066306**

Law-8036:

If both (**q**= [C-P]/X= Quick or Acid Test Ratio= 1.01 times), (**D**= Dividend Paid= 2,370 millions USD), (**s**= [S/S']-1= Sales Growth= 5.00%), (**c**= C/X= Current Ratio= 1.20 times), (**p**= 360 P/V= Procured Inventory Days= 240 Days), (**t**= T/B= Tax Rate= 30.00%), (**v**= V/S= Variable Portion= 19.00%), (**i**= I/S= Interest Portion= 1.00%), (**S**= Sales or Revenues= 51,294 millions USD), (**Q**= Quoted Longterm Liabilities= 35,351 millions USD), (**F**= Fixed Cost= 29,740 millions USD), (**M**= Margin of Contribution= 41,548 millions USD), (**U**= Utilized or Starting Capital= 64,720 millions USD), and (**I**= L/E= Leverage or Gearing Ratio= 99.00%), then it's (**S'**= Sales Past), must be:

$$\mathbf{S'} = 360[\mathbf{c\text{-}q}](\mathbf{I}\{\mathbf{U}+[\mathbf{M\text{-}F\text{-}Si}][1\text{-}\mathbf{t}]\text{-}\mathbf{D}\}\text{-}\mathbf{Q})$$
$$/\{\mathbf{vp}[1+\mathbf{s}]\}$$
$$- 360[1.2000\text{-}1.0100](0.9900$$
$$\{64,720+[41,548\text{-}29,750$$
$$-51,294*0.0100][1\text{-}0.3000]$$
$$-2,370\}\text{-}35,351)/\{0.1900$$
$$*240[1+0.0500]\}$$
$$= \underline{48,851} \text{ millions USD}$$

Corporate IFRS-GAAP (B/S-I/S),ISBN-13: **978-1721066308**, ISBN-10: **1721066306**

Law-8037:

If both (**q**= [C-P]/X= Quick or Acid Test Ratio= 1.01 times), (**D**= Dividend Paid= 2,370 millions USD), (**s**= [S/S']-1= Sales Growth= 5.00%), (**c**= C/X= Current Ratio= 1.20 times), (**p**= 360 P/V= Procured Inventory Days= 240 Days), (**t**= T/B= Tax Rate= 30.00%), (**S'**= Sales of Past Year= 48,851 millions USD), (**i**= I/S= Interest Portion= 1.00%), (**S**= Sales or Revenues= 51,294 millions USD), (**Q**= Quoted Longterm Liabilities= 35,351 millions USD), (**F**= Fixed Cost= 29,740 millions USD), (**M**= Margin of Contribution= 41,548 millions USD), (**U**= Utilized or Starting Capital= 64,720 millions USD), and (**I**= L/E= Leverage or Gearing Ratio= 99.00%), then it's (**v**= Variable Portion planned), is:

$$v= 360[c\text{-}q](I\,\{U+[M\text{-}F\text{-}Si][1\text{-}t]\text{-}D\}\text{-}Q)$$
$$/\{S'p[1+s]\}$$
$$= 360[1.2000\text{-}1.0100](0.9900$$
$$\{64,720+[41,548\text{-}29,750$$
$$-51,294*0.0100][1\text{-}0.3000]$$
$$-2,370\}\text{-}35,351)/\{48,851$$
$$*240[1+0.0500]\}$$
$$= 19.00\%$$

Corporate IFRS-GAAP (B/S-I/S),ISBN-13: **978-1721066308**, ISBN-10: **1721066306**

Law-8038:

If both (q= [C-P]/X= Quick or Acid Test Ratio= 1.01 times), (D= Dividend Paid= 2,370 millions USD), (s= [S/S']-1= Sales Growth= 5.00%), (c= C/X= Current Ratio= 1.20 times), (v= V/S= Variable Portion= 19.00%), (t= T/B= Tax Rate= 30.00%), (S'= Sales of Past Year= 48,851 millions USD), (i= I/S= Interest Portion= 1.00%), (S= Sales or Revenues= 51,294 millions USD), (Q= Quoted Longterm Liabilities= 35,351 millions USD), (F= Fixed Cost= 29,740 millions USD), (M= Margin of Contribution= 41,548 millions USD), (U= Utilized or Starting Capital= 64,720 millions USD), and (l= L/E= Leverage or Gearing Ratio= 99.00%), then it's (p= Procured Inventory Days planned), is:

$$p= 360[c\text{-}q](l\,\{U+[M\text{-}F\text{-}Si][1\text{-}t]\text{-}D\}\text{-}Q)$$
$$/\{S'v[1+s]\}$$
$$= 360[1.2000\text{-}1.0100](0.9900$$
$$\{64,720+[41,548\text{-}29,750$$
$$-51,294*0.0100][1\text{-}0.3000]$$
$$-2,370\}\text{-}35,351)/\{48,851$$
$$*0.1900[1+0.0500]\}$$
$$= \underline{240}\ \text{days}$$

Corporate IFRS-GAAP (B/S-I/S),ISBN-13: **978-1721066308**, ISBN-10: **1721066306**

Law-8039:

If both (**q**= [C-P]/X= Quick or Acid Test Ratio= 1.01
times), (**D**= Dividend Paid= 2,370 millions USD),
(**p**= 360 P/V= Procured Inventory Days= 240 Days),
(**c**= C/X= Current Ratio= 1.20 times), (**v**= V/S=
Variable Portion= 19.00%), (**t**= T/B= Tax Rate=
30.00%), (**$'**= Sales of Past Year= 48,851 millions
USD), (**i**= I/S= Interest Portion= 1.00%), (**$**= Sales or
Revenues= 51,294 millions USD), (**Q**= Quoted
Longterm Liabilities= 35,351 millions USD), (**F**=
Fixed Cost= 29,740 millions USD), (**M**= Margin of
Contribution= 41,548 millions USD), (**U**= Utilized or
Starting Capital= 64,720 millions USD), and (**l**=
L/E= Leverage or Gearing Ratio= 99.00%), then it's
(**s**= Sales Growth planned), is:

$$s= 360[\textbf{c-q}](\textbf{l} \{\textbf{U}+[\textbf{M-F-Si}][1-\textbf{t}]-\textbf{D}\}-\textbf{Q})$$
$$/[\textbf{\$'vp}]-1$$
$$= 360[1.2000-1.0100](0.9900$$
$$\{64,720+[41,548-29,750$$
$$-51,294*0.0100][1-0.3000]$$
$$-2,370\}-35,351)/[48,851$$
$$*0.1900*240]-1$$
$$= 5.00\%$$

Law-8040:

If both (**q**= [C-P]/X= Quick or Acid Test Ratio= 1.01 times), (**D**= Dividend Paid= 2,370 millions USD), (**p**= 360 P/V= Procured Inventory Days= 240 Days), (**s**= [S/S']-1= Sales Growth= 5.00%), (**v**= V/S= Variable Portion= 19.00%), (**t**= T/B= Tax Rate= 30.00%), (**S'**= Sales of Past Year= 48,851 millions USD), (**i**= I/S= Interest Portion= 1.00%), (**S**= Sales or Revenues= 51,294 millions USD), (**Q**= Quoted Longterm Liabilities= 35,351 millions USD), (**F**= Fixed Cost= 29,740 millions USD), (**M**= Margin of Contribution= 41,548 millions USD), (**U**= Utilized or Starting Capital= 64,720 millions USD), and (**l**= L/E= Leverage or Gearing Ratio= 99.00%), then it's (**c**= Current Ratio planned), is:

$$c= q+S'vp[1+s]/[360(l\{U+[M-F-Si]$$
$$[1-t]-D\}-Q)]$$
$$= 1.0100+48,851*0.1900*240$$
$$[1+0.0500]/[360(0.9900$$
$$\{64,720+[41,548-29,750$$
$$-51,294*0.0100][1-0.3000]$$
$$-2,370\}-35,351)]$$
$$= 1.20 \text{ times}$$

Corporate IFRS-GAAP (B/S-I/S),ISBN-13: **978-1721066308**, ISBN-10: **1721066306**

Law-8041:

If both (**c**= C/X= Current Ratio= 1.20 times), (**D**= Dividend Paid= 2,370 millions USD), (**p**= 360 P/V= Procured Inventory Days= 240 Days), (**s**= [S/S']-1= Sales Growth= 5.00%), (**v**= V/S= Variable Portion= 19.00%), (**t**= T/B= Tax Rate= 30.00%), (**S'**= Sales of Past Year= 48,851 millions USD), (**i**= I/S= Interest Portion= 1.00%), (**S**= Sales or Revenues= 51,294 millions USD), (**Q**= Quoted Longterm Liabilities= 35,351 millions USD), (**F**= Fixed Cost= 29,740 millions USD), (**M**= Margin of Contribution= 41,548 millions USD), (**U**= Utilized or Starting Capital= 64,720 millions USD), and (**l**= L/E= Leverage or Gearing Ratio= 99.00%), then it's (**q**= Quick or Acid Test Ratio planned), is:

$$q= c\text{-}S'vp[1+s]/[360(l \{U+[M\text{-}F\text{-}Si] [1\text{-}t]\text{-}D\}\text{-}Q)]$$

$$= 1.2000\text{-}48,851*0.1900*240 [1+0.0500]/[360(0.9900 \{64,720+[41,548\text{-}29,750 \text{-}51,294*0.0100][1\text{-}0.3000] \text{-}2,370\}\text{-}35,351)]$$

$$= 1.01 \text{ times}$$

Corporate IFRS-GAAP (B/S-I/S),ISBN-13: **978-1721066308**, ISBN-10: **1721066306**

<u>Law-8042</u>:

If both (**d**= D/A= Dividend Portion or Payout= 30.00%), (**A**= After Tax Income= 7,899 millions USD), (**t**= T/B= Tax Rate= 30.00%), (**i**= I/S= Interest Portion= 1.00%), (**S**= Sales or Revenues= 51,294 millions USD), (**F**= Fixed Cost= 29,740 millions USD), (**X**= Xpress or Current Debt= 34,196 millions USD), (**M**= Margin of Contribution= 41,548 millions USD), (**U**= Utilized or Starting Capital= 64,720 millions USD), and (**I**= L/E= Leverage or Gearing Ratio= 99.00%), then it's (**Q**= Quoted Longterm Liabilities planned), is:

$$Q = I\{U+[M-F-Si][1-t]-Ad\}-X$$
$$= 0.9900\{64,720+[41,548-29,750$$
$$-51,294*0.0100][1-0.3000]$$
$$-7,899*0.3000\}-34,196$$
$$= \underline{35,351} \text{ millions USD}$$

Corporate IFRS-GAAP (B/S-I/S),ISBN-13: **978-1721066308**, ISBN-10: **1721066306**

Law-8043:

If both (**d**= D/A= Dividend Portion or Payout=
30.00%), (**A**= After Tax Income= 7,899 millions
USD), (**t**= T/B= Tax Rate= 30.00%), (**i**= I/S= Interest
Portion= 1.00%), (**$**= Sales or Revenues= 51,294
millions USD), (**F**= Fixed Cost= 29,740 millions
USD), (**X**= Xpress or Current Debt= 34,196 millions
USD), (**M**= Margin of Contribution= 41,548 millions
USD), (**U**= Utilized or Starting Capital= 64,720
millions USD), and (**Q**= Quoted Longterm
Liabilities= 35,351 millions USD), then it's (**I** =
Leverage or Gearing Ratio planned), is:

$$I = [\mathbf{Q}+\mathbf{X}]/\{\mathbf{U}+[\mathbf{M}-\mathbf{F}-\mathbf{Si}][1-\mathbf{t}]-\mathbf{Ad}\}$$
$$= [35,351+34,196]/\{64,720$$
$$+[41,548-29,750-51,294$$
$$*0.0100][1-0.3000]$$
$$-7,899*0.3000\}$$
$$= 99.00\%$$

Corporate IFRS-GAAP (B/S-I/S),ISBN-13: **978-1721066308**, ISBN-10: **1721066306**

Law-8044:

If both (**d**= D/A= Dividend Portion or Payout= 30.00%), (**A**= After Tax Income= 7,899 millions USD), (**t**= T/B= Tax Rate= 30.00%), (**i**= I/S= Interest Portion= 1.00%), (**S**= Sales or Revenues= 51,294 millions USD), (**F**= Fixed Cost= 29,740 millions USD), (**X**= Xpress or Current Debt= 34,196 millions USD), (**M**= Margin of Contribution= 41,548 millions USD), (**l**= L/E= Leverage or Gearing Ratio= 99.00%), and (**Q**= Quoted Longterm Liabilities= 35,351 millions USD), then it's (**U**= Utilized or Starting Capital planned), is:

$$U= Ad+[Q+X]/l -[M-F-Si][1-t]$$
$$= 7,899*0.3000+[35,351$$
$$+34,196]/0.9900-[41,548$$
$$-29,750-51,294*0.0100]$$
$$[1-0.3000]$$
$$= \underline{64,720} \text{ millions USD}$$

Corporate IFRS-GAAP (B/S-I/S),ISBN-13: **978-1721066308**, ISBN-10: **1721066306**

Law-8045:

If both (**d**= D/A= Dividend Portion or Payout= 30.00%), (**A**= After Tax Income= 7,899 millions USD), (**t**= T/B= Tax Rate= 30.00%), (**i**= I/S= Interest Portion= 1.00%), (**S**= Sales or Revenues= 51,294 millions USD), (**F**= Fixed Cost= 29,740 millions USD), (**X**= Xpress or Current Debt= 34,196 millions USD), (**U**= Utilized or Starting Capital= 64,720 millions USD), (**l**= L/E= Leverage or Gearing Ratio= 99.00%), and (**Q**= Quoted Longterm Liabilities= 35,351 millions USD), then it's (**M**= Margin of Contribution planned), is:

$$M= F+Si+\{Ad+[Q+X]/l-U\}/[1-t]$$
$$= 29,750+51,294*0.0100+\{7,899$$
$$*0.3000+[35,351+34,196]$$
$$/0.9900-64,720\}/[1-0.3000]$$
$$= 41,548 \text{ millions USD}$$

Corporate IFRS-GAAP (B/S-I/S),ISBN-13: **978-1721066308**, ISBN-10: **1721066306**

<u>Law-8046</u>:

If both (**d**= D/A= Dividend Portion or Payout= <u>30.00%</u>), (**A**= After Tax Income= <u>7,899</u> millions USD), (**t**= T/B= Tax Rate= <u>30.00%</u>), (**i**= I/S= Interest Portion= <u>1.00%</u>), (**S**= Sales or Revenues= <u>51,294</u> millions USD), (**M**= Margin of Contribution= <u>41,548</u> millions USD), (**X**= Xpress or Current Debt= <u>34,196</u> millions USD), (**U**= Utilized or Starting Capital= <u>64,720</u> millions USD), (**l**= L/E= Leverage or Gearing Ratio= <u>99.00%</u>), and (**Q**= Quoted Longterm Liabilities= <u>35,351</u> millions USD), then it's (**F**= Fixed Cost planned), is :

$$F= M\text{-}Si\text{-}\{Ad+[Q+X]/l\text{-}U\}/[1\text{-}t]$$
$$= 41,548\text{-}51,294*0.0100\text{-}\{7,899$$
$$*0.3000+[35,351+34,196]$$
$$/0.9900\text{-}64,720\}/[1\text{-}0.3000]$$
$$= \underline{29,750}\ \text{millions USD}$$

Corporate IFRS-GAAP (B/S-I/S),ISBN-13: **978-1721066308**, ISBN-10: **1721066306**

<u>Law-8047</u>:

If both (**d**= D/A= Dividend Portion or Payout= 30.00%), (**A**= After Tax Income= <u>7,899</u> millions USD), (**t**= T/B= Tax Rate= <u>30.00%</u>), (**i**= I/S= Interest Portion= <u>1.00%</u>), (**F**= Fixed Cost= <u>29,740</u> millions USD), (**M**= Margin of Contribution= <u>41,548</u> millions USD), (**X**= Xpress or Current Debt= <u>34,196</u> millions USD), (**U**= Utilized or Starting Capital= <u>64,720</u> millions USD), (**I**= L/E= Leverage or Gearing Ratio= <u>99.00%</u>), and (**Q**= Quoted Longterm Liabilities= <u>35,351</u> millions USD), then it's (**$**= Sales or Revenues planned), is:

$$\$= (M\text{-}F\text{-}\{Ad+[Q+X]/I\text{-}U\}/[1\text{-}t])/i$$
$$= (41,548\text{-}29,750\text{-}\{7,899*0.3000$$
$$+[35,351+34,196]/0.9900$$
$$-64,720\}/[1\text{-}0.3000])/0.0100$$
$$= \underline{51,294} \text{ millions USD}$$

Corporate IFRS-GAAP (B/S-I/S),ISBN-13: **978-1721066308**, ISBN-10: **1721066306**

Law-8048:

If both (**d**= D/A= Dividend Portion or Payout=
30.00%), (**A**= After Tax Income= 7,899 millions
USD), (**t**= T/B= Tax Rate= 30.00%), (**S**= Sales or
Revenues= 51,294 millions USD), (**F**= Fixed Cost=
29,740 millions USD), (**M**= Margin of Contribution=
41,548 millions USD), (**X**= Xpress or Current Debt=
34,196 millions USD), (**U**= Utilized or Starting
Capital= 64,720 millions USD), (**I**= L/E= Leverage
or Gearing Ratio= 99.00%), and (**Q**= Quoted
Longterm Liabilities= 35,351 millions USD), then
it's (**i**= Interest Portion planned), is:

$$i= (M-F-\{Ad+[Q+X]/I-U\}/[1-t])/S$$
$$= (41,548-29,750-\{7,899*0.3000$$
$$+[35,351+34,196]/0.9900$$
$$-64,720\}/[1-0.3000])$$
$$/51,294$$
$$= 1.00\%$$

Corporate IFRS-GAAP (B/S-I/S),ISBN-13: **978-1721066308**, ISBN-10: **1721066306**

Law-8049:

If both (**d**= D/A= Dividend Portion or Payout= 30.00%), (**A**= After Tax Income= 7,899 millions USD), (**i**= I/S= Interest Portion= 1.00%), (**$**= Sales or Revenues= 51,294 millions USD), (**F**= Fixed Cost= 29,740 millions USD), (**M**= Margin of Contribution= 41,548 millions USD), (**X**= Xpress or Current Debt= 34,196 millions USD), (**U**= Utilized or Starting Capital= 64,720 millions USD), (**I**= L/E= Leverage or Gearing Ratio= 99.00%), and (**Q**= Quoted Longterm Liabilities= 35,351 millions USD), then it's (**t**= Tax Rate planned), is:

$$t= 1-\{Ad+[Q+X]/I-U\}/[M-F-Si]$$
$$= 1-\{7,899*0.3000+[35,351+34,196]$$
$$/0.9900-64,720\}/[41,548$$
$$-29,750-51,294*0.0100]$$
$$= 30.00\%$$

Corporate IFRS-GAAP (B/S-I/S),ISBN-13: **978-1721066308**, ISBN-10: **1721066306**

Law-8050:

If both (d= D/A= Dividend Portion or Payout= 30.00%), (t= T/B= Tax Rate= 30.00%), (i= I/S= Interest Portion= 1.00%), (S= Sales or Revenues= 51,294 millions USD), (F= Fixed Cost= 29,740 millions USD), (M= Margin of Contribution= 41,548 millions USD), (X= Xpress or Current Debt= 34,196 millions USD), (U= Utilized or Starting Capital= 64,720 millions USD), (I= L/E= Leverage or Gearing Ratio= 99.00%), and (Q= Quoted Longterm Liabilities= 35,351 millions USD), then it's (A= After Tax Income planned), is:

$$A = \{U+[M-F-Si][1-t]-[Q+X]/I\}/d$$
$$= \{64,720+[41,548-29,750-51,294$$
$$*0.0100][1-0.3000]-[35,351$$
$$+34,196]/0.9900\}/0.3000$$
$$= \underline{7,899} \text{ millions USD}$$

Corporate IFRS-GAAP (B/S-I/S),ISBN-13: **978-1721066308**, ISBN-10: **1721066306**

<u>Law-8051</u>:

If both (**A**= After Tax Income= <u>7,899</u> millions USD), (**t**= T/B= Tax Rate= <u>30.00%</u>), (**i**= I/S= Interest Portion= <u>1.00%</u>), (**S**= Sales or Revenues= <u>51,294</u> millions USD), (**F**= Fixed Cost= <u>29,740</u> millions USD), (**M**= Margin of Contribution= <u>41,548</u> millions USD), (**X**= Xpress or Current Debt= <u>34,196</u> millions USD), (**U**= Utilized or Starting Capital= <u>64,720</u> millions USD), (**I**= L/E= Leverage or Gearing Ratio= <u>99.00%</u>), and (**Q**= Quoted Longterm Liabilities= <u>35,351</u> millions USD), then it's (**d**= Dividend Portion or Payout planned), is:

$$d= \{U+[M-F-Si][1-t]-[Q+X]/I\}/A$$
$$= \{64,720+[41,548-29,750-51,294$$
$$*0.0100][1-0.3000]-[35,351$$
$$+34,196]/0.9900\}/7,899$$
$$= \underline{30.00\%}$$

Corporate IFRS-GAAP (B/S-I/S),ISBN-13: **978-1721066308**, ISBN-10: **1721066306**

Law-8052:

If both (**A**= After Tax Income= <u>7,899</u> millions USD), (**t**= T/B= Tax Rate= <u>30.00%</u>), (**i**= I/S= Interest Portion= <u>1.00%</u>), (**S**= Sales or Revenues= <u>51,294</u> millions USD), (**F**= Fixed Cost= <u>29,740</u> millions USD), (**M**= Margin of Contribution= <u>41,548</u> millions USD), (**d**= D/A= Dividend Portion or Payout= <u>30.00%</u>), (**U**= Utilized or Starting Capital= <u>64,720</u> millions USD), (**l**= L/E= Leverage or Gearing Ratio= <u>99.00%</u>), and (**Q**= Quoted Longterm Liabilities= <u>35,351</u> millions USD), then it's (**X**= Xpress or Current Debt planned), is:

$$X= l\{U+[M-F-Si][1-t]-Ad\}-Q$$
$$= 0.9900\{64,720+[41,548-29,750$$
$$-51,294*0.0100][1-0.3000]$$
$$-7,899*0.3000\}-35,351$$
$$= \underline{34,196} \text{ millions USD}$$

Corporate IFRS-GAAP (B/S-I/S),ISBN-13: **978-1721066308**, ISBN-10: **1721066306**

Law-8053:

If both (**A**= After Tax Income= 7,899 millions USD), (**t**= T/B= Tax Rate= 30.00%), (**P**= Procured Inventories= 6,497 millions USD), (**q**= [C-P]/X= Quick or Acid Test Ratio= 1.01 times), (**i**= I/S= Interest Portion= 1.00%), (**S**= Sales or Revenues= 51,294 millions USD), (**F**= Fixed Cost= 29,740 millions USD), (**M**= Margin of Contribution= 41,548 millions USD), (**d**= D/A= Dividend Portion or Payout= 30.00%), (**U**= Utilized or Starting Capital= 64,720 millions USD), (**l**= L/E= Leverage or Gearing Ratio= 99.00%), and (**c**= C/X= Current Ratio= 1.20 times), then it's (**Q**= Quoted Longterm Liabilities planned), is:

$$Q = l\{U+[M-F-Si][1-t]-Ad\}-P/[c-q]$$
$$= 0.9900\{64,720+[41,548-29,750$$
$$-51,294*0.0100][1-0.3000]$$
$$-7,899*0.3000\}-6,497$$
$$/[1.2000-1.0100]\}$$
$$= 35,351 \text{ millions USD}$$

Corporate IFRS-GAAP (B/S-I/S),ISBN-13: **978-1721066308**, ISBN-10: **1721066306**

Law-8054:

If both (**A**= After Tax Income= 7,899 millions USD), (**t**= T/B= Tax Rate= 30.00%), (**P**= Procured Inventories= 6,497 millions USD), (**q**= [C-P]/X= Quick or Acid Test Ratio= 1.01 times), (**i**= I/S= Interest Portion= 1.00%), (**S**= Sales or Revenues= 51,294 millions USD), (**F**= Fixed Cost= 29,740 millions USD), (**M**= Margin of Contribution= 41,548 millions USD), (**d**= D/A= Dividend Portion or Payout= 30.00%), (**U**= Utilized or Starting Capital= 64,720 millions USD), (**Q**= Quoted Longterm Liabilities= 35,351 millions USD), and (**c**= C/X= Current Ratio= 1.20 times), then it's (**l** = Leverage or Gearing Ratio planned), is:

$$l = \{Q+P/[c-q]\}/\{U+[M-F-Si][1-t]-Ad\}$$
$$= \{35,351+6,497/[1.2000-1.0100]\}$$
$$/\{64,720+[41,548-29,750$$
$$-51,294*0.0100][1-0.3000]$$
$$-7,899*0.3000\}$$
$$= 99.00\%$$

Law-8055:

If both (**A**= After Tax Income= 7,899 millions USD), (**t**= T/B= Tax Rate= 30.00%), (**P**= Procured Inventories= 6,497 millions USD), (**q**= [C-P]/X= Quick or Acid Test Ratio= 1.01 times), (**i**= I/S= Interest Portion= 1.00%), (**S**= Sales or Revenues= 51,294 millions USD), (**F**= Fixed Cost= 29,740 millions USD), (**M**= Margin of Contribution= 41,548 millions USD), (**d**= D/A= Dividend Portion or Payout= 30.00%), (**l**= L/E= Leverage or Gearing Ratio= 99.00%), (**Q**= Quoted Longterm Liabilities= 35,351 millions USD), and (**c**= C/X= Current Ratio= 1.20 times), then it's (**U**= Utilized or Starting Capital planned), is:

$$U= Ad+\{Q+P/[c-q]\}/l-[M-F-Si][1-t]$$

$$= 7,899*0.3000+\{35,351+6,497$$
$$/[1.2000-1.0100]\}/0.9900$$
$$-[41,548-29,750-51,294$$
$$*0.0100][1-0.3000]$$

$$= 64,720 \text{ millions USD}$$

Corporate IFRS-GAAP (B/S-I/S),ISBN-13: **978-1721066308**, ISBN-10: **1721066306**

Law-8056:

If both (**A**= After Tax Income= 7,899 millions USD), (**t**= T/B= Tax Rate= 30.00%), (**P**= Procured Inventories= 6,497 millions USD), (**q**= [C-P]/X= Quick or Acid Test Ratio= 1.01 times), (**i**= I/S= Interest Portion= 1.00%), (**U**= Utilized or Starting Capital= 64,720 millions USD), (**S**= Sales or Revenues= 51,294 millions USD), (**F**= Fixed Cost= 29,740 millions USD), (**d**= D/A= Dividend Portion or Payout= 30.00%), (**I**= L/E= Leverage or Gearing Ratio= 99.00%), (**Q**= Quoted Longterm Liabilities= 35,351 millions USD), and (**c**= C/X= Current Ratio= 1.20 times), then it's (**M**= Margin of Contribution planned), is:

$$M = F+Si+(Ad+\{Q+P/[c-q]\}/I-U)/[1-t]$$
$$= 29,750+51,294*0.0100+(7,899$$
$$*0.3000+\{35,351+6,497$$
$$/[1.2000-1.0100]\}/0.9900$$
$$-64,720)/[1-0.3000]$$
$$= 41,548 \text{ millions USD}$$

Corporate IFRS-GAAP (B/S-I/S),ISBN-13: **978-1721066308**, ISBN-10: **1721066306**

Law-8057:

If both (**A**= After Tax Income= 7,899 millions USD), (**t**= T/B= Tax Rate= 30.00%), (**P**= Procured Inventories= 6,497 millions USD), (**q**= [C-P]/X= Quick or Acid Test Ratio= 1.01 times), (**i**= I/S= Interest Portion= 1.00%), (**U**= Utilized or Starting Capital= 64,720 millions USD), (**$**= Sales or Revenues= 51,294 millions USD), (**M**= Margin of Contribution= 41,548 millions USD), (**d**= D/A= Dividend Portion or Payout= 30.00%), (**/**= L/E= Leverage or Gearing Ratio= 99.00%), (**Q**= Quoted Longterm Liabilities= 35,351 millions USD), and (**c**= C/X= Current Ratio= 1.20 times), then it's (**F**= Fixed Cost planned), is:

$$F= M-\$i-(Ad+\{Q+P/[c-q]\}//-U)/[1-t]$$
$$= 41,548-51,294*0.0100-(7,899$$
$$*0.3000+\{35,351+6,497$$
$$/[1.2000-1.0100]\}/0.9900$$
$$-64,720)/[1-0.3000]$$
$$= 29,750 \text{ millions USD}$$

Corporate IFRS-GAAP (B/S-I/S),ISBN-13: **978-1721066308**, ISBN-10: **1721066306**

Law-8058:

If both (**A**= After Tax Income= 7,899 millions USD),
(**t**= T/B= Tax Rate= 30.00%), (**P**= Procured
Inventories= 6,497 millions USD), (**q**= [C-P]/X=
Quick or Acid Test Ratio= 1.01 times), (**i**= I/S=
Interest Portion= 1.00%), (**U**= Utilized or Starting
Capital= 64,720 millions USD), (**F**= Fixed Cost=
29,740 millions USD), (**M**= Margin of Contribution=
41,548 millions USD), (**d**= D/A= Dividend Portion or
Payout= 30.00%), (**I**= L/E= Leverage or Gearing
Ratio= 99.00%), (**Q**= Quoted Longterm Liabilities=
35,351 millions USD), and (**c**= C/X= Current Ratio=
1.20 times), then it's (**S**= Sales or Revenues planned),
is:

$$S= [M-F-(Ad+\{Q+P/[c-q]\}/I-U)/[1-t]]/i$$
$$= [41,548-29,750-(7,899*0.3000$$
$$+\{35,351+6,497/[1.2000$$
$$-1.0100]\}/0.9900\ 64,720)$$
$$/[1-0.3000]]/0.0100$$
$$= \underline{51,294}\ \text{millions USD}$$

Corporate IFRS-GAAP (B/S-I/S),ISBN-13: **978-1721066308**, ISBN-10: **1721066306**

Law-8059:

If both (**A**= After Tax Income= 7,899 millions USD),
(**t**= T/B= Tax Rate= 30.00%), (**P**= Procured
Inventories= 6,497 millions USD), (**q**= [C-P]/X=
Quick or Acid Test Ratio= 1.01 times), (**S**= Sales or
Revenues= 51,294 millions USD), (**U**= Utilized or
Starting Capital= 64,720 millions USD), (**F**= Fixed
Cost= 29,740 millions USD), (**M**= Margin of
Contribution= 41,548 millions USD), (**d**= D/A=
Dividend Portion or Payout= 30.00%), (**I**= L/E=
Leverage or Gearing Ratio= 99.00%), (**Q**= Quoted
Longterm Liabilities= 35,351 millions USD), and
(**c**= C/X= Current Ratio= 1.20 times), then it's (**i**=
Interest Portion planned), is:

$$i= [M-F-(Ad+\{Q+P/[c-q]\}/I-U)/[1-t]]/S$$
$$= [41,548-29,750-(7,899*0.3000$$
$$+\{35,351+6,497/[1.2000$$
$$-1.0100]\}/0.9900-64,720)$$
$$/[1-0.3000]]/51,294$$
$$= 1.00\%$$

Corporate IFRS-GAAP (B/S-I/S),ISBN-13: **978-1721066308**, ISBN-10: **1721066306**

Law-8060:

If both (**A**= After Tax Income= 7,899 millions USD), (**i**= I/S= Interest Portion= 1.00%), (**P**= Procured Inventories= 6,497 millions USD), (**q**= [C-P]/X= Quick or Acid Test Ratio= 1.01 times), (**S**= Sales or Revenues= 51,294 millions USD), (**U**= Utilized or Starting Capital= 64,720 millions USD), (**F**= Fixed Cost= 29,740 millions USD), (**M**= Margin of Contribution= 41,548 millions USD), (**d**= D/A= Dividend Portion or Payout= 30.00%), (**l**= L/E= Leverage or Gearing Ratio= 99.00%), (**Q**= Quoted Longterm Liabilities= 35,351 millions USD), and (**c**= C/X= Current Ratio= 1.20 times), then it's (**t**= Tax Rate planned), is:

$$t= 1-(Ad+\{Q+P/[c-q]\}/l-U)/[M-F-Si]$$
$$= 1-(7,899*0.3000+\{35,351+6,497$$
$$/[1.2000-1.0100]\}/0.9900$$
$$64,720)/[41,548-29,750$$
$$-51,294*0.0100]$$
$$= 30.00\%$$

`

Corporate IFRS-GAAP (B/S-I/S),ISBN-13: **978-1721066308**, ISBN-10: **1721066306**

Law-8061:

If both (**t**= T/B= Tax Rate= 30.00%), (**i**= I/S= Interest Portion= 1.00%), (**P**= Procured Inventories= 6,497 millions USD), (**q**= [C-P]/X= Quick or Acid Test Ratio= 1.01 times), (**S**= Sales or Revenues= 51,294 millions USD), (**U**= Utilized or Starting Capital= 64,720 millions USD), (**F**= Fixed Cost= 29,740 millions USD), (**M**= Margin of Contribution= 41,548 millions USD), (**d**= D/A= Dividend Portion or Payout= 30.00%), (**l**= L/E= Leverage or Gearing Ratio= 99.00%), (**Q**= Quoted Longterm Liabilities= 35,351 millions USD), and (**c**= C/X= Current Ratio= 1.20 times), then it's (**A**= After Tax Income planned), is:

$$A= (U+[M-F-Si][1-t]-\{Q+P/[c-q]\}/l\ )/d$$
$$= (64,720+[41,548-29,750-51,294$$
$$*0.0100][1-0.3000]-\{35,351$$
$$+6,497/[1.2000-1.0100]\}$$
$$/0.9900)/0.3000$$
$$= \underline{7,899}\ \text{millions USD}$$

Corporate IFRS-GAAP (B/S-I/S),ISBN-13: **978-1721066308**, ISBN-10: **1721066306**

Law-8062:

If both (**t**= T/B= Tax Rate= 30.00%), (**i**= I/S= Interest Portion= 1.00%), (**P**= Procured Inventories= 6,497 millions USD), (**q**= [C-P]/X= Quick or Acid Test Ratio= 1.01 times), (**$**= Sales or Revenues= 51,294 millions USD), (**U**= Utilized or Starting Capital= 64,720 millions USD), (**F**= Fixed Cost= 29,740 millions USD), (**M**= Margin of Contribution= 41,548 millions USD), (**A**= After Tax Income= 7,899 millions USD), (**l**= L/E= Leverage or Gearing Ratio= 99.00%), (**Q**= Quoted Longterm Liabilities= 35,351 millions USD), and (**c**= C/X= Current Ratio= 1.20 times), then it's (**d**= Dividend Portion or payout planned), is:

$$d= (U+[M-F-Si][1-t]-\{Q+P/[c-q]\}/l\ )/A$$
$$= (64,720+[41,548-29,750-51,294$$
$$*0.0100][1-0.3000]-\{35,351$$
$$+6,497/[1.2000-1.0100]\}$$
$$/0.9900)/7,899$$
$$= 30.00\%$$

Corporate IFRS-GAAP (B/S-I/S),ISBN-13: **978-1721066308**, ISBN-10: **1721066306**

Law-8063:

If both (**t**= T/B= Tax Rate= 30.00%), (**i**= I/S= Interest Portion= 1.00%), (**d**= D/A= Dividend Portion or Payout= 30.00%), (**q**= [C-P]/X= Quick or Acid Test Ratio= 1.01 times), (**$**= Sales or Revenues= 51,294 millions USD), (**U**= Utilized or Starting Capital= 64,720 millions USD), (**F**= Fixed Cost= 29,740 millions USD), (**M**= Margin of Contribution= 41,548 millions USD), (**A**= After Tax Income= 7,899 millions USD), (**⌐**= L/E= Leverage or Gearing Ratio= 99.00%), (**Q**= Quoted Longterm Liabilities= 35,351 millions USD), and (**c**= C/X= Current Ratio= 1.20 times), then it's (**P**= Procured Inventories planned), is:

$$P= [c\text{-}q](⌐ \{U+[M\text{-}F\text{-}Si][1\text{-}t]\text{-}Ad\}\text{-}Q)$$
$$= [1.2000\text{-}1.0100](0.9900\{64,720$$
$$+[41,548\text{-}29,750\text{-}51,294$$
$$*0.0100][1\text{-}0.3000]\text{-}7,899$$
$$*0.3000\}\text{-}35,351)$$
$$= \underline{6,497} \text{ millions USD}$$

Corporate IFRS-GAAP (B/S-I/S),ISBN-13: **978-1721066308**, ISBN-10: **1721066306**

Law-8064:

If both (**t**= T/B= Tax Rate= 30.00%), (**i**= I/S= Interest Portion= 1.00%), (**d**= D/A= Dividend Portion or Payout= 30.00%), (**q**= [C-P]/X= Quick or Acid Test Ratio= 1.01 times), (**S**= Sales or Revenues= 51,294 millions USD), (**U**= Utilized or Starting Capital= 64,720 millions USD), (**F**= Fixed Cost= 29,740 millions USD), (**M**= Margin of Contribution= 41,548 millions USD), (**A**= After Tax Income= 7,899 millions USD), (**I**= L/E= Leverage or Gearing Ratio= 99.00%), (**Q**= Quoted Longterm Liabilities= 35,351 millions USD), and (**P**= Procured Inventories= 6,497 millions USD), then it's (**c**= Current Ratio planned), is:

$$c= q+P/(I \{U+[M-F-Si][1-t]-Ad\}-Q)$$
$$= 1.0100+6,497/(0.9900\{64,720$$
$$+[41,548-29,750-51,294$$
$$*0,0100][1-0.3000]-7,899$$
$$*0.3000\}-35,351)$$
$$= 1.20 \text{ times}$$

Corporate IFRS-GAAP (B/S-I/S),ISBN-13: **978-1721066308**, ISBN-10: **1721066306**

Law-8065:

> If both (**t**= T/B= Tax Rate= 30.00%), (**i**= I/S= Interest
> Portion= 1.00%), (**d**= D/A= Dividend Portion or
> Payout= 30.00%), (**c**= C/X= Current Ratio= 1.20
> times), (**S**= Sales or Revenues= 51,294 millions
> USD), (**U**= Utilized or Starting Capital= 64,720
> millions USD), (**F**= Fixed Cost= 29,740 millions
> USD), (**M**= Margin of Contribution= 41,548 millions
> USD), (**A**= After Tax Income= 7,899 millions USD),
> (**l**= L/E= Leverage or Gearing Ratio= 99.00%), (**Q**=
> Quoted Longterm Liabilities= 35,351 millions USD),
> and (**P**= Procured Inventories= 6,497 millions USD),
> then it's (**q**= Quick or Acid Test Ratio planned), is:

$$q = c\text{-}P/(l\ \{U+[M\text{-}F\text{-}Si][1\text{-}t]\text{-}Ad\}\text{-}Q)$$

$$= 1.2000\text{-}6,497/(0.9900\{64,720$$
$$+[41,548\text{-}29,750\text{-}51,294$$
$$*0.0100][1\text{-}0.3000]\text{-}7,899$$
$$*0.3000\}\text{-}35,351)$$

$$= 1.01\ \text{times}$$

Corporate IFRS-GAAP (B/S-I/S),ISBN-13: **978-1721066308**, ISBN-10: **1721066306**

<u>Law-8066</u>:

If both (**t**= T/B= Tax Rate= <u>30.00%</u>), (**i**= I/S= Interest Portion= <u>1.00%</u>), (**V**= Variable Cost= <u>9,746</u> millions USD), (**d**= D/A= Dividend Portion or Payout= <u>30.00%</u>), (**c**= C/X= Current Ratio= <u>1.20</u> times), (**S**= Sales or Revenues= <u>51,294</u> millions USD), (**U**= Utilized or Starting Capital= <u>64,720</u> millions USD), (**F**= Fixed Cost= <u>29,740</u> millions USD), (**M**= Margin of Contribution= <u>41,548</u> millions USD), (**A**= After Tax Income= <u>7,899</u> millions USD), (**l**= L/E= Leverage or Gearing Ratio= <u>99.00%</u>), (**q**= [C-P]/X= Quick or Acid Test Ratio= <u>1.01</u> times), and (**p**= 360 P/V= Procured Inventory Days= <u>240</u> Days), then it's (**Q**= Quoted Longterm Liabilities planned), is:

$$Q = l \{U+[M-F-Si][1-t]-Ad\}-Vp/\{360[c-q]\}$$
$$= 0.9900\{64,720+[41,548-29,750$$
$$-51,294*0.0100][1-0.3000]$$
$$-7,899*0.3000\}-9,746$$
$$*240/\{360[1.2000-1.0100]\}$$
$$= \underline{35,351} \text{ millions USD}$$

Corporate IFRS-GAAP (B/S-I/S),ISBN-13: **978-1721066308**, ISBN-10: **1721066306**

Law-8067:

If both (t= T/B= Tax Rate= 30.00%), (i= I/S= Interest Portion= 1.00%), (V= Variable Cost= 9,746 millions USD), (d= D/A= Dividend Portion or Payout= 30.00%), (c= C/X= Current Ratio= 1.20 times), (S= Sales or Revenues= 51,294 millions USD), (U= Utilized or Starting Capital= 64,720 millions USD), (F= Fixed Cost= 29,740 millions USD), (M= Margin of Contribution= 41,548 millions USD), (A= After Tax Income= 7,899 millions USD), (Q= Quoted Longterm Liabilities= 35,351 millions USD), (q= Quick or Acid Test Ratio= [C-P]/X= 1.01 times), and (p= 360 P/V= Procured Inventory Days= 240 Days), then it's (I = Leverage or Gearing Ratio planned), is:

$$I = (Q+Vp/\{360[c-q]\})$$
$$/\{U+[M-F-Si][1-t]-Ad\}$$
$$= (35,351+9,746*240/\{360[1.2000$$
$$-1.0100]\})/\{64,720+[41,548$$
$$-29,750-51,294*0.0100]$$
$$[1-0.3000]-7,899*0.3000\}$$
$$= 99.00\%$$

Corporate IFRS-GAAP (B/S-I/S),ISBN-13: **978-1721066308**, ISBN-10: **1721066306**

Law-8068:

If both (**t**= T/B= Tax Rate= 30.00%), (**i**= I/S= Interest Portion= 1.00%), (**V**= Variable Cost= 9,746 millions USD), (**d**= D/A= Dividend Portion or Payout= 30.00%), (**c**= C/X= Current Ratio= 1.20 times), (**$**= Sales or Revenues= 51,294 millions USD), (**/**= L/E= Leverage or Gearing Ratio= 99.00%), (**F**= Fixed Cost= 29,740 millions USD), (**M**= Margin of Contribution= 41,548 millions USD), (**A**= After Tax Income= 7,899 millions USD), (**Q**= Quoted Longterm Liabilities= 35,351 millions USD), (**q**= Quick or Acid Test Ratio= [C-P]/X= 1.01 times), and (**p**= 360 P/V= Procured Inventory Days= 240 Days), then it's (**U**= Utilized or Starting Capital planned), is:

$$U = Ad+(Q+Vp/\{360[c-q]\})/I-[M-F-\$i][1-t]$$
$$= 7{,}899*0.3000+(35{,}351+9{,}746$$
$$*240/\{360[1.2000-1.0100]\})$$
$$/0.9900-[41{,}548-29{,}750$$
$$-51{,}294*0.0100][1-0.3000]$$
$$= \underline{64{,}720} \text{ millions USD}$$

Corporate IFRS-GAAP (B/S-I/S),ISBN-13: **978-1721066308**, ISBN-10: **1721066306**

Law-8069:

If both (**t**= T/B= Tax Rate= 30.00%), (**i**= I/S= Interest Portion= 1.00%), (**V**= Variable Cost= 9,746 millions USD), (**d**= D/A= Dividend Portion or Payout= 30.00%), (**c**= C/X= Current Ratio= 1.20 times), (**S**= Sales or Revenues= 51,294 millions USD), (**I**= L/E= Leverage or Gearing Ratio= 99.00%), (**F**= Fixed Cost= 29,740 millions USD), (**U**= Utilized or Starting Capital= 64,720 millions USD), (**A**= After Tax Income= 7,899 millions USD), (**Q**= Quoted Longterm Liabilities= 35,351 millions USD), (**q**= Quick or Acid Test Ratio= [C-P]/X= 1.01 times), and (**p**= 360 P/V= Procured Inventory Days= 240 Days), then it's (**M**= Margin of Contribution planned), is:

$$M= F+Si+[Ad+(Q+Vp/\{360[c-q]\})/I-U]$$
$$/[1-t]$$

$$= 29,750+51,294*0.0100+[7,899$$
$$*0.3000+(35,351+9,746$$
$$*240/\{360[1.2000-1.0100]\})$$
$$/0.9900-64,720]/[1-0.3000]$$

$$= 41,548 \text{ millions USD}$$

Corporate IFRS-GAAP (B/S-I/S),ISBN-13: **978-1721066308**, ISBN-10: **1721066306**

Law-8070:

If both (**t**= T/B= Tax Rate= 30.00%), (**i**= I/S= Interest Portion= 1.00%), (**V**= Variable Cost= 9,746 millions USD), (**d**= D/A= Dividend Portion or Payout= 30.00%), (**c**= C/X= Current Ratio= 1.20 times), (**S**= Sales or Revenues= 51,294 millions USD), (**I**= L/E= Leverage or Gearing Ratio= 99.00%), (**M**= Margin of Contribution= 41,548 millions USD), (**U**= Utilized or Starting Capital= 64,720 millions USD), (**A**= After Tax Income= 7,899 millions USD), (**Q**= Quoted Longterm Liabilities= 35,351 millions USD), (**q**= Quick or Acid Test Ratio= [C-P]/X= 1.01 times), and (**p**= 360 P/V= Procured Inventory Days= 240 Days), then it's (**F**= Fixed Cost planned), is:

$$F= M\text{-}Si\text{-}[Ad+(Q+Vp/\{360[c\text{-}q]\})/I\text{-}U]$$
$$/[1\text{-}t]$$
$$= 41,548\text{-}51,294*0.0100\text{-}[7,899$$
$$*0.3000 +(35,351+9,746$$
$$*240/\{360[1.2000\text{-}1.0100]\})$$
$$/0.9900\text{-}64,720]/[1\text{-}0.3000]$$
$$= 29,750 \text{ millions USD}$$

Corporate IFRS-GAAP (B/S-I/S),ISBN-13: **978-1721066308**, ISBN-10: **1721066306**

Law-8071:

If both (t= T/B= Tax Rate= 30.00%), (i= I/S= Interest Portion= 1.00%), (V= Variable Cost= 9,746 millions USD), (d= D/A= Dividend Portion or Payout= 30.00%), (c= C/X= Current Ratio= 1.20 times), (F= Fixed Cost= 29,740 millions USD), (l= L/E= Leverage or Gearing Ratio= 99.00%), (M= Margin of Contribution= 41,548 millions USD), (U= Utilized or Starting Capital= 64,720 millions USD), (A= After Tax Income= 7,899 millions USD), (Q= Quoted Longterm Liabilities= 35,351 millions USD), (q= Quick or Acid Test Ratio= [C-P]/X= 1.01 times), and (p= 360 P/V= Procured Inventory Days= 240 Days), then it's (S= Sales or Revenues planned), is:

$$S= \{M-F-[Ad+(Q+Vp/\{360[c-q]\})/l-U]/[1-t]\}/i$$

$$= \{41,548-29,750-[7,899*0.3000$$
$$+(35,351+9,746*240/\{360$$
$$[1.2000-1.0100]\})/0.9900$$
$$-64,720]/[1-0.3000]\}/0.0100$$

$$= 51,294 \text{ millions USD}$$

Corporate IFRS-GAAP (B/S-I/S),ISBN-13: **978-1721066308**, ISBN-10: **1721066306**

Law-8072:

If both (**t**= T/B= Tax Rate= 30.00%), (**$**= Sales or Revenues= 51,294 millions USD), (**V**= Variable Cost= 9,746 millions USD), (**d**= D/A= Dividend Portion or Payout= 30.00%), (**c**= C/X= Current Ratio= 1.20 times), (**F**= Fixed Cost= 29,740 millions USD), (**l**= L/E= Leverage or Gearing Ratio= 99.00%), (**M**= Margin of Contribution= 41,548 millions USD), (**U**= Utilized or Starting Capital= 64,720 millions USD), (**A**= After Tax Income= 7,899 millions USD), (**Q**= Quoted Longterm Liabilities= 35,351 millions USD), (**q**= [C-P]/X= Quick or Acid Test Ratio= 1.01 times), and (**p**= 360 P/V= Procured Inventory Days= 240 Days), then it's (**i**= Interest Portion planned), is:

$$i= \{M-F-[Ad+(Q+Vp/\{360[c-q]\})/l-U]/[1-t]\}/\$$$

$$= \{41,548-29,750-[7,899*0.3000$$
$$+(35,351+9,746*240/\{360$$
$$[1.2000-1.0100]\})/0.9900$$
$$-64,720]/[1-0.3000]\}/51,294$$

$$= 1.00\%$$

Corporate IFRS-GAAP (B/S-I/S),ISBN-13: **978-1721066308**, ISBN-10: **1721066306**

Law-8073:

If both (**i**= I/S= Interest Portion= 1.00%), (**S**= Sales or Revenues= 51,294 millions USD), (**V**= Variable Cost= 9,746 millions USD), (**d**= D/A= Dividend Portion or Payout= 30.00%), (**c**= C/X= Current Ratio= 1.20 times), (**F**= Fixed Cost= 29,740 millions USD), (**l**= L/E= Leverage or Gearing Ratio= 99.00%), (**M**= Margin of Contribution= 41,548 millions USD), (**U**= Utilized or Starting Capital= 64,720 millions USD), (**A**= After Tax Income= 7,899 millions USD), (**Q**= Quoted Longterm Liabilities= 35,351 millions USD), (**q**= [C-P]/X= Quick or Acid Test Ratio= 1.01 times), and (**p**= 360 P/V= Procured Inventory Days= 240 Days), then it's (**t**= Tax Rate planned), is:

$$t= 1-[Ad+(Q+Vp/\{360[c-q]\})/l-U]$$
$$/[M-F-Si]$$
$$= 1-[7,899*0.3000 +(35,351+9,746$$
$$*240/\{360[1.2000-1.0100]\})$$
$$/0.9900-64,720]/[41,548$$
$$-29,750-51,294*0.0100]$$
$$= 30.00\%$$

Corporate IFRS-GAAP (B/S-I/S),ISBN-13: **978-1721066308**, ISBN-10: **1721066306**

Law-8074:

If both (**i**= I/S= Interest Portion= 1.00%), (**$**= Sales or Revenues= 51,294 millions USD), (**V**= Variable Cost= 9,746 millions USD), (**d**= D/A= Dividend Portion or Payout= 30.00%), (**c**= C/X= Current Ratio= 1.20 times), (**F**= Fixed Cost= 29,740 millions USD), (**I**= L/E= Leverage or Gearing Ratio= 99.00%), (**M**= Margin of Contribution= 41,548 millions USD), (**U**= Utilized or Starting Capital= 64,720 millions USD), (**t**= T/B= Tax Rate= 30.00%), (**Q**= Quoted Longterm Liabilities= 35,351 millions USD), (**q**= Quick or Acid Test Ratio= [C-P]/X= 1.01 times), and (**p**= 360 P/V= Procured Inventory Days= 240 Days), then it's (**A**= After Tax Income planned), is:

$$A= [U+[M-F-Si][1-t]$$
$$-(Q+Vp/\{360[c-q]\})/I]/d$$
$$= [64,720+[41,548-29,750-51,294$$
$$*0.0100][1-0.3000]-(35,351$$
$$+9,746*240/\{360[1.2000$$
$$-1.0100]\})/0.9900]/0.3000$$
$$= 7,899 \text{ millions USD}$$

Corporate IFRS-GAAP (B/S-I/S),ISBN-13: **978-1721066308**, ISBN-10: **1721066306**

Law-8075:

If both (**i**= I/S= Interest Portion= 1.00%), (**$**= Sales or Revenues= 51,294 millions USD), (**V**= Variable Cost= 9,746 millions USD), (**A**= After Tax Income= 7,899 millions USD), (**c**= C/X= Current Ratio= 1.20 times), (**F**= Fixed Cost= 29,740 millions USD), (**I**= L/E= Leverage or Gearing Ratio= 99.00%), (**M**= Margin of Contribution= 41,548 millions USD), (**U**= Utilized or Starting Capital= 64,720 millions USD), (**t**= T/B= Tax Rate= 30.00%), (**Q**= Quoted Longterm Liabilities= 35,351 millions USD), (**q**= [C-P]/X= Quick or Acid Test Ratio= 1.01 times), and (**p**= 360 P/V= Procured Inventory Days= 240 Days), then it's (**d**= Dividend Portion or Payout planned), is:

$$d = [U+[M\text{-}F\text{-}Si][1\text{-}t]$$
$$-(Q+Vp/\{360[c\text{-}q]\})/I\,]/A$$
$$= [64{,}720+[41{,}548\text{-}29{,}750\text{-}51{,}294$$
$$*0.0100][1\text{-}0.3000]\text{-}(35{,}351$$
$$+9{,}746*240/\{360[1.2000$$
$$-1.0100]\})/0.9900]/7{,}899$$

$$= 30.00\%$$

Corporate IFRS-GAAP (B/S-I/S),ISBN-13: **978-1721066308**, ISBN-10: **1721066306**

Law-8076:

If both (**i**= I/S= Interest Portion= 1.00%), (**$**= Sales or Revenues= 51,294 millions USD), (**d**= D/A= Dividend Portion or Payout= 30.00%), (**A**= After Tax Income= 7,899 millions USD), (**c**= C/X= Current Ratio= 1.20 times), (**F**= Fixed Cost= 29,740 millions USD), (**l**= L/E= Leverage or Gearing Ratio= 99.00%), (**M**= Margin of Contribution= 41,548 millions USD), (**U**= Utilized or Starting Capital= 64,720 millions USD), (**t**= T/B= Tax Rate= 30.00%), (**Q**= Quoted Longterm Liabilities= 35,351 millions USD), (**q**= Quick or Acid Test Ratio= [C-P]/X= 1.01 times), and (**p**= 360 P/V= Procured Inventory Days= 240 Days), then it's (**V**= Variable Cost planned), is:

$$V= [360[c\text{-}q](l \{U+[M\text{-}F\text{-}Si][1\text{-}t]\text{-}Ad\}\text{-}Q)]/p$$

$$= [360[1.2000\text{-}1.0100]\}\,)/0.9900$$
$$\{64,720+[41,548\text{-}29,750$$
$$-51,294*0.0100][1\text{-}0.3000]$$
$$-7,899*0.3000\}$$
$$-35,351)]/240$$

$$= 9,746 \text{ millions USD}$$

Corporate IFRS-GAAP (B/S-I/S),ISBN-13: **978-1721066308**, ISBN-10: **1721066306**

Law-8077:

If both (**i**= I/S= Interest Portion= <u>1.00%</u>), (**$**= Sales or Revenues= <u>51,294</u> millions USD), (**d**= D/A= Dividend Portion or Payout= <u>30.00%</u>), (**A**= After Tax Income= <u>7,899</u> millions USD), (**c**= C/X= Current Ratio= <u>1.20</u> times), (**F**= Fixed Cost= <u>29,740</u> millions USD), (**l**= L/E= Leverage or Gearing Ratio= <u>99.00%</u>), (**M**= Margin of Contribution= <u>41,548</u> millions USD), (**U**= Utilized or Starting Capital= <u>64,720</u> millions USD), (**t**= T/B= Tax Rate= <u>30.00%</u>), (**Q**= Quoted Longterm Liabilities= <u>35,351</u> millions USD), (**q**= Quick or Acid Test Ratio= [C-P]/X= <u>1.01</u> times), and (**V**= Variable Cost= <u>9,746</u> millions USD), then it's (**p**= Procured Inventory Days planned), is:

$$p= [360[\mathbf{c}\text{-}\mathbf{q}](\mathbf{l}\{\mathbf{U}+[\mathbf{M}\text{-}\mathbf{F}\text{-}\mathbf{Si}][1\text{-}\mathbf{t}]\text{-}\mathbf{Ad}\}\text{-}\mathbf{Q})]/\mathbf{V}$$

$$= [360[1.2000\text{-}1.0100]\})/0.9900$$
$$\{64,720+[41,548\text{-}29,750$$
$$-51,294*0.0100][1\text{-}0.3000]$$
$$-7,899*0.3000\}$$
$$-35,351)]/9,746$$

$$= \underline{240} \text{ days}$$

Corporate IFRS-GAAP (B/S-I/S),ISBN-13: **978-1721066308**, ISBN-10: **1721066306**

Law-8078:

If both (**i**= I/S= Interest Portion= 1.00%), (**S**= Sales or Revenues= 51,294 millions USD), (**d**= D/A= Dividend Portion or Payout= 30.00%), (**A**= After Tax Income= 7,899 millions USD), (**p**= 360 P/V= Procured Inventory Days= 240 Days), (**F**= Fixed Cost= 29,740 millions USD), (**l**= L/E= Leverage or Gearing Ratio= 99.00%), (**M**= Margin of Contribution= 41,548 millions USD), (**U**= Utilized or Starting Capital= 64,720 millions USD), (**t**= T/B= Tax Rate= 30.00%), (**Q**= Quoted Longterm Liabilities= 35,351 millions USD), (**q**= [C-P]/X= Quick or Acid Test Ratio= 1.01 times), and (**V**= Variable Cost= 9,746 millions USD), then it's (**c**= Current Ratio planned), is:

$$c= q+Vp/[360(l\{U+[M-F-Si][1-t]-Ad\}-Q)]$$
$$= 1.0100+9,746*240/[360(0.9900$$
$$\{64,720+[41,548-29,750$$
$$-51,294*0.0100][1-0.3000]$$
$$-7,899*0.3000\}-35,351)]$$
$$= 1.20 \text{ times}$$

Corporate IFRS-GAAP (B/S-I/S),ISBN-13: **978-1721066308**, ISBN-10: **1721066306**

Law-8079:

If both (**i**= I/S= Interest Portion= 1.00%), (**S**= Sales or Revenues= 51,294 millions USD), (**d**= D/A= Dividend Portion or Payout= 30.00%), (**A**= After Tax Income= 7,899 millions USD), (**p**= 360P/V= Procured Inventory Days= 240 Days), (**F**= Fixed Cost= 29,740 millions USD), (**l**= L/E= Leverage or Gearing Ratio= 99.00%), (**M**= Margin of Contribution= 41,548 millions USD), (**U**= Utilized or Starting Capital= 64,720 millions USD), (**t**= T/B= Tax Rate= 30.00%), (**Q**= Quoted Longterm Liabilities= 35,351 millions USD), (**c**= C/X= Current Ratio= 1.20 times), and (**V**= Variable Cost= 9,746 millions USD), then it's (**q**= Quick or Acid Test Ratio planned), is:

$$q= c\text{-}Vp/[360(\textbf{\textit{l}} \{U+[M\text{-}F\text{-}Si][1\text{-}t]\text{-}Ad\}\text{-}Q)]$$

$$= 1.2000\text{-}9,746*240/[360(0.9900$$
$$\{64,720+[41,548\text{-}29,750$$
$$\text{-}51,294*0.0100][1\text{-}0.3000]$$
$$\text{-}7,899*0.3000\}\text{-}35,351)]$$

$$= 1.01 \text{ times}$$

Corporate IFRS-GAAP (B/S-I/S),ISBN-13: **978-1721066308**, ISBN-10: **1721066306**

Law-8080:

If both (**i**= I/S= Interest Portion= 1.00%), (**S**= Sales or Revenues= 51,294 millions USD), (**d**= D/A= Dividend Portion or Payout= 30.00%), (**A**= After Tax Income= 7,899 millions USD), (**p**= 360 P/V= Procured Inventory Days= 240 Days), (**F**= Fixed Cost= 29,740 millions USD), (**I**= L/E= Leverage or Gearing Ratio= 99.00%), (**M**= Margin of Contribution= 41,548 millions USD), (**U**= Utilized or Starting Capital= 64,720 millions USD), (**t**= T/B= Tax Rate= 30.00%), (**q**= [C-P]/X= Quick or Acid Test Ratio= 1.01 times), (**c**= C/X= Current Ratio= 1.20 times), and (**v**= V/S= Variable Portion= 19.00%), then it's (**Q**= Quoted Longterm Liabilities planned), is:

$$Q= I\{U+[M\text{-}F\text{-}Si][1\text{-}t]\text{-}Ad\}\text{-}Svp/\{360[c\text{-}q]\}$$
$$= 0.9900\{64,720+[41,548\text{-}29,750$$
$$-51,294*0.0100][1\text{-}0.3000]$$
$$-7,899*0.3000\}\text{-}51,294$$
$$*0.1900*240/\{360[1.2000$$
$$-1.0100]\}$$
$$= 35,351 \text{ millions USD}$$

Corporate IFRS-GAAP (B/S-I/S),ISBN-13: **978-1721066308**, ISBN-10: **1721066306**

Law-8081:

If both (**i**= I/S= Interest Portion= 1.00%), (**S**= Sales or Revenues= 51,294 millions USD), (**d**= D/A= Dividend Portion or Payout= 30.00%), (**A**= After Tax Income= 7,899 millions USD), (**p**= 360 P/V= Procured Inventory Days= 240 Days), (**F**= Fixed Cost= 29,740 millions USD), (**Q**= Quoted Longterm Liabilities= 35,351 millions USD), (**M**= Margin of Contribution= 41,548 millions USD), (**U**= Utilized or Starting Capital= 64,720 millions USD), (**t**= T/B= Tax Rate= 30.00%), (**q**= [C-P]/X= Quick or Acid Test Ratio= 1.01 times), (**c**= C/X= Current Ratio= 1.20 times), and (**v**= V/S= Variable Portion= 19.00%), then it's (**I** = Leverage or Gearing Ratio planned), is:

$$I = (\mathbf{Q} + \mathbf{Svp}/\{360[\mathbf{c}\text{-}\mathbf{q}]\})$$
$$/\{\mathbf{U} + [\mathbf{M}\text{-}\mathbf{F}\text{-}\mathbf{Si}][1\text{-}\mathbf{t}]\text{-}\mathbf{Ad}\}$$
$$= (35,351 + 51,294*0.1900*240/\{360$$
$$[1.2000\text{-}1.0100]\})/\{64,720$$
$$+[41,548\text{-}29,750\text{-}51,294$$
$$*0.0100][1\text{-}0.3000]$$
$$-7,899*0.3000\}$$
$$= 99.00\%$$

Law-8082:

If both (**i**= I/S= Interest Portion= 1.00%), (**S**= Sales or
Revenues= 51,294 millions USD), (**d**= D/A=
Dividend Portion or Payout= 30.00%), (**A**= After Tax
Income= 7,899 millions USD), (**p**= 360 P/V=
Procured Inventory Days= 240 Days), (**F**= Fixed
Cost= 29,740 millions USD), (**Q**= Quoted Longterm
Liabilities= 35,351 millions USD), (**M**= Margin of
Contribution= 41,548 millions USD), (**I**= L/E=
Leverage or Gearing Ratio= 99.00%), (**t**= T/B= Tax
Rate= 30.00%), (**q**= [C-P]/X= Quick or Acid Test
Ratio= 1.01 times), (**c**= C/X= Current Ratio= 1.20
times), and (**v**= V/S= Variable Portion= 19.00%),
then it's (**U**= Utilized or Starting Capital planned), is:

$$U= Ad+(Q+Svp/\{360[c-q]\})/I-[M-F-Si][1-t]$$

$$
\begin{aligned}
&= 7{,}899*0.3000+(35{,}351+51{,}294 \\
&\quad *0.1900*240/\{360[1.2000 \\
&\quad -1.0100]\})/0.9900-[41{,}548 \\
&\quad -29{,}750-51{,}294*0.0100] \\
&\quad [1-0.3000]
\end{aligned}
$$

$$= 64{,}720 \text{ millions USD}$$

Corporate IFRS-GAAP (B/S-I/S),ISBN-13: **978-1721066308**, ISBN-10: **1721066306**

Law-8083:

If both (**i**= I/S= Interest Portion= 1.00%), (**$**= Sales or Revenues= 51,294 millions USD), (**d**= D/A= Dividend Portion or Payout= 30.00%), (**A**= After Tax Income= 7,899 millions USD), (**p**= 360 P/V= Procured Inventory Days= 240 Days), (**F**= Fixed Cost= 29,740 millions USD), (**Q**= Quoted Longterm Liabilities= 35,351 millions USD), (**U**= Utilized or Starting Capital= 64,720 millions USD), (**I**= L/E= Leverage or Gearing Ratio= 99.00%), (**t**= T/B= Tax Rate= 30.00%), (**q**= [C-P]/X= Quick or Acid Test Ratio= 1.01 times), (**c**= C/X= Current Ratio= 1.20 times), and (**v**= V/S= Variable Portion= 19.00%), then it's (**M**= Margin of Contribution planned), is:

$$M= F+\$i+[Ad+(Q+\$vp/\{360[c-q]\})/I-U]$$
$$/[1-t]$$
$$= 29,750+51,294*0.0100+[7,899$$
$$*0.3000+(35,351+51,294$$
$$*0.1900*240/\{360[1.2000$$
$$-1.0100]\})/0.9900-64,720]$$
$$/[1-0.3000]$$
$$= 41,548 \text{ millions USD}$$

Corporate IFRS-GAAP (B/S-I/S),ISBN-13: **978-1721066308**, ISBN-10: **1721066306**

Law-8084:

If both (**i**= I/S= Interest Portion= 1.00%), (**S**= Sales or Revenues= 51,294 millions USD), (**d**= D/A= Dividend Portion or Payout= 30.00%), (**A**= After Tax Income= 7,899 millions USD), (**p**= 360 P/V= Procured Inventory Days= 240 Days), (**M**= Margin of Contribution= 41,548 millions USD), (**Q**= Quoted Longterm Liabilities= 35,351 millions USD), (**U**= Utilized or Starting Capital= 64,720 millions USD), (**I**= L/E= Leverage or Gearing Ratio= 99.00%), (**t**= T/B= Tax Rate= 30.00%), (**q**= [C-P]/X= Quick or Acid Test Ratio= 1.01 times), (**c**= C/X= Current Ratio= 1.20 times), and (**v**= V/S= Variable Portion= 19.00%), then it's (**F**= Fixed Cost planned), is:

$$F = M\text{-}Si\text{-}[Ad+(Q+Svp/\{360[c\text{-}q]\})/I\text{-}U]$$
$$/[1\text{-}t]$$
$$= 41,548\text{-}51,294*0.0100\text{-}[7,899$$
$$*0,3000+(35,351+51,294$$
$$*0.1900*240/\{360[1.2000$$
$$-1.0100]\})/0.9900\text{-}64,720]$$
$$/[1\text{-}0.3000]$$
$$= 29,750 \text{ millions USD}$$

Corporate IFRS-GAAP (B/S-I/S),ISBN-13: **978-1721066308**, ISBN-10: **1721066306**

Law-8085:

If both (**i**= I/S= Interest Portion= 1.00%), (**F**= Fixed Cost= 29,740 millions USD), (**d**= D/A=Dividend Portion or Payout= 30.00%), (**A**= After Tax Income= 7,899 millions USD), (**p**= 360P/V= Procured Inventory Days= 240 Days), (**M**= Margin of Contribution= 41,548 millions USD), (**Q**= Quoted Longterm Liabilities= 35,351 millions USD), (**U**= Utilized or Starting Capital= 64,720 millions USD), (**⌐**= L/E= Leverage or Gearing Ratio= 99.00%), (**t**= T/B= Tax Rate= 30.00%), (**q**= [C-P]/X= Quick or Acid Test Ratio= 1.01 times), (**c**= C/X= Current Ratio= 1.20 times), and (**v**= V/S= Variable Portion= 19.00%), then it's (**$**= Sales or Revenues planned), is :

$$\textbf{S= \{M-F-[Ad+(Q+Svp}/\{360[\textbf{c-q}]\})/\textbf{⌐ -U]}$$
$$/[1\textbf{-t}]\}/\textbf{i}$$
$$= \{41,548-29,750-[7,899*0.3000$$
$$+(35,351+51,294*0.1900$$
$$*240/\{360[1.2000-1.0100]\})$$
$$/0.9900-64,720]$$
$$/[1-0.3000]\}/0.0100$$
$$= 51,294 \text{ millions USD}$$

Corporate IFRS-GAAP (B/S-I/S),ISBN-13: **978-1721066308**, ISBN-10: **1721066306**

Law-8086:

If both (**$**= Sales or Revenues= 51,294 millions
USD), (**F**= Fixed Cost= 29,740 millions USD), (**d**=
D/A= Dividend Portion or Payout= 30.00%), (**A**=
After Tax Income= 7,899 millions USD), (**p**= 360
P/V= Procured Inventory Days= 240 Days), (**M**=
Margin of Contribution= 41,548 millions USD), (**Q**=
Quoted Longterm Liabilities= 35,351 millions USD),
(**U**= Utilized or Starting Capital= 64,720 millions
USD), (**/**= L/E= Leverage or Gearing Ratio=
99.00%), (**t**= T/B= Tax Rate= 30.00%), (**q**= [C-
P]/X= Quick or Acid Test Ratio= 1.01 times), (**c**=
C/X= Current Ratio= 1.20 times), and (**v**= V/S=
Variable Portion= 19.00%), then it's (**i**= Interest
Portion planned), is:

$$i= \{M\text{-}F\text{-}[Ad+(Q+Svp/\{360[c\text{-}q]\})/I\text{-}U]$$
$$/[1\text{-}t]\}/\$$$
$$= \{41,548\text{-}29,750\text{-}[7,899*0.3000$$
$$+(35,351+51,294*0.1900$$
$$*240/\{360[1.2000\text{-}1.0100]\})$$
$$/0.9900\text{-}64,720]$$
$$/[1\text{-}0.3000]\}/51,294$$
$$= 1.00\%$$

Corporate IFRS-GAAP (B/S-I/S),ISBN-13: **978-1721066308**, ISBN-10: **1721066306**

Law-8087:

If both (**$**= Sales or Revenues= 51,294 millions
USD), (**F**= Fixed Cost= 29,740 millions USD), (**d**=
D/A= Dividend Portion or Payout= 30.00%), (**A**=
After Tax Income= 7,899 millions USD), (**p**= 360
P/V= Procured Inventory Days= 240 Days), (**M**=
Margin of Contribution= 41,548 millions USD), (**Q**=
Quoted Longterm Liabilities= 35,351 millions USD),
(**U**= Utilized or Starting Capital= 64,720 millions
USD), (**/**= L/E= Leverage or Gearing Ratio=
99.00%), (**i**= I/S= Interest Portion= 1.00%), (**q**= [C-
P]/X= Quick or Acid Test Ratio= 1.01 times), (**c**=
C/X= Current Ratio= 1.20 times), and (**v**= V/S=
Variable Portion= 19.00%), then it's (**t**= Tax Rate
planned), is:

$$t= 1-[\mathbf{Ad}+(\mathbf{Q}+\mathbf{Svp}/\{360[\mathbf{c}-\mathbf{q}]\})/\mathbf{/}-\mathbf{U}]$$
$$/[\mathbf{M}-\mathbf{F}-\mathbf{Si}]$$
$$= 1-[7,899*0.3000+(35,351+51,294$$
$$*0.1900*240/\{360[1.2000$$
$$-1.0100]\})/0.9900-64,720]$$
$$/[41,548-29,750$$
$$-51,294*0.0100]$$
$$= 30.00\%$$

Corporate IFRS-GAAP (B/S-I/S),ISBN-13: **978-1721066308**, ISBN-10: **1721066306**

Law-8088:

If both (S= Sales or Revenues= 51,294 millions USD), (F= Fixed Cost= 29,740 millions USD), (d= D/A= Dividend Portion or Payout= 30.00%), (t= T/B= Tax Rate= 30.00%), (p= 360 P/V= Procured Inventory Days= 240 Days), (M= Margin of Contribution= 41,548 millions USD), (Q= Quoted Longterm Liabilities= 35,351 millions USD), (U= Utilized or Starting Capital= 64,720 millions USD), (I= L/E= Leverage or Gearing Ratio= 99.00%), (i= I/S= Interest Portion= 1.00%), (q= [C-P]/X= Quick or Acid Test Ratio= 1.01 times), (c= C/X= Current Ratio= 1.20 times), and (v= V/S= Variable Portion= 19.00%), then it's (A= After Tax Income planned), is:

$$A= [\{U+[M-F-Si][1-t]\}$$
$$-(Q+Svp/\{360[c-q]\})/I]/d$$
$$= [\{64,720+[41,548-29,750-51,294$$
$$*0.0100][1-0.3000]\}-(35,351$$
$$+51,294*0.1900*240/\{360$$
$$[1.2000-1.0100]\})$$
$$/0.9900]/0.3000$$
$$= 7,899 \text{ millions USD}$$

Corporate IFRS-GAAP (B/S-I/S),ISBN-13: **978-1721066308**, ISBN-10: **1721066306**

Law-8089:

If both (**$**= Sales or Revenues= 51,294 millions USD), (**F**= Fixed Cost= 29,740 millions USD), (**A**= After Tax Income= 7,899 millions USD), (**t**= T/B= Tax Rate= 30.00%), (**p**= 360 P/V= Procured Inventory Days= 240 Days), (**M**= Margin of Contribution= 41,548 millions USD), (**Q**= Quoted Longterm Liabilities= 35,351 millions USD), (**U**= Utilized or Starting Capital= 64,720 millions USD), (**l**= L/E= Leverage or Gearing Ratio= 99.00%), (**i**= I/S= Interest Portion= 1.00%), (**q**= [C-P]/X= Quick or Acid Test Ratio= 1.01 times), (**c**= C/X= Current Ratio= 1.20 times), and (**v**= V/S= Variable Portion= 19.00%), then it's (**d**= Dividend Portion or Payout planned), is :

$$d= [\{U+[M-F-Si][1-t]\}$$
$$-(Q+Svp/\{360[c-q]\})/l]/A$$
$$= [\{64,720+[41,548-29,750-51,294$$
$$*0.0100][1-0.3000]\}$$
$$-(35,351+51,294*0.1900$$
$$*240/\{360[1.2000$$
$$-1.0100]\})/0.9900]/7,899$$
$$= 30.00\%$$

Corporate IFRS-GAAP (B/S-I/S),ISBN-13: **978-1721066308**, ISBN-10: **1721066306**

Law-8090:

If both (**$**= Sales or Revenues= 51,294 millions USD), (**F**= Fixed Cost= 29,740 millions USD), (**A**= After Tax Income= 7,899 millions USD), (**t**= T/B= Tax Rate= 30.00%), (**p**= 360 P/V= Procured Inventory Days= 240 Days), (**M**= Margin of Contribution= 41,548 millions USD), (**Q**= Quoted Longterm Liabilities= 35,351 millions USD), (**U**= Utilized or Starting Capital= 64,720 millions USD), (**⌿**= L/E= Leverage or Gearing Ratio= 99.00%), (**i**= I/S= Interest Portion= 1.00%), (**q**= [C-P]/X= Quick or Acid Test Ratio= 1.01 times), (**c**= C/X= Current Ratio= 1.20 times), and (**d**= D/A= Dividend Portion or Payout= 30.00%), then it's (**v**= Variable Portion planned), is:

$$v= [360[c\text{-}q](⌿\{U+[M\text{-}F\text{-}Si][1\text{-}t]\text{-}Ad\}\text{-}Q)]$$
$$/[\$p]$$

$$= [360[1.2000\text{-}1.0100](0.9900$$
$$\{64,720+[41,548\text{-}29,750$$
$$\text{-}51,294*0.0100][1\text{-}0.3000]$$
$$\text{-}7,899*0.3000\}\text{-}35,351)]$$
$$/[51,294*240]$$

$$= 19.00\%$$

Corporate IFRS-GAAP (B/S-I/S),ISBN-13: **978-1721066308**, ISBN-10: **1721066306**

Law-8091:

If both (**S**= Sales or Revenues= 51,294 millions USD), (**F**= Fixed Cost= 29,740 millions USD), (**A**= After Tax Income= 7,899 millions USD), (**t**= T/B= Tax Rate= 30.00%), (**v**= V/S= Variable Portion= 19.00%), (**M**= Margin of Contribution= 41,548 millions USD), (**Q**= Quoted Longterm Liabilities= 35,351 millions USD), (**U**= Utilized or Starting Capital= 64,720 millions USD), (**l**= L/E= Leverage or Gearing Ratio= 99.00%), (**i**= I/S= Interest Portion= 1.00%), (**q**= Quick or Acid Test Ratio= [C-P]/X= 1.01 times), (**c**= C/X= Current Ratio= 1.20 times), and (**d**= D/A= Dividend Portion or Payout= 30.00%), then it's (**p**= Procured Inventory Days planned), is

$$p= [360[c-q](l \{U+[M-F-Si][1-t]-Ad\}-Q)]$$
$$/[Sv]$$
$$= [360[1.2000-1.0100](0.9900$$
$$\{64,720+[41,548-29,750$$
$$-51,294*0.0100][1-0.3000]$$
$$-7,899*0.3000\}-35,351)]$$
$$/[51,294*0.1900]$$
$$= \underline{240} \text{ days}$$

Corporate IFRS-GAAP (B/S-I/S),ISBN-13: **978-1721066308**, ISBN-10: **1721066306**

Law-8092:

If both (**$**= Sales or Revenues= 51,294 millions
USD), (**F**= Fixed Cost= 29,740 millions USD), (**A**=
After Tax Income= 7,899 millions USD), (**t**= T/B=
Tax Rate= 30.00%), (**v**= V/S= Variable Portion=
19.00%), (**M**= Margin of Contribution= 41,548
millions USD), (**Q**= Quoted Longterm Liabilities=
35,351 millions USD), (**U**= Utilized or Starting
Capital= 64,720 millions USD), (**l**= L/E= Leverage
or Gearing Ratio= 99.00%), (**i**= I/S= Interest Portion=
1.00%), (**q**= Quick or Acid Test Ratio= [C-P]/X=
1.01 times), (**p**= 360 P/V= Procured Inventory Days=
240 Days), and (**d**= D/A= Dividend Portion or
Payout= 30.00%), then it's (**c**= Current Ratio
planned), is:

$$c= q+Svp/[360(l\{U+[M-F-Si][1-t]-Ad\}-Q)]$$
$$= 1.0100+51,294*0.1900*240/[360$$
$$(0.9900\{64,720+[41,548$$
$$-29,750-51,294*0.0100]$$
$$[1-0.3000]-7,899*0.3000\}$$
$$-35,351)]$$

$$= 1.20 \text{ times}$$

Corporate IFRS-GAAP (B/S-I/S),ISBN-13: **978-1721066308**, ISBN-10: **1721066306**

Law-8093:

If both (**$**= Sales or Revenues= 51,294 millions
USD), (**F**= Fixed Cost= 29,740 millions USD), (**A**=
After Tax Income= 7,899 millions USD), (**t**= T/B=
Tax Rate= 30.00%), (**v**= V/S= Variable Portion=
19.00%), (**M**= Margin of Contribution= 41,548
millions USD), (**Q**= Quoted Longterm Liabilities=
35,351 millions USD), (**U**= Utilized or Starting
Capital= 64,720 millions USD), (**l**= L/E= Leverage
or Gearing Ratio= 99.00%), (**i**= I/S= Interest Portion=
1.00%), (**c**= C/X= Current Ratio= 1.20 times), (**p**=
360 P/V= Procured Inventory Days= 240 Days), and
(**d**= D/A= Dividend Portion or Payout= 30.00%),
then it's (**q**= Quick or Acid Test Ratio planned), is:

$$q= c\text{-}Svp/[360(l\ \{U+[M\text{-}F\text{-}Si][1\text{-}t]\text{-}Ad\}\text{-}Q)]$$
$$= 1.2000\text{-}51{,}294*0.1900*240/[360$$
$$(0.9900\{64{,}720+[41{,}548$$
$$-29{,}750\text{-}51{,}294*0.0100]$$
$$[1\text{-}0.3000]\text{-}7{,}899*0.3000\}$$
$$-35{,}351)]$$
$$= \underline{1.01}\ times$$

Corporate IFRS-GAAP (B/S-I/S),ISBN-13: **978-1721066308**, ISBN-10: **1721066306**

Law-8094:

If both (**$**= Sales or Revenues= 51,294 millions
USD), (**$'**= Sales of Past Year= 48,851 millions
USD), (**s**= [S/S']-1= Sales Growth= 5.00%), (**F**=
Fixed Cost= 29,740 millions USD), (**A**= After Tax
Income= 7,899 millions USD), (**t**= T/B= Tax Rate=
30.00%), (**v**= V/S= Variable Portion= 19.00%), (**M**=
Margin of Contribution= 41,548 millions USD), (**q**=
[C-P]/X= Quick or Acid Test Ratio= 1.01 times), (**U**=
Utilized or Starting Capital= 64,720 millions USD),
(**I**= L/E= Leverage or Gearing Ratio= 99.00%), (**i**=
I/S= Interest Portion= 1.00%), (**c**= C/X= Current
Ratio= 1.20 times), (**p**= 360 P/V= Procured Inventory
Days= 240 Days), and (**d**= D/A= Dividend Portion or
Payout= 30.00%), then it's (**Q**= Quoted Longterm
Liabilities planned), is:

$$Q = I \{U+[M-F-Si][1-t]-Ad\}$$
$$-S'vp[1+s]/\{360[c-q]\}$$
$$= 0.9900\{64,720+[41,548-29,750$$
$$-51,294*0.0100][1-0.3000]$$
$$-7,899*0.3000\}-48,851$$
$$*0.1900*240[1+0.0500]$$
$$/\{360[1.2000-1.0100]\}$$
$$= 35,351 \text{ millions USD}$$

Corporate IFRS-GAAP (B/S-I/S),ISBN-13: **978-1721066308**, ISBN-10: **1721066306**

<u>Law-8095</u>:

If both ($=$ Sales or Revenues= <u>51,294</u> millions USD), ($'=$ Sales of Past Year= <u>48,851</u> millions USD), ($=$ [S/S']-1= Sales Growth= <u>5.00%</u>), (**F**= Fixed Cost= <u>29,740</u> millions USD), (**A**= After Tax Income= <u>7,899</u> millions USD), (**t**= T/B= Tax Rate= <u>30.00%</u>), (**v**= V/S= Variable Portion= <u>19.00%</u>), (**M**= Margin of Contribution= <u>41,548</u> millions USD), (**q**= [C-P]/X= Quick or Acid Test Ratio= <u>1.01</u> times), (**U**= Utilized or Starting Capital= <u>64,720</u> millions USD), (**Q**= Quoted Longterm Liabilities= <u>35,351</u> millions USD), (**i**= I/S= Interest Portion= <u>1.00%</u>), (**c**= C/X= Current Ratio= <u>1.20</u> times), (**p**= 360 P/V= Procured Inventory Days= <u>240</u> Days), and (**d**= D/A= Dividend Portion or Payout= <u>30.00%</u>), then it's (**I** = Leverage or Gearing Ratio planned), is:

$$I = (\mathbf{Q+\$'vp}[1+\mathbf{s}]/\{360[\mathbf{c\text{-}q}]\})$$
$$/\{\mathbf{U+[M\text{-}F\text{-}Si]}[1\text{-}\mathbf{t}]\text{-}\mathbf{Ad}\}$$
$$= (35,351+48,851*0.1900*240$$
$$[1+0.0500]/\{360[1.2000$$
$$-1.0100]\})/\{64,720+[41,548$$
$$-29,750-51,294*0.0100]$$
$$[1-0.3000]-7,899*0.3000\}$$
$$= \underline{99.00\%}$$

Corporate IFRS-GAAP (B/S-I/S),ISBN-13: **978-1721066308**, ISBN-10: **1721066306**

Law-8096:

If both (**$**= Sales or Revenues= 51,294 millions USD), (**$'**= Sales of Past Year= 48,851 millions USD), (**s**= [S/S']-1= Sales Growth= 5.00%), (**F**= Fixed Cost= 29,740 millions USD), (**A**= After Tax Income= 7,899 millions USD), (**t**= T/B= Tax Rate= 30.00%), (**v**= V/S= Variable Portion= 19.00%), (**M**= Margin of Contribution= 41,548 millions USD), (**q**= [C-P]/X= Quick or Acid Test Ratio= 1.01 times), (**l**= L/E= Leverage or Gearing Ratio= 99.00%), (**Q**= Quoted Longterm Liabilities= 35,351 millions USD), (**i**= I/S= Interest Portion= 1.00%), (**c**= C/X= Current Ratio= 1.20 times), (**p**= 360 P/V= Procured Inventory Days= 240 Days), and (**d**= D/A= Dividend Portion or Payout= 30.00%), then it's (**U**= Utilized or Starting Capital planned), is :

$$U= Ad+(Q+S'vp[1+s]/\{360[c-q]\})/l$$
$$-[M-F-Si][1-t]$$

$$= 7,899*0.3000+(35,351+48,851$$
$$*0.1900*240[1+0.0500]$$
$$/\{360[1.2000-1.0100]\})$$
$$/0.9900-[41,548-29,750$$
$$-51,294*0.0100][1-0.3000]$$
$$-7,899*0.3000\}$$

$$= \underline{64,720} \text{ millions USD}$$

Corporate IFRS-GAAP (B/S-I/S),ISBN-13: **978-1721066308**, ISBN-10: **1721066306**

Law-8097:

If both (**$**= Sales or Revenues= 51,294 millions USD), (**$'**= Sales of Past Year= 48,851 millions USD), (**s**= [S/S']-1= Sales Growth= 5.00%), (**F**= Fixed Cost= 29,740 millions USD), (**A**= After Tax Income= 7,899 millions USD), (**t**= T/B= Tax Rate= 30.00%), (**v**= V/S= Variable Portion= 19.00%), (**U**= Utilized or Starting Capital= 64,720 millions USD), (**q**= Quick or Acid Test Ratio= [C-P]/X= 1.01 times), (**l**= L/E= Leverage or Gearing Ratio= 99.00%), (**Q**= Quoted Longterm Liabilities= 35,351 millions USD), (**i**= I/S= Interest Portion= 1.00%), (**c**= C/X= Current Ratio= 1.20 times), (**p**= 360 P/V= Procured Inventory Days= 240 Days), and (**d**= D/A= Dividend Portion or Payout= 30.00%), then it's (**M**= Margin of Contribution planned), is:

$$M= F+Si+[Ad+(Q+S'vp[1+s]$$
$$/\{360[c-q]\})/l-U]/[1-t]$$
$$= 29,750+51,294*0.0100+[7,899$$
$$*0.3000 +(35,351+48,851$$
$$*0.1900*240[1+0.0500]$$
$$/\{360[1.2000-1.0100]\})$$
$$/0.9900-64,720]/[1-0.3000]$$
$$= 41,548 \text{ millions USD}$$

Corporate IFRS-GAAP (B/S-I/S),ISBN-13: **978-1721066308**, ISBN-10: **1721066306**

Law-8098:

If both ($= Sales or Revenues= 51,294 millions USD), ($'= Sales of Past Year= 48,851 millions USD), (s= [S/S']-1= Sales Growth= 5.00%), (M= Margin of Contribution= 41,548 millions USD), (A= After Tax Income= 7,899 millions USD), (t= T/B= Tax Rate= 30.00%), (v= V/S= Variable Portion= 19.00%), (U= Utilized or Starting Capital= 64,720 millions USD), (q= [C-P]/X= Quick or Acid Test Ratio= 1.01 times), (f= L/E= Leverage or Gearing Ratio= 99.00%), (Q= Quoted Longterm Liabilities= 35,351 millions USD), (i= I/S= Interest Portion= 1.00%), (c= C/X= Current Ratio= 1.20 times), (p= 360 P/V= Procured Inventory Days= 240 Days), and (d= D/A= Dividend Portion or Payout= 30.00%), then it's (F= Fixed Cost planned), is:

$$F= M\text{-}Si\text{-}[Ad+(Q+S'vp[1+s]}$$
$$/\{360[c\text{-}q]\})/f\text{-}U]/[1\text{-}t]$$
$$= 41,548\text{-}51,294*0.0100\text{-}[7,899$$
$$*0.3000+(35,351+48,851$$
$$*0.1900*240[1+0.0500]$$
$$/\{360[1.2000\text{-}1.0100]\})$$
$$/0.9900\text{-}64,720]/[1\text{-}0.3000]$$
$$= 29,750 \text{ millions USD}$$

Corporate IFRS-GAAP (B/S-I/S),ISBN-13: **978-1721066308**, ISBN-10: **1721066306**

Law-8099:

If both ($\mathbf{S'}$= Sales of Past Year= 48,851 millions USD), ($\mathbf{s}$= [S/S']-1= Sales Growth= 5.00%), ($\mathbf{M}$= Margin of Contribution= 41,548 millions USD), ($\mathbf{A}$= After Tax Income= 7,899 millions USD), ($\mathbf{t}$= T/B= Tax Rate= 30.00%), ($\mathbf{F}$= Fixed Cost= 29,740 millions USD), ($\mathbf{v}$= V/S= Variable Portion= 19.00%), ($\mathbf{U}$= Utilized or Starting Capital= 64,720 millions USD), ($\mathbf{q}$= [C-P]/X= Quick or Acid Test Ratio= 1.01 times), ($\mathbf{l}$= L/E= Leverage or Gearing Ratio= 99.00%), ($\mathbf{Q}$= Quoted Longterm Liabilities= 35,351 millions USD), ($\mathbf{i}$= I/S= Interest Portion= 1.00%), ($\mathbf{c}$= C/X= Current Ratio= 1.20 times), ($\mathbf{p}$= 360 P/V= Procured Inventory Days= 240 Days), and ($\mathbf{d}$= D/A= Dividend Portion or Payout= 30.00%), then it's ($\mathbf{S}$= Sales or Revenues planned), is:

$$\mathbf{S= \{M\text{-}F\text{-}[Ad+(Q+S'vp}[1+\mathbf{s}]}$$
$$/\{360[\mathbf{c\text{-}q}]\})/\mathbf{l\text{-}U]}/[1\text{-}\mathbf{t}]\}/\mathbf{i}$$

$$= \{41,548\text{-}29,750\text{-}[7,899*0.3000$$
$$+(35,351+48,851*0.1900$$
$$*240[1+0.0500]/\{360$$
$$[1.2000\text{-}1.0100]\})$$
$$/0.9900\text{-}64,720]$$
$$/[1\text{-}0.3000]\}/0.0100$$

$$= 51,294 \text{ millions USD}$$

Corporate IFRS-GAAP (B/S-I/S),ISBN-13: **978-1721066308**, ISBN-10: **1721066306**

Law-8100:

If both (**$'**= Sales of Past Year= 48,851 millions USD), (**s**= [S/S']-1= Sales Growth= 5.00%), (**M**= Margin of Contribution= 41,548 millions USD), (**A**= After Tax Income= 7,899 millions USD), (**t**= T/B= Tax Rate= 30.00%), (**F**= Fixed Cost= 29,740 millions USD), (**v**= V/S= Variable Portion= 19.00%), (**U**= Utilized or Starting Capital= 64,720 millions USD), (**q**= [C-P]/X= Quick or Acid Test Ratio= 1.01 times), (**l**= L/E= Leverage or Gearing Ratio= 99.00%), (**Q**= Quoted Longterm Liabilities= 35,351 millions USD), (**$**= Sales or Revenues= 51,294 millions USD), (**c**= C/X= Current Ratio= 1.20 times), (**p**= 360 P/V= Procured Inventory Days= 240 Days), and (**d**= D/A= Dividend Portion or Payout= 30.00%), then it's (**i**= Interest Portion planned), is:

$$i= \{M-F-[Ad+(Q+\$'vp[1+s]$$
$$/\{360[c-q]\})/I-U]/[1-t]\}/\$$$
$$= \{41,548-29,750-[7,899*0.3000$$
$$+(35,351+48,851*0.1900$$
$$*240[1+0.0500]/\{360$$
$$[1.2000-1.0100]\})$$
$$/0.9900-64,720]$$
$$/[1-0.3000]\}/51,294$$
$$= 1.00\%$$

Corporate IFRS-GAAP (B/S-I/S),ISBN-13: **978-1721066308**, ISBN-10: **1721066306**

Law-8101:

If both (**$'**= Sales of Past Year= <u>48,851</u> millions USD), (**s**= [S/S']-1= Sales Growth= <u>5.00%</u>), (**M**= Margin of Contribution= <u>41,548</u> millions USD), (**A**= After Tax Income= <u>7,899</u> millions USD), (**i**= I/S= Interest Portion= <u>1.00%</u>), (**F**= Fixed Cost= <u>29,740</u> millions USD), (**v**= V/S= Variable Portion= <u>19.00%</u>), (**U**= Utilized or Starting Capital= <u>64,720</u> millions USD), (**q**= [C-P]/X= Quick or Acid Test Ratio= <u>1.01</u> times), (**l**= L/E= Leverage or Gearing Ratio= <u>99.00%</u>), (**Q**= Quoted Longterm Liabilities= <u>35,351</u> millions USD), (**$**= Sales or Revenues= <u>51,294</u> millions USD), (**c**= C/X= Current Ratio= <u>1.20</u> times), (**p**= 360 P/V= Procured Inventory Days= <u>240</u> Days), and (**d**= D/A= Dividend Portion or Payout= <u>30.00%</u>), then it's (**t**= Tax Rate planned), is:

$$t= 1-[\mathbf{Ad}+(\mathbf{Q}+\mathbf{\$'vp}[1+\mathbf{s}]/\{360[\mathbf{c-q}]\})/\mathbf{l}-\mathbf{U}]$$
$$/[\mathbf{M-F-\$i}]$$

$$= 1-[7,899*0.3000+(35,351+48,851$$
$$*0.1900*240[1+0.0500]$$
$$/\{360[1.2000-1.0100]\})$$
$$/0.9900-64,720]/[41,548$$
$$-29,750-51,294*0.0100]$$

$$= \underline{30.00\%}$$

Corporate IFRS-GAAP (B/S-I/S),ISBN-13: **978-1721066308**, ISBN-10: **1721066306**

Law-8102:

If both (**S'**= Sales of Past Year= 48,851 millions USD), (**s**= [S/S']-1= Sales Growth= 5.00%), (**M**= Margin of Contribution= 41,548 millions USD), (**t**= T/B= Tax Rate= 30.00%), (**i**= I/S= Interest Portion= 1.00%), (**F**= Fixed Cost= 29,740 millions USD), (**v**= V/S= Variable Portion= 19.00%), (**U**= Utilized or Starting Capital= 64,720 millions USD), (**q**= Quick or Acid Test Ratio= [C-P]/X= 1.01 times), (**l**= L/E= Leverage or Gearing Ratio= 99.00%), (**Q**= Quoted Longterm Liabilities= 35,351 millions USD), (**S**= Sales or Revenues= 51,294 millions USD), (**c**= C/X= Current Ratio= 1.20 times), (**p**= 360 P/V= Procured Inventory Days= 240 Days), and (**d**= D/A= Dividend Portion or Payout= 30.00%), then it's (**A**= After Tax Income planned), is:

$$\mathbf{A} = [\mathbf{U} + [\mathbf{M} - \mathbf{F} - \mathbf{Si}][1 - \mathbf{t}] - (\mathbf{Q} + \mathbf{S'vp}[1 + \mathbf{s}]$$
$$/\{360[\mathbf{c} - \mathbf{q}]\})/\mathbf{l}]/\mathbf{d}$$

$$= [64,720 + [41,548 - 29,750 - 51,294$$
$$*0.0100][1 - 0.3000] - (35,351$$
$$+48,851*0.1900*240$$
$$[1 + 0.0500]/\{360[1.2000$$
$$-1.0100]\})/0.9900]/0.3000$$

$$= 7,899 \text{ millions USD}$$

Corporate IFRS-GAAP (B/S-I/S),ISBN-13: **978-1721066308**, ISBN-10: **1721066306**

Law-8103:

If both (**S'**= Sales of Past Year= 48,851 millions USD), (**s**= [S/S']-1= Sales Growth= 5.00%), (**M**= Margin of Contribution= 41,548 millions USD), (**t**= T/B= Tax Rate= 30.00%), (**i**= I/S= Interest Portion= 1.00%), (**F**= Fixed Cost= 29,740 millions USD), (**v**= V/S= Variable Portion= 19.00%), (**U**= Utilized or Starting Capital= 64,720 millions USD), (**q**= Quick or Acid Test Ratio= [C-P]/X= 1.01 times), (**l**= L/E= Leverage or Gearing Ratio= 99.00%), (**Q**= Quoted Longterm Liabilities= 35,351 millions USD), (**S**= Sales or Revenues= 51,294 millions USD), (**c**= C/X= Current Ratio= 1.20 times), (**p**= 360 P/V= Procured Inventory Days= 240 Days), and (**A**= After Tax Income= 7,899 millions USD), then it's (**d**= Dividend Portion or Payout planned), is:

$$d= [U+[M-F-Si][1-t]-(Q+S'vp[1+s]$$
$$/\{360[c-q]\})/l]/A$$
$$= [64,720+[41,548-29,750-51,294$$
$$*0.0100][1-0.3000]-(35,351$$
$$+48,851*0.1900*240$$
$$[1+0.0500]/\{360[1.2000$$
$$-1.0100]\})/0.9900]/0.3000$$
$$= 30.00\%$$

Corporate IFRS-GAAP (B/S-I/S),ISBN-13: **978-1721066308**, ISBN-10: **1721066306**

<u>Law-8104</u>:

If both (**d**= D/A= Dividend Portion or Payout= 30.00%), (**s**= [S/S']-1= Sales Growth= 5.00%), (**M**= Margin of Contribution= 41,548 millions USD), (**t**= T/B= Tax Rate= 30.00%), (**i**= I/S= Interest Portion= 1.00%), (**F**= Fixed Cost= 29,740 millions USD), (**v**= V/S= Variable Portion= 19.00%), (**U**= Utilized or Starting Capital= 64,720 millions USD), (**q**= Quick or Acid Test Ratio= [C-P]/X= 1.01 times), (**l**= L/E= Leverage or Gearing Ratio= 99.00%), (**Q**= Quoted Longterm Liabilities= 35,351 millions USD), (**S**= Sales or Revenues= 51,294 millions USD), (**c**= C/X= Current Ratio= 1.20 times), (**p**= 360 P/V= Procured Inventory Days= 240 Days), and (**A**= After Tax Income= 7,899 millions USD), then it's (**S'**= Sales Past), must be:

$$S'= [360[c-q](l\{U+[M-F-Si][1-t]-Ad\}-Q)]$$
$$/\{vp[1+s]\}$$

$$= [360[1.2000-1.0100](0.9900$$
$$\{64,720+[41,548-29,750$$
$$-51,294*0.0100][1-0.3000]$$
$$-7,899*0.3000\}-35,351)]$$
$$/\{0.1900*240[1+0.0500]\}$$

$$= \underline{48,851} \text{ millions USD}$$

Corporate IFRS-GAAP (B/S-I/S),ISBN-13: **978-1721066308**, ISBN-10: **1721066306**

Law-8105:

If both (**d**= D/A= Dividend Portion or Payout=
30.00%), (**s**= [S/S']-1= Sales Growth= 5.00%), (**M**=
Margin of Contribution= 41,548 millions USD), (**t**=
T/B= Tax Rate= 30.00%), (**i**= I/S= Interest Portion=
1.00%), (**F**= Fixed Cost= 29,740 millions USD), (**S'**=
Sales of Past Year= 48,851 millions USD), (**U**=
Utilized or Starting Capital= 64,720 millions USD),
(**q**= [C-P]/X= Quick or Acid Test Ratio= 1.01 times),
(**l**= L/E= Leverage or Gearing Ratio= 99.00%), (**Q**=
Quoted Longterm Liabilities= 35,351 millions USD),
(**S**= Sales or Revenues= 51,294 millions USD), (**c**=
C/X= Current Ratio= 1.20 times), (**p**= 360 P/V=
Procured Inventory Days= 240 Days), and (**A**= After
Tax Income= 7,899 millions USD), then it's (**v**=
Variable Portion planned), is:

$$v= [360[\textbf{c-q}](\textbf{\textit{l}} \{\textbf{U}+[\textbf{M-F-Si}][1\text{-}\textbf{t}]\text{-}\textbf{Ad}\}\text{-}\textbf{Q})]$$
$$/\{\textbf{S'p}[1+\textbf{s}]\}$$
$$= [360[1.2000\text{-}1.0100](0.9900$$
$$\{64,720+[41,548\text{-}29,750$$
$$-51,294*0.0100][1\text{-}0.3000]$$
$$-7,899*0.3000\}\text{-}35,351)]$$
$$/\{48,851*240[1+0.0500]\}$$
$$= \underline{19.00\%}$$

Corporate IFRS-GAAP (B/S-I/S),ISBN-13: **978-1721066308**, ISBN-10: **1721066306**

Law-8106:

If both (**d**= D/A= Dividend Portion or Payout= 30.00%), (**s**= [S/S']-1= Sales Growth= 5.00%), (**M**= Margin of Contribution= 41,548 millions USD), (**t**= T/B= Tax Rate= 30.00%), (**i**= I/S= Interest Portion= 1.00%), (**F**= Fixed Cost= 29,740 millions USD), (**S'**= Sales of Past Year= 48,851 millions USD), (**U**= Utilized or Starting Capital= 64,720 millions USD), (**q**= [C-P]/X= Quick or Acid Test Ratio= 1.01 times), (**l**= L/E= Leverage or Gearing Ratio= 99.00%), (**Q**= Quoted Longterm Liabilities= 35,351 millions USD), (**S**= Sales or Revenues= 51,294 millions USD), (**c**= C/X= Current Ratio= 1.20 times), (**v**= V/S= Variable Portion= 19.00%), and (**A**= After Tax Income= 7,899 millions USD), then it's (**p**= Procured Inventory Days planned), is:

$$p= [360[c-q](l\{U+[M-F-Si][1-t]-Ad\}-Q)]$$
$$/\{S'v[1+s]\}$$
$$= [360[1.2000-1.0100](0.9900$$
$$\{64,720+[41,548-29,750$$
$$-51,294*0.0100][1-0.3000]$$
$$-7,899*0.3000\}-35,351)]$$
$$/\{48,851*0.1900[1+0.0500]\}$$
$$= \underline{240} \text{ days}$$

Corporate IFRS-GAAP (B/S-I/S),ISBN-13: **978-1721066308**, ISBN-10: **1721066306**

Law-8107:

If both (d= D/A= Dividend Portion or Payout= 30.00%), (p= 360 P/V= Procured Inventory Days= 240 Days), (M= Margin of Contribution= 41,548 millions USD), (t= T/B= Tax Rate= 30.00%), (i= I/S= Interest Portion= 1.00%), (F= Fixed Cost= 29,740 millions USD), (S'= Sales of Past Year= 48,851 millions USD), (U= Utilized or Starting Capital= 64,720 millions USD), (q= [C-P]/X= Quick or Acid Test Ratio= 1.01 times), (l= L/E= Leverage or Gearing Ratio= 99.00%), (Q= Quoted Longterm Liabilities= 35,351 millions USD), (S= Sales or Revenues= 51,294 millions USD), (c= C/X= Current Ratio= 1.20 times), (v= V/S= Variable Portion= 19.00%), and (A= After Tax Income= 7,899 millions USD), then it's (s= Sales Growth planned), is:

$$s= [360[c\text{-}q](l\ \{U+[M\text{-}F\text{-}Si][1\text{-}t]\text{-}Ad\}\text{-}Q)]$$
$$/[S'vp]\text{-}1$$
$$= [360[1.2000\text{-}1.0100](0.9900$$
$$\{64,720+[41,548\text{-}29,750$$
$$\text{-}51,294*0.0100][1\text{-}0.3000]$$
$$\text{-}7,899*0.3000\}\text{-}35,351)]$$
$$/[48,851*0.1900*240]\text{-}1$$
$$= 5.00\%$$

Corporate IFRS-GAAP (B/S-I/S),ISBN-13: **978-1721066308**, ISBN-10: **1721066306**

Law-8108:

If both (**d**= D/A= Dividend Portion or Payout= 30.00%), (**p**= 360 P/V= Procured Inventory Days= 240 Days), (**M**= Margin of Contribution= 41,548 millions USD), (**t**= T/B= Tax Rate= 30.00%), (**i**= I/S= Interest Portion= 1.00%), (**F**= Fixed Cost= 29,740 millions USD), (**S'**= Sales of Past Year= 48,851 millions USD), (**U**= Utilized or Starting Capital= 64,720 millions USD), (**q**= [C-P]/X= Quick or Acid Test Ratio= 1.01 times), (**l**= L/E= Leverage or Gearing Ratio= 99.00%), (**Q**= Quoted Longterm Liabilities= 35,351 millions USD), (**S**= Sales or Revenues= 51,294 millions USD), (**s**= [S/S']-1= Sales Growth= 5.00%), (**v**= V/S= Variable Portion= 19.00%), and (**A**= After Tax Income= 7,899 millions USD), then it's (**c**= Current Ratio planned), is:

$$c= q+S'vp[1+s]/[360(\boldsymbol{l}\{U+[M-F-Si]$$
$$[1-t]-\boldsymbol{Ad}\}-\boldsymbol{Q})]$$
$$= 1.0100+48,851*0.1900*240$$
$$[1+0.0500]/[360(0.9900$$
$$\{64,720+[41,548-29,750$$
$$-51,294*0.0100][1-0.3000]$$
$$-7,899*0.3000\}-35,351)]$$
$$= \underline{1.20}\ \text{times}$$

Corporate IFRS-GAAP (B/S-I/S),ISBN-13: **978-1721066308**, ISBN-10: **1721066306**

Law-8109:

If both (**d**= D/A= Dividend Portion or Payout= 30.00%), (**p**= 360 P/V= Procured Inventory Days= 240 Days), (**M**= Margin of Contribution= 41,548 millions USD), (**t**= T/B= Tax Rate= 30.00%), (**i**= I/S= Interest Portion= 1.00%), (**F**= Fixed Cost= 29,740 millions USD),(**S'**= Sales of Past Year= 48,851 millions USD), (**U**= Utilized or Starting Capital= 64,720 millions USD), (**c**= C/X= Current Ratio= 1.20 times), (**Q**= Quoted Longterm Liabilities= 35,351 millions USD), (**S**= Sales or Revenues= 51,294 millions USD), (**s**= [S/S']-1= Sales Growth= 5.00%), (**v**= V/S= Variable Portion= 19.00%), and (**A**= After Tax Income= 7,899 millions USD), then it's (**q**= Quick or Acid Test Ratio planned), is:

$$q= c\text{-}S'vp[1+s]/[360(I\{U+[M\text{-}F\text{-}Si]$$
$$[1\text{-}t]\text{-}Ad\}\text{-}Q)]$$
$$= 1.2000\text{-}48{,}851*0.1900*240$$
$$[1+0.0500]/[360(0.9900$$
$$\{64{,}720+[41{,}548\text{-}29{,}750$$
$$\text{-}51{,}294*0.0100][1\text{-}0.3000]$$
$$\text{-}7{,}899*0.3000\}\text{-}35{,}351)]$$

$$= 1.01 \text{ times}$$

Corporate IFRS-GAAP (B/S-I/S),ISBN-13: **978-1721066308**, ISBN-10: **1721066306**

Law-8110:

If both (**d**= D/A= Dividend Portion or Payout= 30.00%), (**M**= Margin of Contribution= 41,548 millions USD), (**t**= T/B= Tax Rate= 30.00%), (**i**= I/S= Interest Portion= 1.00%), (**F**= Fixed Cost= 29,740 millions USD), (**U**= Utilized or Starting Capital= 64,720 millions USD), (**I**= L/E= Leverage or Gearing Ratio= 99.00%), (**S**= Sales or Revenues= 51,294 millions USD), and (**X**= Xpress or Current Debt= 34,196 millions USD), then it's (**Q**= Quoted Longterm Liabilities planned), is:

$$Q= I \{U+[M-F-Si][1-t][1-d]\}-X$$
$$= 0.9900\{64,720+[41,548-29,750$$
$$-51,294*0.0100][1-0.3000]$$
$$[1-0.3000]\}-34,196$$
$$= \underline{35,351} \text{ millions USD}$$

Law-8111:

If both (**d**= D/A= Dividend Portion or Payout= 30.00%), (**M**= Margin of Contribution= 41,548 millions USD), (**t**= T/B= Tax Rate= 30.00%), (**i**= I/S= Interest Portion= 1.00%), (**F**= Fixed Cost= 29,740 millions USD), (**U**= Utilized or Starting Capital= 64,720 millions USD), (**Q**= Quoted Longterm Liabilities= 35,351 millions USD), (**$**= Sales or Revenues= 51,294 millions USD), and (**X**= Xpress or Current Debt= 34,196 millions USD), then it's (**I** = Leverage or Gearing Ratio planned), is:

$$I = [Q+X]/\{U+[M-F-\$i][1-t][1-d]\}$$
$$= [35,351+34,196]/\{64,720+[41,548$$
$$-29,750-51,294*0.0100]$$
$$[1-0.3000][1-0.3000]\}$$
$$= 99.00\%$$

Corporate IFRS-GAAP (B/S-I/S),ISBN-13: **978-1721066308**, ISBN-10: **1721066306**

Law-8112:

If both (**d**= D/A= Dividend Portion or Payout= 30.00%), (**M**= Margin of Contribution= 41,548 millions USD), (**t**= T/B= Tax Rate= 30.00%), (**i**= I/S= Interest Portion= 1.00%), (**F**= Fixed Cost= 29,740 millions USD), (**U**= Utilized or Starting Capital= 64,720 millions USD), (**Q**= Quoted Longterm Liabilities= 35,351 millions USD), (**S**= Sales or Revenues= 51,294 millions USD), and (**X**= Xpress or Current Debt= 34,196 millions USD), then it's (**U**= Utilized or Starting Capital planned), is:

$$\mathbf{U}= [\mathbf{Q}+\mathbf{X}]/\mathbf{I} -[\mathbf{M}-\mathbf{F}-\mathbf{Si}][1-\mathbf{t}][1-\mathbf{d}]$$
$$= [35,351+34,196]/0.9900-[41,548$$
$$-29,750-51,294*0.0100]$$
$$[1-0.3000][1-0.3000]$$
$$= \underline{64,720} \text{ millions USD}$$

Corporate IFRS-GAAP (B/S-I/S),ISBN-13: **978-1721066308**, ISBN-10: **1721066306**

Law-8113:

If both (**d**= D/A= Dividend Portion or Payout= 30.00%), (**U**= Utilized or Starting Capital= 64,720 millions USD), (**t**= T/B= Tax Rate= 30.00%), (**i**= I/S= Interest Portion= 1.00%), (**F**= Fixed Cost= 29,740 millions USD), (**l**= L/E= Leverage or Gearing Ratio= 99.00%), (**Q**= Quoted Longterm Liabilities= 35,351 millions USD), (**$**= Sales or Revenues= 51,294 millions USD), and (**X**= Xpress or Current Debt= 34,196 millions USD), then it's (**M**= Margin of Contribution planned), is:

$$M= F+\$i+\{[Q+X]/l-U\}/\{[1-t][1-d]\}$$
$$= 29,750+51,294*0.0100+\{[35,351$$
$$+34,196]/0.9900-64,720\}$$
$$/\{[1-0.3000][1-0.3000]\}$$
$$= 41,548 \text{ millions USD}$$

Corporate IFRS-GAAP (B/S-I/S),ISBN-13: **978-1721066308**, ISBN-10: **1721066306**

Law-8114:

If both (**d**= D/A= Dividend Portion or Payout= 30.00%), (**U**= Utilized or Starting Capital= 64,720 millions USD), (**t**= T/B= Tax Rate= 30.00%), (**i**= I/S= Interest Portion= 1.00%), (**F**= Fixed Cost= 29,740 millions USD), (**I**= L/E= Leverage or Gearing Ratio= 99.00%), (**Q**= Quoted Longterm Liabilities= 35,351 millions USD), (**S**= Sales or Revenues= 51,294 millions USD), and (**X**= Xpress or Current Debt= 34,196 millions USD), then it's (**F**= Fixed Cost planned), is:

$$F= M\text{-}Si\text{-}\{[Q+X]/I\text{-}U\}/\{[1\text{-}t][1\text{-}d]\}$$
$$= 41,548\text{-}51,294*0.0100\text{-}\{[35,351$$
$$+34,196]/0.9900\text{-}64,720\}$$
$$/\{[1\text{-}0.3000][1\text{-}0.3000]\}$$
$$= \underline{29,750} \text{ millions USD}$$

Corporate IFRS-GAAP (B/S-I/S),ISBN-13: **978-1721066308**, ISBN-10: **1721066306**

Law-8115:

If both (**d**= D/A= Dividend Portion or Payout= 30.00%), (**U**= Utilized or Starting Capital= 64,720 millions USD), (**M**= Margin of Contribution= 41,548 millions USD), (**t**= T/B= Tax Rate= 30.00%), (**i**= I/S= Interest Portion= 1.00%), (**F**= Fixed Cost= 29,740 millions USD), (**l**= L/E= Leverage or Gearing Ratio= 99.00%), (**Q**= Quoted Longterm Liabilities= 35,351 millions USD), and (**X**= Xpress or Current Debt= 34,196 millions USD), then it's (**S**= Sales or Revenues planned), is:

$$S= (M-F-\{[Q+X]/l-U\}/\{[1-t][1-d]\})/i$$
$$= (41,548-29,750-\{[35,351+34,196]$$
$$/0.9900-64,720\}/\{[1-0.3000]$$
$$[1-0.3000]\})/0.0100$$
$$= \underline{51,294} \text{ millions USD}$$

Corporate IFRS-GAAP (B/S-I/S),ISBN-13: **978-1721066308**, ISBN-10: **1721066306**

Law-8116:

If both (**d**= D/A= Dividend Portion or Payout= 30.00%), (**U**= Utilized or Starting Capital= 64,720 millions USD), (**M**= Margin of Contribution= 41,548 millions USD), (**t**= T/B= Tax Rate= 30.00%), (**$**= Sales or Revenues= 51,294 millions USD), (**F**= Fixed Cost= 29,740 millions USD), (**f**= L/E= Leverage or Gearing Ratio= 99.00%), (**Q**= Quoted Longterm Liabilities= 35,351 millions USD), and (**X**= Xpress or Current Debt= 34,196 millions USD), then it's (**i**= Interest Portion planned), is:

$$i= (M-F-\{[Q+X]/f-U\}/\{[1-t][1-d]\})/\$$$
$$= (41,548-29,750-\{[35,351+34,196]$$
$$/0.9900-64,720\}/\{[1-0.3000]$$
$$[1-0.3000]\})/51,294$$
$$= \underline{1.00\%}$$

Law-8117:

If both (**d**= D/A= Dividend Portion or Payout= 30.00%), (**U**= Utilized or Starting Capital= 64,720 millions USD), (**M**= Margin of Contribution= 41,548 millions USD), (**i**= I/S= Interest Portion= 1.00%), (**S**= Sales or Revenues= 51,294 millions USD), (**F**= Fixed Cost= 29,740 millions USD), (**I**= L/E= Leverage or Gearing Ratio= 99.00%), (**Q**= Quoted Longterm Liabilities= 35,351 millions USD), and (**X**= Xpress or Current Debt= 34,196 millions USD), then it's (**t**= Tax Rate planned), is:

$$t= 1-\{[\mathbf{Q}+\mathbf{X}]/\mathbf{I}-\mathbf{U}\}/\{[1-\mathbf{d}][\mathbf{M}-\mathbf{F}-\mathbf{Si}]\}$$
$$= 1-\{[35,351+34,196]/0.9900-64,720\}$$
$$/\{[1-0.3000][41,548-29,750$$
$$-51,294*0.0100]\}$$
$$= 30.00\%$$

Corporate IFRS-GAAP (B/S-I/S),ISBN-13: **978-1721066308**, ISBN-10: **1721066306**

Law-8118:

If both (**t**= T/B= Tax Rate= 30.00%), (**U**= Utilized or Starting Capital= 64,720 millions USD), (**M**= Margin of Contribution= 41,548 millions USD), (**i**= I/S= Interest Portion= 1.00%), (**$**= Sales or Revenues= 51,294 millions USD), (**F**= Fixed Cost= 29,740 millions USD), (**I**= L/E= Leverage or Gearing Ratio= 99.00%), (**Q**= Quoted Longterm Liabilities= 35,351 millions USD), and (**X**= Xpress or Current Debt= 34,196 millions USD), then it's (**d**= Dividend Portion or Payout planned), is:

$$d= 1-\{[Q+X]/I-U\}/\{[1-t][M-F-\$i]\}$$
$$= 1-\{[35,351+34,196]/0.9900-64,720\}$$
$$/\{[1-0.3000][41,548-29,750$$
$$-51,294*0.0100]\}$$
$$= 30.00\%$$

Corporate IFRS-GAAP (B/S-I/S),ISBN-13: **978-1721066308**, ISBN-10: **1721066306**

<u>Law-8119</u>:

If both (**t**= T/B= Tax Rate= <u>30.00%</u>), (**U**= Utilized or
Starting Capital= <u>64,720</u> millions USD), (**M**= Margin
of Contribution= <u>41,548</u> millions USD), (**i**= I/S=
Interest Portion= <u>1.00%</u>), (**S**= Sales or Revenues=
<u>51,294</u> millions USD), (**F**= Fixed Cost= <u>29,740</u>
millions USD), (**l**= L/E= Leverage or Gearing Ratio=
<u>99.00%</u>), (**Q**= Quoted Longterm Liabilities= <u>35,351</u>
millions USD), and (**d**= D/A= Dividend Portion or
Payout= <u>30.00%</u>), then it's (**X**= Current Debt
planned), is:

$$X = l\{U+[M-F-Si][1-t][1-d]\}-Q$$
$$= 0.9900\{64,720+[41,548-29,750$$
$$-51,294*0.0100][1-0.3000]$$
$$[1-0.3000]\}-35,351$$
$$= \underline{34,196} \text{ millions USD}$$

Corporate IFRS-GAAP (B/S-I/S),ISBN-13: **978-1721066308**, ISBN-10: **1721066306**

<u>Law-8120</u>:

If both (**t**= T/B= Tax Rate= <u>30.00%</u>), (**U**= Utilized or Starting Capital= <u>64,720</u> millions USD), (**M**= Margin of Contribution= <u>41,548</u> millions USD), (**i**= I/S= Interest Portion= <u>1.00%</u>), (**S**= Sales or Revenues= <u>51,294</u> millions USD), (**F**= Fixed Cost= <u>29,740</u> millions USD), (**l**= L/E= Leverage or Gearing Ratio= <u>99.00%</u>), (**q**= [C-P]/X= Quick or Acid Test Ratio= <u>1.01</u> times), (**P**= Procured Inventories= <u>6,497</u> millions USD), (**c**= C/X= Current Ratio= <u>1.20</u> times), and (**d**= D/A= Dividend Portion or Payout= <u>30.00%</u>), then it's (**Q**= Quoted Longterm Liabilities planned), is:

$$Q = \boldsymbol{l}\{U+[M-F-Si][1-t][1-d]\}-P/[c-q]$$
$$= 0.9900\{64,720+[41,548-29,750$$
$$-51,294*0.0100][1-0.3000]$$
$$[1-0.3000]\}-6,497/[1.2000$$
$$-1.0100]$$
$$= \underline{35,351} \text{ millions USD}$$

Corporate IFRS-GAAP (B/S-I/S),ISBN-13: **978-1721066308**, ISBN-10: **1721066306**

Law-8121:

If both (**t**= T/B= Tax Rate= 30.00%), (**U**= Utilized or Starting Capital= 64,720 millions USD), (**M**= Margin of Contribution= 41,548 millions USD), (**i**= I/S= Interest Portion= 1.00%), (**$**= Sales or Revenues= 51,294 millions USD), (**F**= Fixed Cost= 29,740 millions USD), (**Q**= Quoted Longterm Liabilities= 35,351 millions USD), (**q**= [C-P]/X= Quick or Acid Test Ratio= 1.01 times), (**P**= Procured Inventories= 6,497 millions USD), (**c**= C/X= Current Ratio= 1.20 times), and (**d**= D/A= Dividend Portion or Payout= 30.00%), then it's (**l**= L/E= Leverage or Gearing Ratio planned), is:

$$l = \{Q+P/[c-q]\}/\{U+[M-F-\$i][1-t][1-d]\}$$
$$= \{35,351+6,497/[1.2000-1.0100]\}$$
$$/\{64,720+[41,548-29,750$$
$$-51,294*0.0100][1-0.3000]$$
$$[1-0.3000]\}$$
$$= 99.00\%$$

Corporate IFRS-GAAP (B/S-I/S),ISBN-13: **978-1721066308**, ISBN-10: **1721066306**

Law-8122:

If both (**t**= T/B= Tax Rate= 30.00%), (**l**= L/E= Leverage or Gearing Ratio= 99.00%), (**M**= Margin of Contribution= 41,548 millions USD), (**i**= I/S= Interest Portion= 1.00%), (**S**= Sales or Revenues= 51,294 millions USD), (**F**= Fixed Cost= 29,740 millions USD), (**Q**= Quoted Longterm Liabilities= 35,351 millions USD), (**q**= [C-P]/X= Quick or Acid Test Ratio= 1.01 times), (**P**= Procured Inventories= 6,497 millions USD), (**c**= C/X= Current Ratio= 1.20 times), and (**d**= D/A= Dividend Portion or Payout= 30.00%), then it's (**U**= Utilized or Starting Capital planned), is:

$$U= \{Q+P/[c-q]\}/l -[M-F-Si][1-t][1-d]$$
$$= \{35,351+6,497/[1.2000-1.0100]\}$$
$$/0.9900-[41,548-29,750$$
$$-51,294*0.0100][1-0.3000]$$
$$[1\ 0.3000]$$
$$= \underline{64,720} \text{ millions USD}$$

Corporate IFRS-GAAP (B/S-I/S),ISBN-13: **978-1721066308**, ISBN-10: **1721066306**

Law-8123:

If both (**t**= T/B= Tax Rate= 30.00%), (**l**= L/E= Leverage or Gearing Ratio= 99.00%), (**U**= Utilized or Starting Capital= 64,720 millions USD), (**i**= I/S= Interest Portion= 1.00%), (**S**= Sales or Revenues= 51,294 millions USD), (**F**= Fixed Cost= 29,740 millions USD), (**Q**= Quoted Longterm Liabilities= 35,351 millions USD), (**q**= [C-P]/X= Quick or Acid Test Ratio= 1.01 times), (**P**= Procured Inventories= 6,497 millions USD), (**c**= C/X= Current Ratio= 1.20 times), and (**d**= D/A= Dividend Portion or Payout= 30.00%), then it's (**M**= Margin of Contribution planned), is:

$$M= F+Si+(\{Q+P/[c-q]\}/l)/\{[1-t][1-d]\}$$
$$= 29,750+51,294*0.0100+(\{35,351 \\ +6,497/[1.2000-1.0100]\} \\ /0.9900)/\{[1-0.3000] \\ [1-0.3000]\}$$
$$= \underline{41,548} \text{ millions USD}$$

Corporate IFRS-GAAP (B/S-I/S),ISBN-13: **978-1721066308**, ISBN-10: **1721066306**

Law-8124:

If both (**t**= T/B= Tax Rate= 30.00%), (**l**= L/E= Leverage or Gearing Ratio= 99.00%), (**U**= Utilized or Starting Capital= 64,720 millions USD), (**i**= I/S= Interest Portion= 1.00%), (**S**= Sales or Revenues= 51,294 millions USD), (**M**= Margin of Contribution= 41,548 millions USD), (**Q**= Quoted Longterm Liabilities= 35,351 millions USD), (**q**= [C-P]/X= Quick or Acid Test Ratio= 1.01 times), (**P**= Procured Inventories= 6,497 millions USD), (**c**= C/X= Current Ratio= 1.20 times), and (**d**= D/A= Dividend Portion or Payout= 30.00%), then it's (**F**= Fixed Cost planned), is:

$$F= M-Si-(\{Q+P/[c-q]\}/l)/\{[1-t][1-d]\}$$

$$= 41,548-51,294*0.0100-(\{35,351$$
$$+6,497/[1.2000-1.0100]\}$$
$$/0.9900)/\{[1-0.3000]$$
$$[1-0.3000]\}$$

$$= 29,750 \text{ millions USD}$$

Corporate IFRS-GAAP (B/S-I/S),ISBN-13: **978-1721066308**, ISBN-10: **1721066306**

Law-8125:

If both (**t**= T/B= Tax Rate= 30.00%), (**I**= L/E= Leverage or Gearing Ratio= 99.00%), (**U**= Utilized or Starting Capital= 64,720 millions USD), (**i**= I/S= Interest Portion= 1.00%), (**F**= Fixed Cost= 29,740 millions USD), (**M**= Margin of Contribution= 41,548 millions USD), (**Q**= Quoted Longterm Liabilities= 35,351 millions USD), (**q**= [C-P]/X= Quick or Acid Test Ratio= 1.01 times), (**P**= Procured Inventories= 6,497 millions USD), (**c**= C/X= Current Ratio= 1.20 times), and (**d**= D/A= Dividend Portion or Payout= 30.00%), then it's (**$**= Sales or Revenues planned), is:

$$\$= [\text{M-F-}(\{\text{Q+P}/[\text{c-q}]\}/\textbf{I})/\{[1-\textbf{t}][1-\textbf{d}]\}]/\textbf{i}$$
$$= [41,548-29,750-(\{35,351+6,497$$
$$/[1.2000-1.0100]\}/0.9900)$$
$$/\{[1-0.3000][1-0.3000]\}]$$
$$/0.0100$$
$$= \underline{51,294} \text{ millions USD}$$

Corporate IFRS-GAAP (B/S-I/S),ISBN-13: **978-1721066308**, ISBN-10: **1721066306**

Law-8126:

If both (**t**= T/B= Tax Rate= <u>30.00%</u>), (**l**= L/E= Leverage or Gearing Ratio= <u>99.00%</u>), (**U**= Utilized or Starting Capital= <u>64,720</u> millions USD), (**S**= Sales or Revenues= <u>51,294</u> millions USD), (**F**= Fixed Cost= <u>29,740</u> millions USD), (**M**= Margin of Contribution= <u>41,548</u> millions USD), (**Q**= Quoted Longterm Liabilities= <u>35,351</u> millions USD), (**q**= [C-P]/X= Quick or Acid Test Ratio= <u>1.01</u> times), (**P**= Procured Inventories= <u>6,497</u> millions USD), (**c**= C/X= Current Ratio= <u>1.20</u> times), and (**d**= D/A= Dividend Portion or Payout= <u>30.00%</u>), then it's (**i**= Interest Portion planned), is:

$$i= [M-F-(\{Q+P/[c-q]\}/l)/\{[1-t][1-d]\}]/S$$

$$= [41,548-29,750-(\{35,351+6,497$$
$$/[1.2000-1.0100]\}/0.9900)$$
$$/\{[1-0.3000][1-0.3000]\}]$$
$$/51,294$$

$$= \underline{1.00\%}$$

Corporate IFRS-GAAP (B/S-I/S),ISBN-13: **978-1721066308**, ISBN-10: **1721066306**

Law-8127:

If both (**i**= I/S= Interest Portion= 1.00%), (**f**= L/E= Leverage or Gearing Ratio= 99.00%), (**U**= Utilized or Starting Capital= 64,720 millions USD), (**S**= Sales or Revenues= 51,294 millions USD), (**F**= Fixed Cost= 29,740 millions USD), (**M**= Margin of Contribution= 41,548 millions USD), (**Q**= Quoted Longterm Liabilities= 35,351 millions USD), (**q**= [C-P]/X= Quick or Acid Test Ratio= 1.01 times), (**P**= Procured Inventories= 6,497 millions USD), (**c**= C/X= Current Ratio= 1.20 times), and (**d**= D/A= Dividend Portion or Payout= 30.00%), then it's (**t**= Tax Rate planned), is:

$$\mathbf{t}= 1-(\{\mathbf{Q}+\mathbf{P}/[\mathbf{c}-\mathbf{q}]\}/\mathbf{f})/\{[1-\mathbf{d}][\mathbf{M}-\mathbf{F}-\mathbf{Si}]\}$$
$$= 1- (\{35,351+6,497/[1.2000-1.0100]\}$$
$$/0.9900)/\{[1-0.3000][41,548$$
$$-29,750-51,294*0.0100]\}$$
$$= 30.00\%$$

Corporate IFRS-GAAP (B/S-I/S),ISBN-13: **978-1721066308**, ISBN-10: **1721066306**

Law-8128:

If both (**i**= I/S= Interest Portion= 1.00%), (**I**= L/E= Leverage or Gearing Ratio= 99.00%), (**U**= Utilized or Starting Capital= 64,720 millions USD), (**S**= Sales or Revenues= 51,294 millions USD), (**F**= Fixed Cost= 29,740 millions USD), (**M**= Margin of Contribution= 41,548 millions USD), (**Q**= Quoted Longterm Liabilities= 35,351 millions USD), (**q**= [C-P]/X= Quick or Acid Test Ratio= 1.01 times), (**P**= Procured Inventories= 6,497 millions USD), (**c**= C/X= Current Ratio= 1.20 times), and (**t**= T/B= Tax Rate= 30.00%),   then it's (**d**= Dividend Portion or Payout planned), is:

$$\mathbf{d}= 1-(\{\mathbf{Q}+\mathbf{P}/[\mathbf{c}-\mathbf{q}]\}/\mathbf{I})/\{[1-\mathbf{t}][\mathbf{M}-\mathbf{F}-\mathbf{Si}]\}$$
$$= 1- (\{35,351+6,497/[1.2000-1.0100]\}$$
$$/0.9900)/\{[1-0.3000][41,548$$
$$-29,750-51,294*0.0100]\}$$
$$= 30.00\%$$

Corporate IFRS-GAAP (B/S-I/S),ISBN-13: **978-1721066308**, ISBN-10: **1721066306**

Law-8129:

If both (**i**= I/S= Interest Portion= 1.00%), (**f**= L/E= Leverage or Gearing Ratio= 99.00%), (**U**= Utilized or Starting Capital= 64,720 millions USD), (**S**= Sales or Revenues= 51,294 millions USD), (**F**= Fixed Cost= 29,740 millions USD), (**M**= Margin of Contribution= 41,548 millions USD), (**Q**= Quoted Longterm Liabilities= 35,351 millions USD), (**q**= [C-P]/X= Quick or Acid Test Ratio= 1.01 times), (**d**= D/A= Dividend Portion or Payout= 30.00%), (**c**= C/X= Current Ratio= 1.20 times), and (**t**= T/B= Tax Rate= 30.00%), then it's (**P**= Procured Inventories Days planned), is:

$$P= [c-q](I\{U+[M-F-Si][1-t][1-d]\}-Q)$$
$$= [1.2000-1.0100](0.9900\{64,720$$
$$+[41,548-29,750-51,294$$
$$*0.0100][1-0.3000]$$
$$[1-0.3000]\}-35,351)$$
$$= \underline{6,497} \text{ millions USD}$$

Corporate IFRS-GAAP (B/S-I/S),ISBN-13: **978-1721066308**, ISBN-10: **1721066306**

Law-8130:

If both (**i**= I/S= Interest Portion= 1.00%), (**l**= L/E= Leverage or Gearing Ratio= 99.00%), (**U**= Utilized or Starting Capital= 64,720 millions USD), (**S**= Sales or Revenues= 51,294 millions USD), (**F**= Fixed Cost= 29,740 millions USD), (**M**= Margin of Contribution= 41,548 millions USD), (**Q**= Quoted Longterm Liabilities= 35,351 millions USD), (**q**= [C-P]/X= Quick or Acid Test Ratio= 1.01 times), (**d**= D/A= Dividend Portion or Payout= 30.00%), (**P**= Procured Inventories= 6,497 millions USD), and (**t**= T/B= Tax Rate= 30.00%), then it's (**c**= Current Ratio planned), is:

$$c= q+P/(l \{U+[M-F-Si][1-t][1-d]\}-Q)$$
$$= 1.0100+6,497/(0.9900\{64,720$$
$$+[41,548-29,750-51,294$$
$$*0.0100][1-0.3000]$$
$$[1-0.3000]\}-35,351)$$
$$= 1.20 \text{ times}$$

Corporate IFRS-GAAP (B/S-I/S),ISBN-13: **978-1721066308**, ISBN-10: **1721066306**

Law-8131:

If both (**i**= I/S= Interest Portion= 1.00%), (**I**= L/E= Leverage or Gearing Ratio= 99.00%), (**U**= Utilized or Starting Capital= 64,720 millions USD), (**S**= Sales or Revenues= 51,294 millions USD), (**F**= Fixed Cost= 29,740 millions USD), (**M**= Margin of Contribution= 41,548 millions USD), (**Q**= Quoted Longterm Liabilities= 35,351 millions USD), (**c**= C/X= Current Ratio= 1.20 times), (**d**= D/A= Dividend Portion or Payout= 30.00%), (**P**= Procured Inventories= 6,497 millions USD), and (**t**= T/B= Tax Rate= 30.00%), then it's (**q**= Quick or Acid Test Ratio planned), is:

$$q= c\text{-}P/(I \{U+[M\text{-}F\text{-}Si][1\text{-}t][1\text{-}d]\}\text{-}Q)$$
$$= 1.2000\text{-}6,497/(0.9900\{64,720$$
$$+[41,548\text{-}29,750\text{-}51,294$$
$$*0.0100][1\text{-}0.3000]$$
$$[1\text{-}0.3000]\}\text{-}35,351)$$
$$= 1.01 \text{ times}$$

Corporate IFRS-GAAP (B/S-I/S),ISBN-13: **978-1721066308**, ISBN-10: **1721066306**

Law-8132:

If both (**i**= I/S= Interest Portion= <u>1.00%</u>), (**I**= L/E= Leverage or Gearing Ratio= <u>99.00%</u>), (**U**= Utilized or Starting Capital= <u>64,720</u> millions USD), (**S**= Sales or Revenues= <u>51,294</u> millions USD), (**F**= Fixed Cost= <u>29,740</u> millions USD), (**M**= Margin of Contribution= <u>41,548</u> millions USD), (**q**= [C-P]/X= Quick or Acid Test Ratio= <u>1.01</u> times), (**c**= C/X= Current Ratio= <u>1.20</u> times), (**d**= D/A= Dividend Portion or Payout= <u>30.00%</u>), (**p**= 360 P/V= Procured Inventory Days= <u>240</u> Days), (**V**= Variable Cost= <u>9,746</u> millions USD), and (**t**= T/B= Tax Rate= <u>30.00%</u>), then it's (**Q**= Quoted Longterm Liabilities planned), is:

$$Q= I\{U+[M\text{-}F\text{-}Si][1\text{-}t][1\text{-}d]\}\text{-}Vp/\{360[c\text{-}q]\}$$
$$= 0.9900\{64,720+[41,548\text{-}29,750$$
$$-51,294*0.0100][1\text{-}0.3000]$$
$$[1\text{-}0.3000]\}\text{-}9,746*240$$
$$/\{360[1.2000\text{-}1.0100]\}$$
$$= \underline{35,351}\ \text{millions USD}$$

Corporate IFRS-GAAP (B/S-I/S),ISBN-13: **978-1721066308**, ISBN-10: **1721066306**

Law-8133:

If both (**i**= I/S= Interest Portion= 1.00%), (**Q**= Quoted Longterm Liabilities= 35,351 millions USD), (**U**= Utilized or Starting Capital= 64,720 millions USD), (**$**= Sales or Revenues= 51,294 millions USD), (**F**= Fixed Cost= 29,740 millions USD), (**M**= Margin of Contribution= 41,548 millions USD), (**q**= [C-P]/X= Quick or Acid Test Ratio= 1.01 times), (**c**= C/X= Current Ratio= 1.20 times), (**d**= D/A= Dividend Portion or Payout= 30.00%), (**p**= 360 P/V= Procured Inventory Days= 240 Days), (**V**= Variable Cost= 9,746 millions USD), and (**t**= T/B= Tax Rate= 30.00%), then it's (**l** = Leverage or Gearing Ratio planned), is:

$$l = (\mathbf{Q}+\mathbf{Vp}/\{360[\mathbf{c}\text{-}\mathbf{q}]\})$$
$$/\{\mathbf{U}+[\mathbf{M}\text{-}\mathbf{F}\text{-}\mathbf{Si}][1\text{-}\mathbf{t}][1\text{-}\mathbf{d}]\}$$
$$= (35,351+9,746*240/\{360[1.2000$$
$$-1.0100]\})/\{64,720+[41,548$$
$$-29,750-51,294*0.0100]$$
$$[1\text{-}0.3000][1\text{-}0.3000]\}$$
$$= \underline{99.00\%}$$

Corporate IFRS-GAAP (B/S-I/S),ISBN-13: **978-1721066308**, ISBN-10: **1721066306**

Law-8134:

If both (**i**= I/S= Interest Portion= 1.00%), (**l**= L/E= Leverage or Gearing Ratio= 99.00%), (**Q**= Quoted Longterm Liabilities= 35,351 millions USD), (**S**= Sales or Revenues= 51,294 millions USD), (**F**= Fixed Cost= 29,740 millions USD), (**M**= Margin of Contribution= 41,548 millions USD), (**q**= [C-P]/X= Quick or Acid Test Ratio= 1.01 times), (**c**= C/X= Current Ratio= 1.20 times), (**d**= D/A= Dividend Portion or Payout= 30.00%), (**p**= 360 P/V= Procured Inventory Days= 240 Days), (**V**= Variable Cost= 9,746 millions USD), and (**t**= T/B= Tax Rate= 30.00%), then it's (**U**= Utilized or Starting Capital planned), is:

$$\mathbf{U} = (\mathbf{Q}+\mathbf{Vp}/\{360[\mathbf{c}\text{-}\mathbf{q}]\})/\mathbf{l}\text{-}[\mathbf{M}\text{-}\mathbf{F}\text{-}\mathbf{Si}][1\text{-}\mathbf{t}][1\text{-}\mathbf{d}]$$

$$= (35,351+9,746*240/\{360[1.2000$$
$$-1.0100]\})/0.9900-[41,548$$
$$-29,750-51,294*0.0100]$$
$$[1-0.3000][1-0.3000]$$
$$= \underline{64,720} \text{ millions USD}$$

Corporate IFRS-GAAP (B/S-I/S),ISBN-13: **978-1721066308**, ISBN-10: **1721066306**

Law-8135:

If both (**i**= I/S= Interest Portion= 1.00%), (**l**= L/E= Leverage or Gearing Ratio= 99.00%), (**Q**= Quoted Longterm Liabilities= 35,351 millions USD), (**S**= Sales or Revenues= 51,294 millions USD), (**F**= Fixed Cost= 29,740 millions USD), (**U**= Utilized or Starting Capital= 64,720 millions USD), (**q**= [C-P]/X= Quick or Acid Test Ratio= 1.01 times), (**c**= C/X= Current Ratio= 1.20 times), (**d**= D/A= Dividend Portion or Payout= 30.00%), (**p**= 360 P/V= Procured Inventory Days= 240 Days), (**V**= Variable Cost= 9,746 millions USD), and (**t**= T/B= Tax Rate= 30.00%), then it's (**M**= Margin of Contribution planned), is:

$$M= F+Si+[(Q+Vp/\{360[c-q]\})/l-U]$$
$$/\{[1-t][1-d]\}$$
$$= 29,750+51,294*0.0100+[(35,351$$
$$+9,746*240/\{360[1.2000$$
$$-1.0100]\})/0.9900-64,720]$$
$$/\{[1-0.3000][1-0.3000]\}$$
$$= 41,548 \text{ millions USD}$$

Corporate IFRS-GAAP (B/S-I/S),ISBN-13: **978-1721066308**, ISBN-10: **1721066306**

Law-8136:

If both (**i**= I/S= Interest Portion= 1.00%), (**I**= L/E= Leverage or Gearing Ratio= 99.00%), (**Q**= Quoted Longterm Liabilities= 35,351 millions USD), (**S**= Sales or Revenues= 51,294 millions USD), (**M**= Margin of Contribution= 41,548 millions USD), (**U**= Utilized or Starting Capital= 64,720 millions USD), (**q**= [C-P]/X= Quick or Acid Test Ratio= 1.01 times), (**c**= C/X= Current Ratio= 1.20 times), (**d**= D/A= Dividend Portion or Payout= 30.00%), (**p**= 360 P/V= Procured Inventory Days= 240 Days), (**V**= Variable Cost= 9,746 millions USD), and (**t**= T/B= Tax Rate= 30.00%), then it's (**F**= Fixed Cost planned), is:

$$F= M\text{-}Si\text{-}[(Q+Vp/\{360[c\text{-}q]\})/I\text{-}U]$$
$$/\{[1\text{-}t][1\text{-}d]\}$$
$$= 41,548\text{-}51,294*0.0100\text{-}[(35,351$$
$$+9,746*240/\{360[1.2000$$
$$-1.0100]\})/0.9900\text{-}64,720]$$
$$/\{[1\text{-}0.3000][1\text{-}0.3000]\}$$
$$= 29,750 \text{ millions USD}$$

Corporate IFRS-GAAP (B/S-I/S),ISBN-13: **978-1721066308**, ISBN-10: **1721066306**

Law-8137:

If both (i= I/S= Interest Portion= 1.00%), (l= L/E= Leverage or Gearing Ratio= 99.00%), (Q= Quoted Longterm Liabilities= 35,351 millions USD), (F= Fixed Cost= 29,740 millions USD), (M= Margin of Contribution= 41,548 millions USD), (U= Utilized or Starting Capital= 64,720 millions USD), (q= Quick or Acid Test Ratio= [C-P]/X= 1.01 times), (c= C/X= Current Ratio= 1.20 times), (d= D/A= Dividend Portion or Payout= 30.00%), (p= 360 P/V= Procured Inventory Days= 240 Days), (V= Variable Cost= 9,746 millions USD), and (t= T/B= Tax Rate= 30.00%), then it's (S= Sales or Revenues planned), is:

$$S= \{M-F-[(Q+Vp/\{360[c-q]\})/l-U] /\{[1-t][1-d]\}\}/i$$

$$= \{41,548-29,750-[(35,351+9,746 \\ *240/\{360[1.2000-1.0100]\}) \\ /0.9900-64,720]/\{[1-0.3000] \\ [1-0.3000]\}\}/0.0100$$

$$= 51,294 \text{ millions USD}$$

Corporate IFRS-GAAP (B/S-I/S),ISBN-13: **978-1721066308**, ISBN-10: **1721066306**

Law-8138:

If both (**$**= Sales or Revenues= 51,294 millions USD), (**/**= L/E= Leverage or Gearing Ratio= 99.00%), (**Q**= Quoted Longterm Liabilities= 35,351 millions USD), (**F**= Fixed Cost= 29,740 millions USD), (**M**= Margin of Contribution= 41,548 millions USD), (**U**= Utilized or Starting Capital= 64,720 millions USD), (**q**= Quick or Acid Test Ratio= [C-P]/X= 1.01 times), (**c**= C/X= Current Ratio= 1.20 times), (**d**= D/A= Dividend Portion or Payout= 30.00%), (**p**= 360 P/V= Procured Inventory Days= 240 Days), (**V**= Variable Cost= 9,746 millions USD), and (**t**= T/B= Tax Rate= 30.00%), then it's (**i**= Interest Portion planned), is:

$$i= \{M\text{-}F\text{-}[(Q+Vp/\{360[c\text{-}q]\})//\text{-}U]$$
$$/\{[1\text{-}t][1\text{-}d]\}\}/\$$$
$$= \{41,548\text{-}29,750\text{-}[(35,351+9,746$$
$$*240/\{360[1.2000\text{-}1.0100]\})$$
$$/0.9900\text{-}64,720]/\{[1\text{-}0.3000]$$
$$[1\text{-}0.3000]\}\}/51,294$$
$$= 1.00\%$$

Corporate IFRS-GAAP (B/S-I/S),ISBN-13: **978-1721066308**, ISBN-10: **1721066306**

Law-8139:

If both (**$**= Sales or Revenues= 51,294 millions USD), (**/**= L/E= Leverage or Gearing Ratio= 99.00%), (**Q**= Quoted Longterm Liabilities= 35,351 millions USD), (**F**= Fixed Cost= 29,740 millions USD), (**M**= Margin of Contribution= 41,548 millions USD), (**U**= Utilized or Starting Capital= 64,720 millions USD), (**q**= Quick or Acid Test Ratio= [C-P]/X= 1.01 times), (**c**= C/X= Current Ratio= 1.20 times), (**d**= D/A= Dividend Portion or Payout= 30.00%), (**p**= 360 P/V= Procured Inventory Days= 240 Days), (**V**= Variable Cost= 9,746 millions USD), and (**i**= I/S= Interest Portion= 1.00%), then it's (**t**= Tax Rate planned), is:

$$t= 1-[(Q+Vp/\{360[c-q]\})/I-U]$$
$$/\{[1-d][M-F-Si]\}$$
$$= 1-[(35,351+9,746*240/\{360[1.2000$$
$$-1.0100]\})/0.9900-64,720]$$
$$/\{[1-0.3000][41,548-29,750$$
$$-51,294*0.0100]\}$$
$$= 30.00\%$$

Law-8140:

If both (**$**= Sales or Revenues= 51,294 millions
USD), (**I**= L/E= Leverage or Gearing Ratio=
99.00%), (**Q**= Quoted Longterm Liabilities= 35,351
millions USD), (**F**= Fixed Cost= 29,740 millions
USD), (**M**= Margin of Contribution= 41,548 millions
USD), (**U**= Utilized or Starting Capital= 64,720
millions USD), (**q**= [C-P]/X= Quick or Acid Test
Ratio= 1.01 times), (**c**= C/X= Current Ratio= 1.20
times), (**t**= T/B= Tax Rate= 30.00%), (**p**= 360 P/V=
Procured Inventory Days= 240 Days), (**V**= Variable
Cost= 9,746 millions USD), and (**i**= I/S= Interest
Portion= 1.00%), then it's (**d**= Dividend Portion or
Payout planned), is:

$$d= 1-[(Q+Vp/\{360[c-q]\})/I-U]$$
$$/\{[1-t][M-F-Si]\}$$
$$= 1-[(35,351+9,746*240/\{360[1.2000$$
$$-1.0100]\})/0.9900-64,720]$$
$$/\{[1-0.3000][41,548-29,750$$
$$-51,294*0.0100]\}$$
$$= 30.00\%$$

Corporate IFRS-GAAP (B/S-I/S),ISBN-13: **978-1721066308**, ISBN-10: **1721066306**

<u>Law-8141</u>:

If both (**$**= Sales or Revenues= <u>51,294</u> millions USD), (**/**= L/E= Leverage or Gearing Ratio= <u>99.00%</u>), (**Q**= Quoted Longterm Liabilities= <u>35,351</u> millions USD), (**F**= Fixed Cost= <u>29,740</u> millions USD), (**M**= Margin of Contribution= <u>41,548</u> millions USD), (**U**= Utilized or Starting Capital= <u>64,720</u> millions USD), (**q**= Quick or Acid Test Ratio= [C-P]/X= <u>1.01</u> times), (**c**= C/X= Current Ratio= <u>1.20</u> times), (**t**= T/B= Tax Rate= <u>30.00%</u>), (**p**= 360 P/V= Procured Inventory Days= <u>240</u> Days), (**d**= D/A= Dividend Portion or Payout= <u>30.00%</u>), and (**i**= I/S= Interest Portion= <u>1.00%</u>), then it's (**V**= Variable Cost planned), is:

$$\mathbf{V}= [360[\mathbf{c}\text{-}\mathbf{q}](\mathbf{/} \{\mathbf{U}+\{[\mathbf{M}\text{-}\mathbf{F}\text{-}\mathbf{Si}]$$
$$[1\text{-}\mathbf{t}][1\text{-}\mathbf{d}]\}\text{-}\mathbf{Q})]/\mathbf{p}$$
$$= [360[1.2000\text{-}1.0100](0.9900\{64,720$$
$$+[41,548\text{-}29,750\text{-}51,294$$
$$*0.0100][1\text{-}0.3000]$$
$$[1\text{-}0.3000]\}\text{-}35,351)]/240$$
$$= \underline{9,745} \text{ millions USD}$$

Corporate IFRS-GAAP (B/S-I/S),ISBN-13: **978-1721066308**, ISBN-10: **1721066306**

Law-8142:

   If both (**$**= Sales or Revenues= 51,294 millions
USD), (**I**= L/E= Leverage or Gearing Ratio=
99.00%), (**Q**= Quoted Longterm Liabilities= 35,351
millions USD), (**F**= Fixed Cost= 29,740 millions
USD), (**M**= Margin of Contribution= 41,548 millions
USD), (**U**= Utilized or Starting Capital= 64,720
millions USD), (**q**= Quick or Acid Test Ratio= [C-
P]/X= 1.01 times), (**c**= C/X= Current Ratio= 1.20
times), (**t**= T/B= Tax Rate= 30.00%), (**V**= Variable
Cost= 9,746 millions USD), (**d**= D/A= Dividend
Portion or Payout= 30.00%), and (**i**= I/S= Interest
Portion= 1.00%), then it's (**p**= Procured Inventory
Days planned), is:

$$p= [360[c\text{-}q](I\,\{U+\{[M\text{-}F\text{-}Si]$$
$$[1\text{-}t][1\text{-}d]\}\text{-}Q)]/V$$
$$= [360[1.2000\text{-}1.0100](0.9900$$
$$\{64,720+[41,548\text{-}29,750$$
$$\text{-}51,294*0.0100][1\text{-}0.3000]$$
$$[1\text{-}0.3000]\}\text{-}35,351)]/9,745$$

$$= \underline{240}\ \text{days}$$

Corporate IFRS-GAAP (B/S-I/S),ISBN-13: **978-1721066308**, ISBN-10: **1721066306**

<u>Law-8143</u>:

If both (**\$**= Sales or Revenues= <u>51,294</u> millions USD), (**/**= L/E= Leverage or Gearing Ratio= <u>99.00%</u>), (**Q**= Quoted Longterm Liabilities= <u>35,351</u> millions USD), (**F**= Fixed Cost= <u>29,740</u> millions USD), (**M**= Margin of Contribution= <u>41,548</u> millions USD), (**U**= Utilized or Starting Capital= <u>64,720</u> millions USD), (**q**= Quick or Acid Test Ratio= [C-P]/X= <u>1.01</u> times), (**p**= 360 P/V= Procured Inventory Days= <u>240</u> Days), (**t**= T/B= Tax Rate= <u>30.00%</u>), (**V**= Variable Cost= <u>9,746</u> millions USD), (**d**= D/A= Dividend Portion or Payout= <u>30.00%</u>), and (**i**= I/S= Interest Portion= <u>1.00%</u>), then it's (**c**= Current Ratio planned), is:

$$c= q+Vp/[360(I \{U+\{[M-F-\$i] [1-t][1-d]\}-Q)]$$

$$= 1.0100+9,746*240/[360(0.9900 \{64,720+[41,548-29,750 -51,294*0.0100][1-0.3000] [1-0.3000]\}-35,351)]$$

$$= \underline{1.20} \text{ times}$$

Corporate IFRS-GAAP (B/S-I/S),ISBN-13: **978-1721066308**, ISBN-10: **1721066306**

Law-8144:

If both (**$**= Sales or Revenues= 51,294 millions
USD), (**I**= L/E= Leverage or Gearing Ratio=
99.00%), (**Q**= Quoted Longterm Liabilities= 35,351
millions USD), (**F**= Fixed Cost= 29,740 millions
USD), (**M**= Margin of Contribution= 41,548 millions
USD), (**U**= Utilized or Starting Capital= 64,720
millions USD), (**c**= C/X= Current Ratio= 1.20 times),
(**p**= 360 P/V= Procured Inventory Days= 240 Days),
(**t**= T/B= Tax Rate= 30.00%), (**V**= Variable Cost=
9,746 millions USD), (**d**= D/A= Dividend Portion or
Payout= 30.00%), and (**i**= I/S= Interest Portion=
1.00%), then it's (**q**= Quick or Acid Test Ratio
planned), is:

$$q= c\text{-}Vp/[360(I \{U+\{[M\text{-}F\text{-}Si]$$
$$[1\text{-}t][1\text{-}d]\}\text{-}Q)]$$
$$= 1.2000\text{-}9,746*240/[360(0.9900$$
$$\{64,720+[41,548\text{-}29,750$$
$$-51,294*0.0100][1\text{-}0.3000]$$
$$[1\text{-}0.3000]\}\text{-}35,351)]$$
$$= \underline{1.01} \text{ times}$$

Corporate IFRS-GAAP (B/S-I/S),ISBN-13: **978-1721066308**, ISBN-10: **1721066306**

<u>Law-8145</u>:

If both (**$**= Sales or Revenues= <u>51,294</u> millions USD), (**I**= L/E= Leverage or Gearing Ratio= <u>99.00%</u>), (**q**= [C-P]/X= Quick or Acid Test Ratio= <u>1.01</u> times), (**F**= Fixed Cost= <u>29,740</u> millions USD), (**M**= Margin of Contribution= <u>41,548</u> millions USD), (**U**= Utilized or Starting Capital= <u>64,720</u> millions USD), (**c**= C/X= Current Ratio= <u>1.20</u> times), (**p**= 360 P/V= Procured Inventory Days= <u>240</u> Days), (**t**= T/B= Tax Rate= <u>30.00%</u>), (**v**= V/S= Variable Portion= <u>19.00%</u>), (**d**= D/A= Dividend Portion or Payout= <u>30.00%</u>), and (**i**= I/S= Interest Portion= <u>1.00%</u>), then it's (**Q**= Quoted Longterm Liabilities planned), is:

$$Q= I\{U+\{[M\text{-}F\text{-}Si][1\text{-}t][1\text{-}d]\}$$
$$-Svp/\{360[c\text{-}q]\}$$
$$= 0.9900\{64{,}720+[41{,}548\text{-}29{,}750$$
$$-51{,}294*0.0100][1\text{-}0.3000]$$
$$[1\text{-}0.3000]\}\text{-}51{,}294*0.1900$$
$$*240/\{360[1.2000\text{-}1.0100]\}$$
$$= \underline{35{,}351} \text{ millions USD}$$

Corporate IFRS-GAAP (B/S-I/S),ISBN-13: **978-1721066308**, ISBN-10: **1721066306**

Law-8146:

If both (**$**= Sales or Revenues= 51,294 millions USD), (**Q**= Quoted Longterm Liabilities= 35,351 millions USD), (**q**= [C-P]/X= Quick or Acid Test Ratio= 1.01 times), (**F**= Fixed Cost= 29,740 millions USD), (**M**= Margin of Contribution= 41,548 millions USD), (**U**= Utilized or Starting Capital= 64,720 millions USD), (**c**= C/X= Current Ratio= 1.20 times), (**p**= 360 P/V= Procured Inventory Days= 240 Days), (**t**= T/B= Tax Rate= 30.00%), (**v**= V/S= Variable Portion= 19.00%), (**d**= D/A= Dividend Portion or Payout= 30.00%), and (**i**= I/S= Interest Portion= 1.00%), then it's (**I** = Leverage or Gearing Ratio planned), is:

$$I = (Q+Svp/\{360[c-q]\})/\{U+\{[M-F-Si][1-t][1-d]\}$$
$$= (35,351+51,294*0.1900*240$$
$$/\{360[1.2000-1.0100]\})$$
$$/\{64,720+[41,548-29,750$$
$$-51,294*0.0100][1-0.3000]$$
$$[1-0.3000]\}$$
$$= \underline{99.00\%}$$

Corporate IFRS-GAAP (B/S-I/S),ISBN-13: **978-1721066308**, ISBN-10: **1721066306**

<u>Law-8147</u>:

If both (**$**= Sales or Revenues= <u>51,294</u> millions USD), (**Q**= Quoted Longterm Liabilities= <u>35,351</u> millions USD), (**q**= [C-P]/X= Quick or Acid Test Ratio= <u>1.01</u> times), (**F**= Fixed Cost= <u>29,740</u> millions USD), (**M**= Margin of Contribution= <u>41,548</u> millions USD), (**I**= L/E= Leverage or Gearing Ratio= <u>99.00%</u>), (**c**= C/X= Current Ratio= <u>1.20</u> times), (**p**= 360 P/V= Procured Inventory Days= <u>240</u> Days), (**t**= T/B= Tax Rate= <u>30.00%</u>), (**v**= V/S= Variable Portion= <u>19.00%</u>), (**d**= D/A= Dividend Portion or Payout= <u>30.00%</u>), and (**i**= I/S= Interest Portion= <u>1.00%</u>), then it's (**U**= Utilized or Starting Capital planned), is:

$$U= (Q+\$vp/\{360[c-q]\})/I$$
$$-[M-F-\$i][1-t][1-d]$$
$$= (35,351+51,294*0.1900*240$$
$$/\{360[1.2000-1.0100]\})$$
$$/0.9900-[41,548-29,750$$
$$-51,294*0.0100][1-0.3000]$$
$$[1-0.3000]$$
$$= \underline{64,720} \text{ millions USD}$$

Corporate IFRS-GAAP (B/S-I/S),ISBN-13: **978-1721066308**, ISBN-10: **1721066306**

Law-8148:

If both ($= Sales or Revenues= 51,294 millions USD), (**Q**= Quoted Longterm Liabilities= 35,351 millions USD), (**q**= [C-P]/X= Quick or Acid Test Ratio= 1.01 times), (**F**= Fixed Cost= 29,740 millions USD), (**U**= Utilized or Starting Capital= 64,720 millions USD), (**I**= L/E= Leverage or Gearing Ratio= 99.00%), (**c**= C/X= Current Ratio= 1.20 times), (**p**= 360 P/V= Procured Inventory Days= 240 Days), (**t**= T/B= Tax Rate= 30.00%), (**v**= V/S= Variable Portion= 19.00%), (**d**= D/A= Dividend Portion or Payout= 30.00%), and (**i**= I/S= Interest Portion= 1.00%), then it's (**M**= Margin of Contribution planned), is:

$$M= F+\$i+[(Q+\$vp/\{360[c-q]\})/I-U]$$
$$/\{[1-t][1-d]\}$$

$$= 29,750+51,294*0.0100 +[(35,351$$
$$+51,294*0.1900*240/\{360$$
$$[1.2000-1.0100]\})/0.9900$$
$$-64,720]/\{[1-0.3000]$$
$$[1-0.3000]\}$$

$$= 41,548 \text{ millions USD}$$

Corporate IFRS-GAAP (B/S-I/S),ISBN-13: **978-1721066308**, ISBN-10: **1721066306**

Law-8149:

If both (**S**= Sales or Revenues= 51,294 millions USD), (**Q**= Quoted Longterm Liabilities= 35,351 millions USD), (**q**= [C-P]/X= Quick or Acid Test Ratio= 1.01 times), (**M**= Margin of Contribution= 41,548 millions USD), (**U**= Utilized or Starting Capital= 64,720 millions USD), (**I**= L/E= Leverage or Gearing Ratio= 99.00%), (**c**= C/X= Current Ratio= 1.20 times), (**p**= 360 P/V= Procured Inventory Days= 240 Days), (**t**= T/B= Tax Rate= 30.00%), (**v**= V/S= Variable Portion= 19.00%), (**d**= D/A= Dividend Portion or Payout= 30.00%), and (**i**= I/S= Interest Portion= 1.00%), then it's (**F**= Fixed Cost planned), is:

$$F= M\text{-}Si\text{-}[(Q+Svp/\{360[c\text{-}q]\})/I\text{-}U]$$
$$/\{[1\text{-}t][1\text{-}d]\}$$
$$= 41,548\text{-}51,294*0.0100\text{-}[(35,351$$
$$+51,294*0.1900*240/\{360$$
$$[1.2000\text{-}1.0100]\})/0.9900$$
$$-64,720]/\{[1\text{-}0.3000]$$
$$[1\text{-}0.3000]\}$$
$$= \underline{29,750} \text{ millions USD}$$

Corporate IFRS-GAAP (B/S-I/S),ISBN-13: **978-1721066308**, ISBN-10: **1721066306**

Law-8150:

If both (**F**= Fixed Cost= 29,740 millions USD), (**Q**= Quoted Longterm Liabilities= 35,351 millions USD), (**q**= [C-P]/X= Quick or Acid Test Ratio= 1.01 times), (**M**= Margin of Contribution= 41,548 millions USD), (**U**= Utilized or Starting Capital= 64,720 millions USD), (**I**= L/E= Leverage or Gearing Ratio= 99.00%), (**c**= C/X= Current Ratio= 1.20 times), (**p**= 360 P/V= Procured Inventory Days= 240 Days), (**t**= T/B= Tax Rate= 30.00%), (**v**= V/S= Variable Portion= 19.00%), (**d**= D/A= Dividend Portion or Payout= 30.00%), and (**i**= I/S= Interest Portion= 1.00%), then it's (**S**= Sales or Revenues planned), is:

$$S= \{M\text{-}F\text{-}[(Q+Svp/\{360[c\text{-}q]\})/I\text{-}U]$$
$$/\{[1\text{-}t][1\text{-}d]\}\}/i$$
$$= \{41,548\text{-}29,750\text{-}[(35,351+51,294$$
$${}^{*}0.1900{}^{*}240/\{360[1.2000$$
$$\text{-}1.0100]\})/0.9900\text{-}64,720]$$
$$/\{[1\text{-}0.3000][1\text{-}0.3000]\}\}$$
$$/0.0100$$
$$= 51,294 \text{ millions USD}$$

Corporate IFRS-GAAP (B/S-I/S),ISBN-13: **978-1721066308**, ISBN-10: **1721066306**

Law-8151:

If both (**F**= Fixed Cost= <u>29,740</u> millions USD), (**Q**= Quoted Longterm Liabilities= <u>35,351</u> millions USD), (**q**= [C-P]/X= Quick or Acid Test Ratio= <u>1.01</u> times), (**M**= Margin of Contribution= <u>41,548</u> millions USD), (**U**= Utilized or Starting Capital= <u>64,720</u> millions USD), (**l**= L/E= Leverage or Gearing Ratio= <u>99.00%</u>), (**c**= C/X= Current Ratio= <u>1.20</u> times), (**p**= 360 P/V= Procured Inventory Days= <u>240</u> Days), (**t**= T/B= Tax Rate= <u>30.00%</u>), (**v**= V/S= Variable Portion= <u>19.00%</u>), (**d**= D/A= Dividend Portion or Payout= <u>30.00%</u>), and (**S**= Sales or Revenues= <u>51,294</u> millions USD), then it's (**i**= Interest Portion planned), is:

$$i= \{M\text{-}F\text{-}[(Q\text{+}Svp/\{360[c\text{-}q]\})/l\text{-}U]$$
$$/\{[1\text{-}t][1\text{-}d]\}\}/S$$
$$= \{41,548\text{-}29,750\text{-}[(35,351\text{+}51,294$$
$$*0.1900*240/\{360[1.2000$$
$$\text{-}1.0100]\})/0.9900\text{-}64,720]$$
$$/\{[1\text{-}0.3000][1\text{-}0.3000]\}\}$$
$$/51,294$$

$$= \underline{1.00\%}$$

Corporate IFRS-GAAP (B/S-I/S),ISBN-13: **978-1721066308**, ISBN-10: **1721066306**

Law-8152:

If both (**F**= Fixed Cost= 29,740 millions USD), (**Q**= Quoted Longterm Liabilities= 35,351 millions USD), (**q**= [C-P]/X= Quick or Acid Test Ratio= 1.01 times), (**M**= Margin of Contribution= 41,548 millions USD), (**U**= Utilized or Starting Capital= 64,720 millions USD), (**l**= L/E= Leverage or Gearing Ratio= 99.00%), (**c**= C/X= Current Ratio= 1.20 times), (**p**= 360 P/V= Procured Inventory Days= 240 Days), (**i**= I/S= Interest Portion= 1.00%), (**v**= V/S= Variable Portion= 19.00%), (**d**= D/A= Dividend Portion or Payout= 30.00%), and (**S**= Sales or Revenues= 51,294 millions USD), then it's (**t**= Tax Rate planned), is:

$$t= 1-[(Q+Svp/\{360[c-q]\})/l -U]$$
$$/\{[1-d][M-F-Si]\}$$
$$= 1-[(35,351+51,294*0.1900*240$$
$$/\{360[1.2000-1.0100]\})$$
$$/0.9900-64,720]/\{[1-0.3000]$$
$$[41,548-29,750$$
$$-51,294*0.0100]\}$$
$$= 30.00\%$$

Corporate IFRS-GAAP (B/S-I/S),ISBN-13: **978-1721066308**, ISBN-10: **1721066306**

Law-8153:

If both (**F**= Fixed Cost= 29,740 millions USD), (**Q**= Quoted Longterm Liabilities= 35,351 millions USD), (**q**= [C-P]/X= Quick or Acid Test Ratio= 1.01 times), (**M**= Margin of Contribution= 41,548 millions USD), (**U**= Utilized or Starting Capital= 64,720 millions USD), (**I**= L/E= Leverage or Gearing Ratio= 99.00%), (**c**= C/X= Current Ratio= 1.20 times), (**p**= 360 P/V= Procured Inventory Days= 240 Days), (**i**= I/S= Interest Portion= 1.00%), (**v**= V/S= Variable Portion= 19.00%), (**t**= T/B= Tax Rate= 30.00%), and (**S**= Sales or Revenues= 51,294 millions USD), then it's (**d**= Dividend Portion or Payout planned), is:

$$d= 1-[(Q+Svp/\{360[c-q]\})/I-U]$$
$$/\{1-t\}[M-F-Si]\}$$
$$= 1-[(35,351+51,294*0.1900*240$$
$$/\{360[1.2000-1.0100]\})$$
$$/0.9900-64,720]/\{[1-0.3000]$$
$$[41,548-29,750$$
$$-51,294*0.0100]\}$$
$$= 30.00\%$$

Corporate IFRS-GAAP (B/S-I/S),ISBN-13: **978-1721066308**, ISBN-10: **1721066306**

Law-8154:

If both (**F**= Fixed Cost= 29,740 millions USD), (**Q**= Quoted Longterm Liabilities= 35,351 millions USD), (**q**= [C-P]/X= Quick or Acid Test Ratio= 1.01 times), (**M**= Margin of Contribution= 41,548 millions USD), (**U**= Utilized or Starting Capital= 64,720 millions USD), (**l**= L/E= Leverage or Gearing Ratio= 99.00%), (**c**= C/X= Current Ratio= 1.20 times), (**p**= 360 P/V= Procured Inventory Days= 240 Days), (**i**= I/S= Interest Portion= 1.00%), (**d**= D/A= Dividend Portion or Payout= 30.00%), (**t**= T/B= Tax Rate= 30.00%), and (**$**= Sales or Revenues= 51,294 millions USD), then it's (**v**= Variable Portion planned), is:

$$v= 360[\mathbf{c}\text{-}\mathbf{q}](\mathbf{l}\ \{\mathbf{U}+[\mathbf{M}\text{-}\mathbf{F}\text{-}\$\mathbf{i}][1\text{-}\mathbf{t}][1\text{-}\mathbf{d}]\}\text{-}\mathbf{Q})$$
$$/[\$\mathbf{p}]$$

$$= 360[1.2000\text{-}1.0100](0.9900$$
$$\{64,720+[41,548\text{-}29,750$$
$$-51,294*0.0100][1\text{-}0.3000]$$
$$[1\text{-}0.3000]\}\text{-}35,351)$$
$$/[0.1900*240]$$

$$= 51,294 \text{ millions USD}$$

Corporate IFRS-GAAP (B/S-I/S),ISBN-13: 978-1721066308, ISBN-10: 1721066306

Law-8155:

If both (**F**= Fixed Cost= 29,740 millions USD), (**Q**= Quoted Longterm Liabilities= 35,351 millions USD), (**q**= [C-P]/X= Quick or Acid Test Ratio= 1.01 times), (**M**= Margin of Contribution= 41,548 millions USD), (**U**= Utilized or Starting Capital= 64,720 millions USD), (**l**= L/E= Leverage or Gearing Ratio= 99.00%), (**c**= C/X= Current Ratio= 1.20 times), (**v**= V/S= Variable Portion= 19.00%), (**i**= I/S= Interest Portion= 1.00%), (**d**= D/A= Dividend Portion or Payout= 30.00%), (**t**= T/B= Tax Rate= 30.00%), and (**$**= Sales or Revenues= 51,294 millions USD), then it's (**p**= Procured Inventory Days planned), is:

$$p= 360[\mathbf{c}\text{-}\mathbf{q}](\mathbf{l}\ \{\mathbf{U}+[\mathbf{M}\text{-}\mathbf{F}\text{-}\mathbf{\$i}][1\text{-}\mathbf{t}][1\text{-}\mathbf{d}]\}\text{-}\mathbf{Q})$$
$$/[\mathbf{\$v}]$$

$$= 360[1.2000\text{-}1.0100](0.9900$$
$$\{64,720+[41,548\text{-}29,750$$
$$\text{-}51,294*0.0100][1\text{-}0.3000]$$
$$[1\text{-}0.3000]\}\text{-}35,351)$$
$$/[51,294*240]$$

$$= \underline{240}\ days$$

Law-8156:

If both (**F**= Fixed Cost= 29,740 millions USD), (**Q**= Quoted Longterm Liabilities= 35,351 millions USD), (**q**= [C-P]/X= Quick or Acid Test Ratio= 1.01 times), (**M**= Margin of Contribution= 41,548 millions USD), (**U**= Utilized or Starting Capital= 64,720 millions USD), (**f**= L/E= Leverage or Gearing Ratio= 99.00%), (**p**= 360 P/V= Procured Inventory Days= 240 Days), (**v**= V/S= Variable Portion= 19.00%), (**i**= I/S= Interest Portion= 1.00%), (**d**= D/A= Dividend Portion or Payout= 30.00%), (**t**= T/B= Tax Rate= 30.00%), and (**S**= Sales or Revenues= 51,294 millions USD), then it's (**c**= Current Ratio planned), is:

$$c= q+Svp/[360(f\{U+[M-F-Si]$$
$$[1-t][1-d]\}-Q)]$$
$$= 1.0100+51,294*0.1900*240$$
$$/[360(0.9900\{64,720$$
$$+[41,548-29,750-51,294$$
$$*0.0100][1-0.3000]$$
$$[1-0.3000]\}-35,351)]$$
$$= \underline{1.20} \text{ times}$$

Corporate IFRS-GAAP (B/S-I/S),ISBN-13: **978-1721066308**, ISBN-10: **1721066306**

Law-8157:

If both (**F**= Fixed Cost= 29,740 millions USD), (**Q**= Quoted Longterm Liabilities= 35,351 millions USD), (**c**= C/X= Current Ratio= 1.20 times), (**M**= Margin of Contribution= 41,548 millions USD), (**U**= Utilized or Starting Capital= 64,720 millions USD), (**l**= L/E= Leverage or Gearing Ratio= 99.00%), (**p**= 360 P/V= Procured Inventory Days= 240 Days), (**v**= V/S= Variable Portion= 19.00%), (**i**= I/S= Interest Portion= 1.00%), (**d**= D/A= Dividend Portion or Payout= 30.00%), (**t**= T/B= Tax Rate= 30.00%), and (**S**= Sales or Revenues= 51,294 millions USD), then it's (**q**= Quick or Acid Ratio planned), is:

$$q= c\text{-}Svp/[360(l\ \{U+[M\text{-}F\text{-}Si]$$
$$[1\text{-}t][1\text{-}d]\}\text{-}Q)]$$
$$= 1.2000\text{-}51,294*0.1900*240$$
$$/[360(0.9900\{64,720$$
$$+[41,548\text{-}29,750\text{-}51,294$$
$$*0.0100][1\text{-}0.3000]$$
$$[1\text{-}0.3000]\}\text{-}35,351)]$$
$$= 1.01 \text{ times}$$

Corporate IFRS-GAAP (B/S-I/S),ISBN-13: **978-1721066308**, ISBN-10: **1721066306**

Law-8158:

If both (**F**= Fixed Cost= <u>29,740</u> millions USD), (**q**= [C-P]/X= Quick or Acid Test Ratio= <u>1.01</u> times), (**S'**= Sales of Past Year= <u>48,851</u> millions USD), (**s**= [S/S']-1= Sales Growth= <u>5.00%</u>), (**c**= C/X= Current Ratio= <u>1.20</u> times), (**M**= Margin of Contribution= <u>41,548</u> millions USD), (**U**= Utilized or Starting Capital= <u>64,720</u> millions USD), (**F**= L/E= Leverage or Gearing Ratio= <u>99.00%</u>), (**p**= 360 P/V= Procured Inventory Days= <u>240</u> Days), (**v**= V/S= Variable Portion= <u>19.00%</u>), (**i**= I/S= Interest Portion= <u>1.00%</u>), (**d**= D/A= Dividend Portion or Payout= <u>30.00%</u>), (**t**= T/B= Tax Rate= <u>30.00%</u>), and (**S**= Sales or Revenues= <u>51,294</u> millions USD), then it's (**Q**= Quoted Longterm Liabilities planned), is:

$$Q = I\{U+[M-F-Si][1-t][1-d]\}$$
$$-S'vp[1+s]/\{360[c-q]\}$$
$$= 0.9900\{64,720+[41,548-29,750$$
$$-51,294*0.0100][1-0.3000]$$
$$[1-0.3000]\}-48,851*0.1900$$
$$*240[1+0.0500]$$
$$/\{360[1.2000-1.0100]\}$$
$$= \underline{35,351} \text{ millions USD}$$

Corporate IFRS-GAAP (B/S-I/S),ISBN-13: **978-1721066308**, ISBN-10: **1721066306**

Law-8159:

If both (**F**= Fixed Cost= 29,740 millions USD), (**q**= [C-P]/X= Quick or Acid Test Ratio= 1.01 times), (**S'**= Sales of Past Year= 48,851 millions USD), (**s**= [S/S']-1= Sales Growth= 5.00%), (**c**= C/X= Current Ratio= 1.20 times), (**M**= Margin of Contribution= 41,548 millions USD), (**U**= Utilized or Starting Capital= 64,720 millions USD), (**Q**= Quoted Longterm Liabilities= 35,351 millions USD), (**p**= 360 P/V= Procured Inventory Days= 240 Days), (**v**= V/S= Variable Portion= 19.00%), (**i**= I/S= Interest Portion= 1.00%), (**d**= D/A= Dividend Portion or Payout= 30.00%), (**t**= T/B= Tax Rate= 30.00%), and (**S**= Sales or Revenues= 51,294 millions USD), then it's (**I** = Leverage or Gearing Ratio planned), is:

$$I = (Q+S'vp[1+s]/\{360[c-q]\})$$
$$/\{U+[M-F-Si][1-t][1-d]\}$$
$$= (35,351+48,851*0.1900*240$$
$$[1+0.0500]/\{360[1.2000$$
$$-1.0100]\})/\{64,720+[41,548$$
$$-29,750-51,294*0.0100]$$
$$[1-0.3000][1-0.3000]\}$$
$$= \underline{99.00\%}$$

Corporate IFRS-GAAP (B/S-I/S),ISBN-13: **978-1721066308**, ISBN-10: **1721066306**

Law-8160:

  If both (**F**= Fixed Cost= 29,740 millions USD), (**q**= [C-P]/X= Quick or Acid Test Ratio= 1.01 times), (**S'**= Sales of Past Year= 48,851 millions USD), (**s**= [S/S']-1= Sales Growth= 5.00%), (**c**= C/X= Current Ratio= 1.20 times), (**M**= Margin of Contribution= 41,548 millions USD), (**I**= L/E= Leverage or Gearing Ratio= 99.00%), (**Q**= Quoted Longterm Liabilities= 35,351 millions USD), (**p**= 360 P/V= Procured Inventory Days= 240 Days), (**v**= V/S= Variable Portion= 19.00%), (**i**= I/S= Interest Portion= 1.00%), (**d**= D/A= Dividend Portion or Payout= 30.00%), (**t**= T/B= Tax Rate= 30.00%), and (**S**= Sales or Revenues= 51,294 millions USD), then it's (**U**= Utilized or Starting Capital planned), is:

$$U= (Q+S'vp[1+s]/\{360[c-q]\})/I$$
$$-[M-F-Si][1-t][1-d]$$
$$= (35,351+48,851*0.1900*240$$
$$[1+0.0500]/\{360[1.2000$$
$$-1.0100]\})/0.9900-[41,548$$
$$-29,750-51,294*0.0100]$$
$$[1-0.3000][1-0.3000]\}$$
$$= \underline{64,720} \text{ millions USD}$$

Corporate IFRS-GAAP (B/S-I/S),ISBN-13: **978-1721066308**, ISBN-10: **1721066306**

Law-8161:

If both (**F**= Fixed Cost= <u>29,740</u> millions USD), (**q**= [C-P]/X= Quick or Acid Test Ratio= <u>1.01</u> times), (**S'**= Sales of Past Year= <u>48,851</u> millions USD), (**s**= [S/S']-1= Sales Growth= <u>5.00%</u>), (**c**= C/X= Current Ratio= <u>1.20</u> times), (**U**= Utilized or Starting Capital= <u>64,720</u> millions USD), (**l**= L/E= Leverage or Gearing Ratio= <u>99.00%</u>), (**Q**= Quoted Longterm Liabilities= <u>35,351</u> millions USD), (**p**= 360 P/V= Procured Inventory Days= <u>240</u> Days), (**v**= V/S= Variable Portion= <u>19.00%</u>), (**i**= I/S= Interest Portion= <u>1.00%</u>), (**d**= D/A= Dividend Portion or Payout= <u>30.00%</u>), (**t**= T/B= Tax Rate= <u>30.00%</u>), and (**S**= Sales or Revenues= <u>51,294</u> millions USD), then it's (**M**= Margin of Contribution planned), is:

$$M= F+Si+[(Q+S'vp[1+s]/\{360[c-q]\})/l-U]/\{[1-t][1-d]\}$$

$$= 29,750+51,294*0.0100 +[(35,351$$
$$+48,851*0.1900*240$$
$$[1+0.0500]/\{360[1.2000$$
$$-1.0100]\})/0.9900-64,720]$$
$$/\{[1-0.3000][1-0.3000]\}$$

$$= \underline{41,548} \text{ millions USD}$$

Corporate IFRS-GAAP (B/S-I/S),ISBN-13: **978-1721066308**, ISBN-10: **1721066306**

Law-8162:

If both (**M**= Margin of Contribution= 41,548 millions USD), (**q**= [C-P]/X= Quick or Acid Test Ratio= 1.01 times), (**S'**= Sales of Past Year= 48,851 millions USD), (**s**= [S/S']-1= Sales Growth= 5.00%), (**c**= C/X= Current Ratio= 1.20 times), (**U**= Utilized or Starting Capital= 64,720 millions USD), (**l**= L/E= Leverage or Gearing Ratio= 99.00%), (**Q**= Quoted Longterm Liabilities= 35,351 millions USD), (**p**= 360 P/V= Procured Inventory Days= 240 Days), (**v**= V/S= Variable Portion= 19.00%), (**i**= I/S= Interest Portion= 1.00%), (**d**= D/A= Dividend Portion or Payout= 30.00%), (**t**= T/B= Tax Rate= 30.00%), and (**S**= Sales or Revenues= 51,294 millions USD), then it's (**F**= Fixed Cost planned), is:

$$F= M\text{-}Si\text{-}[(Q\text{+}S'vp[1\text{+}s]/\{360[c\text{-}q]\})/l\text{-}U]$$
$$/\{[1\text{-}t][1\text{-}d]\}$$
$$= 41,548\text{-}51,294*0.0100\text{-}[(35,351$$
$$+48,851*0.1900*240$$
$$[1+0.0500]/\{360[1.2000$$
$$-1.0100]\})/0.9900\text{-}64,720]$$
$$/\{[1\text{-}0.3000][1\text{-}0.3000]\}$$
$$= 29,750 \text{ millions USD}$$

Corporate IFRS-GAAP (B/S-I/S),ISBN-13: **978-1721066308**, ISBN-10: **1721066306**

<u>Law-8163</u>:

If both (**M**= Margin of Contribution= 41,548 millions USD), (**q**= [C-P]/X= Quick or Acid Test Ratio= 1.01 times), (**S'**= Sales of Past Year= 48,851 millions USD), (**s**= [S/S']-1= Sales Growth= 5.00%), (**c**= C/X= Current Ratio= 1.20 times), (**U**= Utilized or Starting Capital= 64,720 millions USD), (**l**= L/E= Leverage or Gearing Ratio= 99.00%), (**Q**= Quoted Longterm Liabilities= 35,351 millions USD), (**p**= 360 P/V= Procured Inventory Days= 240 Days), (**v**= V/S= Variable Portion= 19.00%), (**i**= I/S= Interest Portion= 1.00%), (**d**= D/A= Dividend Portion or Payout= 30.00%), (**t**= T/B= Tax Rate= 30.00%), and (**F**= Fixed Cost= 29,740 millions USD), then it's (**S**= Sales or Revenues planned), is:

$$S= \{M\text{-}F\text{-}[(Q+S'vp[1+s]/\{360[c\text{-}q]\})/l\text{-}U]$$
$$/\{[1\text{-}t][1\text{-}d]\}\}/i$$
$$= \{41,548\text{-}29,750\text{-}[(35,351+48,851$$
$$*0.1900*240[1+0.0500]$$
$$/\{360[1.2000\text{-}1.0100]\})$$
$$/0.9900\text{ -}64,720]/\{[1\text{-}0.3000]$$
$$[1\text{-}0.3000]\}\}/0.0100$$
$$= \underline{51,294}\text{ millions USD}$$

Corporate IFRS-GAAP (B/S-I/S),ISBN-13: **978-1721066308**, ISBN-10: **1721066306**

Law-8164:

If both (**M**= Margin of Contribution= 41,548 millions USD), (**q**= [C-P]/X= Quick or Acid Test Ratio= 1.01 times), (**S'**= Sales of Past Year= 48,851 millions USD), (**s**= [S/S']-1= Sales Growth= 5.00%), (**c**= C/X= Current Ratio= 1.20 times), (**U**= Utilized or Starting Capital= 64,720 millions USD), (**l**= L/E= Leverage or Gearing Ratio= 99.00%), (**Q**= Quoted Longterm Liabilities= 35,351 millions USD), (**p**= 360 P/V= Procured Inventory Days= 240 Days), (**v**= V/S= Variable Portion= 19.00%), (**S**= Sales or Revenues= 51,294 millions USD), (**d**= D/A= Dividend Portion or Payout= 30.00%), (**t**= T/B= Tax Rate= 30.00%), and (**F**= Fixed Cost= 29,740 millions USD), then it's (**i**= Interest Portion planned), is:

$$i= \{M\text{-}F\text{-}[(Q+S'vp[1+s]/\{360[c\text{-}q]\})/l\text{-}U] /\{[1\text{-}t][1\text{-}d]\}\}/S$$

$$= \{41,548\text{-}29,750\text{-}[(35,351+48,851 \\ *0.1900*240[1+0.0500] \\ /\{360[1.2000\text{-}1.0100]\}) \\ /0.9900\text{-}64,720]/\{[1\text{-}0.3000] \\ [1\text{-}0.3000]\}\}/51,294$$

$$= 1.00\%$$

Corporate IFRS-GAAP (B/S-I/S),ISBN-13: **978-1721066308**, ISBN-10: **1721066306**

<u>Law-8165</u>:

If both (**M**= Margin of Contribution= <u>41,548</u> millions USD), (**q**= [C-P]/X= Quick or Acid Test Ratio= <u>1.01</u> times), (**S'**= Sales of Past Year= <u>48,851</u> millions USD), (**s**= [S/S']-1= Sales Growth= <u>5.00%</u>), (**c**= C/X= Current Ratio= <u>1.20</u> times), (**U**= Utilized or Starting Capital= <u>64,720</u> millions USD), (**l**= L/E= Leverage or Gearing Ratio= <u>99.00%</u>), (**Q**= Quoted Longterm Liabilities= <u>35,351</u> millions USD), (**p**= 360 P/V= Procured Inventory Days= <u>240</u> Days), (**v**= V/S= Variable Portion= <u>19.00%</u>), (**S**= Sales or Revenues= <u>51,294</u> millions USD), (**d**= D/A= Dividend Portion or Payout= <u>30.00%</u>), (**i**= I/S= Interest Portion= <u>1.00%</u>), and (**F**= Fixed Cost= <u>29,740</u> millions USD), then it's (**t**= Tax Rate planned), is:

$$t= 1-[(Q+S'vp[1+s]/\{360[c-q]\})/l-U]$$
$$/\{[1-d][M-F-Si]\}$$
$$= 1-[(35,351+48,851*0.1900*240$$
$$[1+0.0500]/\{360[1.2000$$
$$-1.0100]\})/0.9900-64,720]$$
$$/\{[1-0.3000][41,548-29,750$$
$$-51,294*0.0100]\}$$
$$= \underline{30.00\%}$$

Corporate IFRS-GAAP (B/S-I/S),ISBN-13: **978-1721066308**, ISBN-10: **1721066306**

Law-8166:

If both (**M**= Margin of Contribution= 41,548 millions
USD), (**q**= [C-P]/X= Quick or Acid Test Ratio= 1.01
times), (**S'**= Sales of Past Year= 48,851 millions
USD), (**s**= [S/S']-1= Sales Growth= 5.00%), (**c**=
C/X= Current Ratio= 1.20 times), (**U**= Utilized or
Starting Capital= 64,720 millions USD), (**l**= L/E=
Leverage or Gearing Ratio= 99.00%), (**Q**= Quoted
Longterm Liabilities= 35,351 millions USD), (**p**=
360 P/V= Procured Inventory Days= 240 Days), (**v**=
V/S= Variable Portion= 19.00%), (**S**= Sales or
Revenues= 51,294 millions USD), (**t**= T/B= Tax
Rate= 30.00%), (**i**= I/S= Interest Portion= 1.00%),
and (**F**= Fixed Cost= 29,740 millions USD), then it's
(**d**= Dividend Portion or Payout planned), is:

$$d= 1-[(Q+S'vp[1+s]/\{360[c-q]\})/l-U]$$
$$/\{[1-t][M-F-Si]\}$$
$$= 1-[(35,351+48,851*0.1900*240$$
$$[1+0.0500]/\{360[1.2000$$
$$-1.0100]\})/0.9900-64,720]$$
$$/\{[1-0.3000][41,548-29,750$$
$$-51,294*0.0100]\}$$
$$= 30.00\%$$

Corporate IFRS-GAAP (B/S-I/S),ISBN-13: **978-1721066308**, ISBN-10: **1721066306**

Law-8167:

If both (**M**= Margin of Contribution= 41,548 millions USD), (**q**= [C-P]/X= Quick or Acid Test Ratio= 1.01 times), (**d**= D/A= Dividend Portion or Payout= 30.00%), (**s**= [S/S']-1= Sales Growth= 5.00%), (**c**= C/X= Current Ratio= 1.20 times), (**U**= Utilized or Starting Capital= 64,720 millions USD), (**l**= L/E= Leverage or Gearing Ratio= 99.00%), (**Q**= Quoted Longterm Liabilities= 35,351 millions USD), (**p**= 360 P/V= Procured Inventory Days= 240 Days), (**v**= V/S= Variable Portion= 19.00%), (**S**= Sales or Revenues= 51,294 millions USD), (**t**= T/B= Tax Rate= 30.00%), (**i**= I/S= Interest Portion= 1.00%), and (**F**= Fixed Cost= 29,740 millions USD), then it's (**S'**= Sales Past), must be:

$$S'= [360[c-q](l \{U+[M-F-Si][1-t][1-d]\}-Q)]$$
$$/\{vp[1+s]\}$$
$$= [360[1.2000-1.0100](0.9900$$
$$\{64,720+[41,548-29,750$$
$$-51,294*0.0100][1-0.3000]$$
$$[1-0.3000]\}- 35,351)]$$
$$/\{0.1900*240[1+0.0500]\}$$
$$= 48,851 \text{ millions USD}$$

Corporate IFRS-GAAP (B/S-I/S),ISBN-13: **978-1721066308**, ISBN-10: **1721066306**

Law-8168:

If both (**M**= Margin of Contribution= 41,548 millions USD), (**q**= [C-P]/X= Quick or Acid Test Ratio= 1.01 times), (**d**= D/A= Dividend Portion or Payout= 30.00%), (**s**= [S/S']-1= Sales Growth= 5.00%), (**c**= C/X= Current Ratio= 1.20 times), (**U**= Utilized or Starting Capital= 64,720 millions USD), (**l**= L/E= Leverage or Gearing Ratio= 99.00%), (**Q**= Quoted Longterm Liabilities= 35,351 millions USD), (**p**= 360 P/V= Procured Inventory Days= 240 Days), (**S'**= Sales of Past Year= 48,851 millions USD), (**S**= Sales or Revenues= 51,294 millions USD), (**t**= T/B= Tax Rate= 30.00%), (**i**= I/S= Interest Portion= 1.00%), and (**F**= Fixed Cost= 29,740 millions USD), then it's (**v**= Variable Portion planned), is:

$$v= [360[\mathbf{c}\text{-}\mathbf{q}](\mathbf{l}\,\{\mathbf{U}+[\mathbf{M}\text{-}\mathbf{F}\text{-}\mathbf{Si}][1\text{-}\mathbf{t}][1\text{-}\mathbf{d}]\}\text{-}\mathbf{Q})]$$
$$/\{\mathbf{S'p}[1+\mathbf{s}]\}$$
$$= [360[1.2000\text{-}1.0100](0.9900$$
$$\{64,720+[41,548\text{-}29,750$$
$$-51,294*0.0100][1\text{-}0.3000]$$
$$[1\text{-}0.3000]\}\text{-}35,351)]/\{48,851$$
$$*240[1+0.0500]\}$$
$$= 19.00\%$$

Corporate IFRS-GAAP (B/S-I/S),ISBN-13: **978-1721066308**, ISBN-10: **1721066306**

Law-8169:

If both (**M**= Margin of Contribution= 41,548 millions USD), (**q**= [C-P]/X= Quick or Acid Test Ratio= 1.01 times), (**d**= D/A= Dividend Portion or Payout= 30.00%), (**s**= [S/S']-1= Sales Growth= 5.00%), (**c**= C/X= Current Ratio= 1.20 times), (**U**= Utilized or Starting Capital= 64,720 millions USD), (**l**= L/E= Leverage or Gearing Ratio= 99.00%), (**Q**= Quoted Longterm Liabilities= 35,351 millions USD), (**v**= V/S= Variable Portion= 19.00%), (**S'**= Sales of Past Year= 48,851 millions USD), (**S**= Sales or Revenues= 51,294 millions USD), (**t**= T/B= Tax Rate= 30.00%), (**i**= I/S= Interest Portion= 1.00%), and (**F**= Fixed Cost= 29,740 millions USD), then it's (**p**= Procured Inventory Days planned), is:

$$p= [360[c\text{-}q](l\ \{U+[M\text{-}F\text{-}Si][1\text{-}t][1\text{-}d]\}\text{-}Q)]$$
$$/\{S'v[1+s]\}$$
$$= [360[1.2000\text{-}1.0100](0.9900$$
$$\{64,720+[41,548\text{-}29,750$$
$$\text{-}51,294*0.0100][1\text{-}0.3000]$$
$$[1\text{-}0.3000]\}\text{-}35,351)]/\{48,851$$
$$*0.1900[1+0.0500]\}$$
$$= \underline{240}\ \text{days}$$

Corporate IFRS-GAAP (B/S-I/S),ISBN-13: **978-1721066308**, ISBN-10: **1721066306**

Law-8170:

If both (**M**= Margin of Contribution= 41,548 millions USD), (**q**= [C-P]/X= Quick or Acid Test Ratio= 1.01 times), (**d**= D/A= Dividend Portion or Payout= 30.00%), (**p**= 360 P/V= Procured Inventory Days= 240 Days), (**c**= C/X= Current Ratio= 1.20 times), (**U**= Utilized or Starting Capital= 64,720 millions USD), (**⌐**= L/E= Leverage or Gearing Ratio= 99.00%), (**Q**= Quoted Longterm Liabilities= 35,351 millions USD), (**v**= V/S= Variable Portion= 19.00%), (**$'**= Sales of Past Year= 48,851 millions USD), (**$**= Sales or Revenues= 51,294 millions USD), (**t**= T/B= Tax Rate= 30.00%), (**i**= I/S= Interest Portion= 1.00%), and (**F**= Fixed Cost= 29,740 millions USD), then it's (**s**= Sales Growth planned), is:

$$s= [360[c\text{-}q](⌐\{U+[M\text{-}F\text{-}\$i][1\text{-}t][1\text{-}d]\}\text{-}Q)]$$
$$/[\$'vp]\ 1$$
$$= [360[1.2000\text{-}1.0100](0.9900$$
$$\{64,720+[41,548\text{-}29,750$$
$$\text{-}51,294*0.0100][1\text{-}0.3000]$$
$$[1\text{-}0.3000]\}\text{-}35,351)]$$
$$/[48,851*0.1900*240]\text{-}1$$
$$= \underline{5.00\%}$$

Corporate IFRS-GAAP (B/S-I/S),ISBN-13: **978-1721066308**, ISBN-10: **1721066306**

<u>Law-8171</u>:

If both (**M**= Margin of Contribution= <u>41,548</u> millions USD), (**q**= [C-P]/X= Quick or Acid Test Ratio= <u>1.01</u> times), (**d**= D/A= Dividend Portion or Payout= <u>30.00%</u>), (**p**= 360 P/V= Procured Inventory Days= <u>240</u> Days), (**s**= [S/S']-1= Sales Growth= <u>5.00%</u>), (**U**= Utilized or Starting Capital= <u>64,720</u> millions USD), (**f**= L/E= Leverage or Gearing Ratio= <u>99.00%</u>), (**Q**= Quoted Longterm Liabilities= <u>35,351</u> millions USD), (**v**= V/S= Variable Portion= <u>19.00%</u>), (**S'**= Sales of Past Year= <u>48,851</u> millions USD), (**S**= Sales or Revenues= <u>51,294</u> millions USD), (**t**= T/B= Tax Rate= <u>30.00%</u>), (**i**= I/S= Interest Portion= <u>1.00%</u>), and (**F**= Fixed Cost= <u>29,740</u> millions USD), then it's (**c**= Current Ratio planned), is:

$$c= q+S'vp[1+s]/[360(f\{U+[M-F-Si]$$
$$[1-t][1-d]\}-Q)]$$
$$= 1.0100+48,851*0.1900*240$$
$$[1+0.0500]/[360(0.9900$$
$$\{64,720+[41,548-29,750$$
$$-51,294*0.0100][1-0.3000]$$
$$[1-0.3000]\}-35,351)]$$
$$= \underline{1.20} \text{ times}$$

Law-8172:

If both (**M**= Margin of Contribution= 41,548 millions USD), (**c**= C/X= Current Ratio= 1.20 times), (**d**= D/A=Dividend Portion or Payout= 30.00%), (**p**= 360 P/V= Procured Inventory Days= 240 Days), (**s**= [S/S']-1= Sales Growth= 5.00%), (**U**= Utilized or Starting Capital= 64,720 millions USD), (**l**= L/E= Leverage or Gearing Ratio= 99.00%), (**Q**= Quoted Longterm Liabilities= 35,351 millions USD), (**v**= V/S= Variable Portion= 19.00%), (**S'**= Sales of Past Year= 48,851 millions USD), (**S**= Sales or Revenues= 51,294 millions USD), (**t**= T/B= Tax Rate= 30.00%), (**i**= I/S= Interest Portion= 1.00%), and (**F**= Fixed Cost= 29,740 millions USD), then it's (**q**= Quick or Acid Test Ratio planned), is:

$$q = c\text{-}S'vp[1+s]/[360(l\{U+[M\text{-}F\text{-}Si]$$
$$[1\text{-}t][1\text{-}d]\}\text{-}Q)]$$
$$= 1.2000\text{-}48,851*0.1900*240$$
$$[1+0.0500]/[360(0.9900$$
$$\{64,720+[41,548\text{-}29,750$$
$$\text{-}51,294*0.0100][1\text{-}0.3000]$$
$$[1\text{-}0.3000]\}\text{-}35,351)]$$
$$= 1.01 \text{ times}$$

Corporate IFRS-GAAP (B/S-I/S),ISBN-13: **978-1721066308**, ISBN-10: **1721066306**

Law-8173:

If both (**M**= Margin of Contribution= 41,548 millions USD), (**D**= Dividend Paid= 2,370 millions USD), (**U**= Utilized or Starting Capital= 64,720 millions USD), (**F**= L/E= Leverage or Gearing Ratio= 99.00%), (**X**= Xpress or Current Debt= 34,196 millions USD), (**S'**= Sales of Past Year= 48,851 millions USD), (**s**= [S/S']-1= Sales Growth= 5.00%), (**T**= Tax Paid= 3,385 millions USD), (**i**= I/S= Interest Portion= 1.00%), and (**F**= Fixed Cost= 29,740 millions USD), then it's (**Q**= Quoted Longterm Liabilities planned), is:

$$Q = F\{U+M-F-S'i[1+s]-T-D\}-X$$
$$= 0.9900\{64,720+41,548-29,750$$
$$-48,851*0.0100[1+0.0500]$$
$$-3,385-2,370\}-34,196$$
$$= 35,351 \text{ millions USD}$$

Corporate IFRS-GAAP (B/S-I/S),ISBN-13: **978-1721066308**, ISBN-10: **1721066306**

Law-8174:

If both (**M**= Margin of Contribution= 41,548 millions USD), (**D**= Dividend Paid= 2,370 millions USD), (**U**= Utilized or Starting Capital= 64,720 millions USD), (**Q**= Quoted Longterm Liabilities= 35,351 millions USD), (**X**= Xpress or Current Debt= 34,196 millions USD), (**S'**= Sales of Past Year= 48,851 millions USD), (**s**= [S/S']-1= Sales Growth= 5.00%), (**T**= Tax Paid= 3,385 millions USD), (**i**= I/S= Interest Portion= 1.00%), and (**F**= Fixed Cost= 29,740 millions USD), then it's (**I** = Leverage or Gearing Ratio planned), is:

$$I = [Q+X]/\{U+M-F-S'i[1+s]-T-D\}$$
$$= [35,351+34,196]/\{64,720+41,548$$
$$-29,750-48,851*0.0100$$
$$[1+0.0500]-3,385-2,370\}$$
$$= 99.00\%$$

Corporate IFRS-GAAP (B/S-I/S),ISBN-13: **978-1721066308**, ISBN-10: **1721066306**

Law-8175:

If both (**M**= Margin of Contribution= 41,548 millions USD), (**D**= Dividend Paid= 2,370 millions USD), (**I**= L/E= Leverage or Gearing Ratio= 99.00%), (**Q**= Quoted Longterm Liabilities= 35,351 millions USD), (**X**= Xpress or Current Debt= 34,196 millions USD), (**S'**= Sales of Past Year= 48,851 millions USD), (**s**= [S/S']-1= Sales Growth= 5.00%), (**T**= Tax Paid= 3,385 millions USD), (**i**= I/S= Interest Portion= 1.00%), and (**F**= Fixed Cost= 29,740 millions USD), then it's (**U**= Utilized or Starting Capital planned), is:

$$U = F + S'i[1+s] + T + D + [Q+X]/I - M$$

$$= 29,750 + 48,851*0.0100[1+0.0500]$$
$$+3,385 + 2,370 + [35,351$$
$$+34,196]/0.9900 - 41,548$$

$$= \underline{64,720} \text{ millions USD}$$

Corporate IFRS-GAAP (B/S-I/S),ISBN-13: **978-1721066308**, ISBN-10: **1721066306**

Law-8176:

If both (**U**= Utilized or Starting Capital= 64,720 millions USD), (**D**= Dividend Paid= 2,370 millions USD), (**I**= L/E= Leverage or Gearing Ratio= 99.00%), (**Q**= Quoted Longterm Liabilities= 35,351 millions USD), (**X**= Xpress or Current Debt= 34,196 millions USD), (**S'**= Sales of Past Year= 48,851 millions USD), (**s**= [S/S']-1= Sales Growth= 5.00%), (**T**= Tax Paid= 3,385 millions USD), (**i**= I/S= Interest Portion= 1.00%), and (**F**= Fixed Cost= 29,740 millions USD), then it's (**M**= Margin of Contribution planned), is:

$$M= F+S'i[1+s]+T+D+[Q+X]/I-U$$

$$= 29,750+48,851*0.0100[1+0.0500]$$
$$+3,385+2,370+[35,351$$
$$+34,196]/0.9900-64,720$$

$$= 41,548 \text{ millions USD}$$

Corporate IFRS-GAAP (B/S-I/S),ISBN-13: **978-1721066308**, ISBN-10: **1721066306**

Law-8177:

If both (**U**= Utilized or Starting Capital= 64,720 millions USD), (**D**= Dividend Paid= 2,370 millions USD), (**I**= L/E= Leverage or Gearing Ratio= 99.00%), (**Q**= Quoted Longterm Liabilities= 35,351 millions USD), (**X**= Xpress or Current Debt= 34,196 millions USD), (**S'**= Sales of Past Year= 48,851 millions USD), (**s**= [S/S']-1= Sales Growth= 5.00%), (**T**= Tax Paid= 3,385 millions USD), (**i**= I/S= Interest Portion= 1.00%), and (**M**= Margin of Contribution= 41,548 millions USD), then it's (**F**= Fixed Cost planned), is:

$$F= U+M-S'i[1+s]-T-D-[Q+X]/I$$
$$= 64,720+41,548-48,851*0.0100$$
$$[1+0.0500]-3,385-2,370$$
$$-[35,351+34,196]/0.9900$$
$$= 29,750 \text{ millions}$$

Corporate IFRS-GAAP (B/S-I/S),ISBN-13: **978-1721066308**, ISBN-10: **1721066306**

Law-8178:

If both (**U**= Utilized or Starting Capital= 64,720 millions USD), (**D**= Dividend Paid= 2,370 millions USD), (**I**= L/E= Leverage or Gearing Ratio= 99.00%), (**Q**= Quoted Longterm Liabilities= 35,351 millions USD), (**X**= Xpress or Current Debt= 34,196 millions USD), (**F**= Fixed Cost= 29,740 millions USD), (**s**= [S/S']-1= Sales Growth= 5.00%), (**T**= Tax Paid= 3,385 millions USD), (**i**= I/S= Interest Portion= 1.00%), and (**M**= Margin of Contribution= 41,548 millions USD), then it's (**S'**= Sales Past), must be:

$$S' = \{U+M-F-T-D-[Q+X]/I\}/\{i[1+s]\}$$
$$= \{64,720+41,548-29,750-3,385-2,370$$
$$-[35,351+34,196]/0.9900\}$$
$$/\{0.0100[1+0.0500]\}$$
$$= \underline{48,851} \text{ millions USD}$$

Corporate IFRS-GAAP (B/S-I/S),ISBN-13: **978-1721066308**, ISBN-10: **1721066306**

Law-8179:

If both (**U**= Utilized or Starting Capital= 64,720 millions USD), (**D**= Dividend Paid= 2,370 millions USD), (**I**= L/E= Leverage or Gearing Ratio= 99.00%), (**Q**= Quoted Longterm Liabilities= 35,351 millions USD), (**X**= Xpress or Current Debt= 34,196 millions USD), (**F**= Fixed Cost= 29,740 millions USD), (**s**= [S/S']-1= Sales Growth= 5.00%), (**T**= Tax Paid= 3,385 millions USD), (**S'**= Sales of Past Year= 48,851 millions USD), and (**M**= Margin of Contribution= 41,548 millions USD), then it's (**i**= Interest Portion planned), is:

$$i= \{U+M-F-T-D-[Q+X]/I\}/\{S'[1+s]\}$$
$$= \{64,720+41,548-29,750-3,385$$
$$-2,370-[35,351+34,196]$$
$$/0.9900\}/\{48,851$$
$$[1+0.0500]\}$$
$$= \underline{1.00\%}$$

Corporate IFRS-GAAP (B/S-I/S),ISBN-13: **978-1721066308**, ISBN-10: **1721066306**

Law-8180:

If both (**U**= Utilized or Starting Capital= 64,720 millions USD), (**D**= Dividend Paid= 2,370 millions USD), (**I**= L/E= Leverage or Gearing Ratio= 99.00%), (**Q**= Quoted Longterm Liabilities= 35,351 millions USD), (**X**= Xpress or Current Debt= 34,196 millions USD), (**F**= Fixed Cost= 29,740 millions USD), (**i**= I/S= Interest Portion= 1.00%), (**T**= Tax Paid= 3,385 millions USD), (**$'**= Sales of Past Year= 48,851 millions USD), and (**M**= Margin of Contribution= 41,548 millions USD), then it's (**s**= Sales Growth planned), is:

$$s= \{U+M-F-T-D-[Q+X]/I\}/[\$'i]-1$$
$$= \{64,720+41,548-29,750-3,385$$
$$-2,370-[35,351+34,196]$$
$$/0.9900\}/[48,851$$
$$*0.0100]-1$$
$$= 5.00\%$$

Corporate IFRS-GAAP (B/S-I/S),ISBN-13: **978-1721066308**, ISBN-10: **1721066306**

Law-8181:

If both (**U**= Utilized or Starting Capital= 64,720 millions USD), (**D**= Dividend Paid= 2,370 millions USD), (**I**= L/E= Leverage or Gearing Ratio= 99.00%), (**Q**= Quoted Longterm Liabilities= 35,351 millions USD), (**X**= Xpress or Current Debt= 34,196 millions USD), (**F**= Fixed Cost= 29,740 millions USD), (**i**= I/S= Interest Portion= 1.00%), (**s**= [S/S']-1= Sales Growth= 5.00%), (**S'**= Sales of Past Year= 48,851 millions USD), and (**M**= Margin of Contribution= 41,548 millions USD), then it's (**T**= Tax Paid planned), is:

$$T= U+M-F-S'i[1+s]-D-[Q+X]/I$$

$$= 64{,}720+41{,}548-29{,}750-48{,}851$$
$$*0.0100[1+0.0500]-2{,}370$$
$$-[35{,}351+34{,}196]/0.9900$$

$$= \underline{3{,}385} \text{ millions USD}$$

Corporate IFRS-GAAP (B/S-I/S),ISBN-13: **978-1721066308**, ISBN-10: **1721066306**

Law-8182:

If both (**U**= Utilized or Starting Capital= 64,720 millions USD), (**T**= Tax Paid= 3,385 millions USD), (**I**= L/E= Leverage or Gearing Ratio= 99.00%), (**Q**= Quoted Longterm Liabilities= 35,351 millions USD), (**X**= Xpress or Current Debt= 34,196 millions USD), (**F**= Fixed Cost= 29,740 millions USD), (**i**= I/S= Interest Portion= 1.00%), (**s**= [S/S']-1= Sales Growth= 5.00%), (**S'**= Sales of Past Year= 48,851 millions USD), and (**M**= Margin of Contribution= 41,548 millions USD), then it's (**D**= Dividend Paid planned), is:

$$D= U+M-F-S'i[1+s]-T-[Q+X]/I$$
$$= 64,720+41,548-29,750-48,851$$
$$*0.0100[1+0.0500]-3,385$$
$$-[35,351+34,196]/0.9900$$
$$= 2,370 \text{ millions USD}$$

Corporate IFRS-GAAP (B/S-I/S),ISBN-13: **978-1721066308**, ISBN-10: **1721066306**

Law-8183:

If both (**U**= Utilized or Starting Capital= 64,720 millions USD), (**T**= Tax Paid= 3,385 millions USD), (**I**= L/E= Leverage or Gearing Ratio= 99.00%), (**Q**= Quoted Longterm Liabilities= 35,351 millions USD), (**D**= Dividend Paid= 2,370 millions USD), (**F**= Fixed Cost= 29,740 millions USD), (**i**= I/S= Interest Portion= 1.00%), (**s**= [S/S']-1= Sales Growth= 5.00%), (**S'**= Sales of Past Year= 48,851 millions USD), and (**M**= Margin of Contribution= 41,548 millions USD), then it's (**X**= Xpress or Current Debt planned), is:

$$X = I\{U+M-F-S'i[1+s]-T-D\}-Q$$
$$= 0.9900\{64,720+41,548-29,750$$
$$-48,851*0.0100[1+0.0500]$$
$$-3,385-2,370\}-35,351$$
$$= \underline{34,196} \text{ millions USD}$$

Corporate IFRS-GAAP (B/S-I/S),ISBN-13: **978-1721066308**, ISBN-10: **1721066306**

Law-8184:

If both (**U**= Utilized or Starting Capital= 64,720 millions USD), (**T**= Tax Paid= 3,385 millions USD), (**I**= L/E= Leverage or Gearing Ratio= 99.00%), (**c**= C/X= Current Ratio= 1.20 times), (**q**= [C-P]/X= Quick or Acid Test Ratio= 1.01 times), (**P**= Procured Inventories= 6,497 millions USD), (**D**= Dividend Paid= 2,370 millions USD), (**F**= Fixed Cost= 29,740 millions USD), (**i**= I/S= Interest Portion= 1.00%), (**s**= [S/S']-1= Sales Growth= 5.00%), (**S'**= Sales of Past Year= 48,851 millions USD), and (**M**= Margin of Contribution= 41,548 millions USD), then it's (**Q**= Quoted Longterm Liabilities planned), is:

$$Q = I\{U+M-F-S'i[1+s]-T-D\}-P/[c-q]$$
$$= 0.9900\{64,720+41,548-29,750$$
$$-48,851*0.0100[1+0.0500]$$
$$-3,385-2,370\}-6,497$$
$$/[1.2000-1.0100]$$
$$= 35,351 \text{ millions USD}$$

Corporate IFRS-GAAP (B/S-I/S),ISBN-13: **978-1721066308**, ISBN-10: **1721066306**

<u>Law-8185</u>:

If both (**U**= Utilized or Starting Capital= <u>64,720</u> millions USD), (**T**= Tax Paid= <u>3,385</u> millions USD), (**Q**= Quoted Longterm Liabilities= <u>35,351</u> millions USD), (**c**= C/X= Current Ratio= <u>1.20</u> times), (**q**= Quick or Acid Test Ratio= [C-P]/X= <u>1.01</u> times), (**P**= Procured Inventories= <u>6,497</u> millions USD), (**D**= Dividend Paid= <u>2,370</u> millions USD), (**F**= Fixed Cost= <u>29,740</u> millions USD), (**i**= I/S= Interest Portion= <u>1.00%</u>), (**s**= [S/S']-1= Sales Growth= <u>5.00%</u>), (**S'**= Sales of Past Year= <u>48,851</u> millions USD), and (**M**= Margin of Contribution= <u>41,548</u> millions USD), then it's (**I** = Leverage or Gearing Ratioplanned), is:

$$I = \{Q+P/[c-q]\}/\{U+M-F-S'i[1+s]-T-D\}$$
$$= \{35,351+6,497/[1.2000-1.0100]\}$$
$$/\{64,720+41,548-29,750$$
$$-48,851*0.0100[1+0.0500]$$
$$-3,385-2,370\}$$
$$= \underline{99.00\%}$$

Corporate IFRS-GAAP (B/S-I/S),ISBN-13: **978-1721066308**, ISBN-10: **1721066306**

Law-8186:

If both (**Q**= Quoted Longterm Liabilities= 35,351
millions USD), (**T**= Tax Paid= 3,385 millions USD),
(**I**= L/E= Leverage or Gearing Ratio= 99.00%), (**c**=
C/X= Current Ratio= 1.20 times), (**q**= [C-P]/X=
Quick or Acid Test Ratio= 1.01 times), (**P**= Procured
Inventories= 6,497 millions USD), (**D**= Dividend
Paid= 2,370 millions USD), (**F**= Fixed Cost= 29,740
millions USD), (**i**= I/S= Interest Portion= 1.00%), (**s**=
[S/S']-1= Sales Growth= 5.00%), (**S'**= Sales of Past
Year= 48,851 millions USD), and (**M**= Margin of
Contribution= 41,548 millions USD), then it's (**U**=
Utilized or Starting Capital planned), is:

$$U = F+S'i[1+s]+T+D+\{Q+P/[c-q]\}/I-M$$
$$= 29,750+48,851*0.0100[1+0.0500]$$
$$+3,385+2,370+\{35,351$$
$$+6,497/[1.2000\ 1.0100]\}$$
$$/0.9900-41,548$$

$$= \underline{64,720}\text{ millions USD}$$

Corporate IFRS-GAAP (B/S-I/S),ISBN-13: **978-1721066308**, ISBN-10: **1721066306**

Law-8187:

If both (**Q**= Quoted Longterm Liabilities= 35,351 millions USD), (**T**= Tax Paid= 3,385 millions USD), (**I**= L/E= Leverage or Gearing Ratio= 99.00%), (**c**= C/X= Current Ratio= 1.20 times), (**q**= [C-P]/X= Quick or Acid Test Ratio= 1.01 times), (**P**= Procured Inventories= 6,497 millions USD), (**D**= Dividend Paid= 2,370 millions USD), (**F**= Fixed Cost= 29,740 millions USD), (**i**= I/S= Interest Portion= 1.00%), (**s**= [S/S']-1= Sales Growth= 5.00%), (**S'**= Sales of Past Year= 48,851 millions USD), and (**U**= Utilized or Starting Capital= 64,720 millions USD), then it's (**M**= Margin of Contribution planned), is:

$$\textbf{M= F+S'i[1+s]+T+D+\{Q+P/[c-q]\}/I-U}$$
$$= 29,750+48,851*0.0100[1+0.0500]$$
$$+3,385+2,370+\{35,351$$
$$+6,497/[1.2000-1.0100]\}$$
$$/0.9900-64,720$$
$$= \underline{41,548} \text{ millions USD}$$

Corporate IFRS-GAAP (B/S-I/S),ISBN-13: **978-1721066308**, ISBN-10: **1721066306**

Law-8188:

If both (**Q**= Quoted Longterm Liabilities= 35,351 millions USD), (**T**= Tax Paid= 3,385 millions USD), (**l**= L/E= Leverage or Gearing Ratio= 99.00%), (**c**= C/X= Current Ratio= 1.20 times), (**q**= [C-P]/X= Quick or Acid Test Ratio= 1.01 times), (**P**= Procured Inventories= 6,497 millions USD), (**D**= Dividend Paid= 2,370 millions USD), (**M**= Margin of Contribution= 41,548 millions USD), (**i**= I/S= Interest Portion= 1.00%), (**s**= [S/S']-1= Sales Growth= 5.00%), (**S'**= Sales of Past Year= 48,851 millions USD), and (**U**= Utilized or Starting Capital= 64,720 millions USD), then it's (**F**= Fixed Cost planned), is:

$$F= U+M-S'i[1+s]-T-D+\{Q+P/[c-q]\}/l$$

$$= 64,720+41,548-48,851*0.0100$$
$$[1+0.0500]-3,385-2,370$$
$$-\{35,351+6,497/[1.2000$$
$$-1.0100]\}/0.9900$$

$$= 29,750 \text{ millions USD}$$

Corporate IFRS-GAAP (B/S-I/S),ISBN-13: **978-1721066308**, ISBN-10: **1721066306**

<u>Law-8189</u>:

If both (**Q**= Quoted Longterm Liabilities= <u>35,351</u> millions USD), (**T**= Tax Paid= <u>3,385</u> millions USD), (**∦**= L/E= Leverage or Gearing Ratio= <u>99.00%</u>), (**c**= C/X= Current Ratio= <u>1.20</u> times), (**q**= [C-P]/X= Quick or Acid Test Ratio= <u>1.01</u> times), (**P**= Procured Inventories= <u>6,497</u> millions USD), (**D**= Dividend Paid= <u>2,370</u> millions USD), (**M**= Margin of Contribution= <u>41,548</u> millions USD), (**i**= I/S= Interest Portion= <u>1.00%</u>), (**s**= [S/S']-1= Sales Growth= <u>5.00%</u>), (**F**= Fixed Cost= <u>29,740</u> millions USD), and (**U**= Utilized or Starting Capital= <u>64,720</u> millions USD), then it's (**S'**= Sales Past), must be:

$$S' = \{U+M-F-T-D+\{Q+P/[c-q]\}/\!\!/\}/\{i[1+s]\}$$
$$= \{64,720+41,548-29,750-3,385$$
$$-2,370-\{35,351+6,497$$
$$/[1.2000-1.0100]\}/0.9900\}$$
$$/\{0.0100[1+0.0500]\}$$
$$= \underline{48,851} \text{ millions USD}$$

Corporate IFRS-GAAP (B/S-I/S),ISBN-13: **978-1721066308**, ISBN-10: **1721066306**

Law-8190:

If both (**Q**= Quoted Longterm Liabilities= 35,351 millions USD), (**T**= Tax Paid= 3,385 millions USD), (**I**= L/E= Leverage or Gearing Ratio= 99.00%), (**c**= C/X= Current Ratio= 1.20 times), (**q**= [C-P]/X= Quick or Acid Test Ratio= 1.01 times), (**P**= Procured Inventories= 6,497 millions USD), (**D**= Dividend Paid= 2,370 millions USD), (**M**= Margin of Contribution= 41,548 millions USD), (**S'**= Sales of Past Year= 48,851 millions USD), (**s**= [S/S']-1= Sales Growth= 5.00%), (**F**= Fixed Cost= 29,740 millions USD), and (**U**= Utilized or Starting Capital= 64,720 millions USD), then it's (**i**= Interest Portion planned), is:

$$i= \{U+M-F-T-D+\{Q+P/[c-q]\}/I\}/\{S'[1+s]\}$$
$$= \{64,720+41,548-29,750-3,385$$
$$-2,370-\{35,351 \mid 6,497$$
$$/[1.2000-1.0100]\}/0.9900\}$$
$$/\{48,851[1+0.0500]\}$$
$$= 1.00\%$$

Corporate IFRS-GAAP (B/S-I/S),ISBN-13: **978-1721066308**, ISBN-10: **1721066306**

<u>Law-8191</u>:

If both (**Q**= Quoted Longterm Liabilities= <u>35,351</u> millions USD), (**T**= Tax Paid= <u>3,385</u> millions USD), (**I**= L/E= Leverage or Gearing Ratio= <u>99.00%</u>), (**c**= C/X= Current Ratio= <u>1.20</u> times), (**q**= [C-P]/X= Quick or Acid Test Ratio= <u>1.01</u> times), (**P**= Procured Inventories= <u>6,497</u> millions USD), (**D**= Dividend Paid= <u>2,370</u> millions USD), (**M**= Margin of Contribution= <u>41,548</u> millions USD), (**S'**= Sales of Past Year= <u>48,851</u> millions USD), (**i**= I/S= Interest Portion= <u>1.00%</u>), (**F**= Fixed Cost= <u>29,740</u> millions USD), and (**U**= Utilized or Starting Capital= <u>64,720</u> millions USD), then it's (**s**= Sales Growth planned), is:

$$s= (U+M-F-T-D+\{Q+P/[c-q]\}/I)/[S'i]-1$$
$$= (64,720+41,548-29,750-3,385$$
$$-2,370-\{35,351+6,497$$
$$/[1.2000-1.0100]\}/0.9900)$$
$$/[48,851*0.0100]-1$$
$$= \underline{5.00\%}$$

Corporate IFRS-GAAP (B/S-I/S),ISBN-13: **978-1721066308**, ISBN-10: **1721066306**

Law-8192:

If both (**Q**= Quoted Longterm Liabilities= 35,351 millions USD), (**s**= [S/S']-1= Sales Growth= 5.00%), (**l**= L/E= Leverage or Gearing Ratio= 99.00%), (**c**= C/X= Current Ratio= 1.20 times), (**q**= [C-P]/X= Quick or Acid Test Ratio= 1.01 times), (**P**= Procured Inventories= 6,497 millions USD), (**D**= Dividend Paid= 2,370 millions USD), (**M**= Margin of Contribution= 41,548 millions USD), (**S'**= Sales of Past Year= 48,851 millions USD), (**i**= I/S= Interest Portion= 1.00%), (**F**= Fixed Cost= 29,740 millions USD), and (**U**= Utilized or Starting Capital= 64,720 millions USD), then it's (**T**= Tax Paid planned), is:

$$T= U+M-F-S'i[1+s]-D+\{Q+P/[c-q]\}/l$$
$$= 64,720+41,548-29,750-48,851$$
$$*0.0100[1+0.0500]-2,370$$
$$-\{35,351+6,497/[1.2000$$
$$-1.0100]\}/0.9900$$
$$= 3,385 \text{ millions USD}$$

Corporate IFRS-GAAP (B/S-I/S),ISBN-13: **978-1721066308**, ISBN-10: **1721066306**

Law-8193:

If both (**Q**= Quoted Longterm Liabilities= 35,351 millions USD), (**s**= [S/S']-1= Sales Growth= 5.00%), (**l**= L/E= Leverage or Gearing Ratio= 99.00%), (**c**= C/X= Current Ratio= 1.20 times), (**q**= [C-P]/X= Quick or Acid Test Ratio= 1.01 times), (**P**= Procured Inventories= 6,497 millions USD), (**T**= Tax Paid= 3,385 millions USD), (**M**= Margin of Contribution= 41,548 millions USD), (**S'**= Sales of Past Year= 48,851 millions USD), (**i**= I/S= Interest Portion= 1.00%), (**F**= Fixed Cost= 29,740 millions USD), and (**U**= Utilized or Starting Capital= 64,720 millions USD), then it's (**D**= Dividend Paid planned), is:

$$D = U+M-F-S'i[1+s]-T+\{Q+P/[c-q]\}/l$$
$$= 64,720+41,548-29,750-48,851$$
$$*0.0100[1+0.0500]-3,385$$
$$-\{35,351+6,497/[1.2000$$
$$-1.0100]\}/0.9900$$
$$= 2,370 \text{ millions USD}$$

Corporate IFRS-GAAP (B/S-I/S),ISBN-13: **978-1721066308**, ISBN-10: **1721066306**

<u>Law-8194</u>:

If both (**Q**= Quoted Longterm Liabilities= <u>35,351</u> millions USD), (**s**= [S/S']-1= Sales Growth= <u>5.00%</u>), (**l**= L/E= Leverage or Gearing Ratio= <u>99.00%</u>), (**c**= C/X= Current Ratio= <u>1.20</u> times), (**q**= [C-P]/X= Quick or Acid Test Ratio= <u>1.01</u> times), (**D**= Dividend Paid= <u>2,370</u> millions USD), (**T**= Tax Paid= <u>3,385</u> millions USD), (**M**= Margin of Contribution= <u>41,548</u> millions USD), (**S'**= Sales of Past Year= <u>48,851</u> millions USD), (**i**= I/S= Interest Portion= <u>1.00%</u>), (**F**= Fixed Cost= <u>29,740</u> millions USD), and (**U**= Utilized or Starting Capital= <u>64,720</u> millions USD), then it's (**P**= Procured Inventory Days planned), is:

$$P= [c-q](l \{U+M-F-S'i[1+s]-T-D\}-Q)$$
$$= [1.2000-1.0100](0.9900\{64,720$$
$$+41,548-29,750-48,851$$
$${}^{ll}0.0100[1+0.0500]-3,385$$
$$-2,370\}-35,351)$$
$$= \underline{6,497} \text{ millions USD}$$

Law-8195:

If both (**Q**= Quoted Longterm Liabilities= 35,351 millions USD), (**s**= [S/S']-1= Sales Growth= 5.00%), (**I**= L/E= Leverage or Gearing Ratio= 99.00%), (**P**= Procured Inventories= 6,497 millions USD), (**q**= [C-P]/X= Quick or Acid Test Ratio= 1.01 times), (**D**= Dividend Paid= 2,370 millions USD), (**T**= Tax Paid= 3,385 millions USD), (**M**= Margin of Contribution= 41,548 millions USD), (**S'**= Sales of Past Year= 48,851 millions USD), (**i**= I/S= Interest Portion= 1.00%), (**F**= Fixed Cost= 29,740 millions USD), and (**U**= Utilized or Starting Capital= 64,720 millions USD), then it's (**c**= Current Ratio planned), is:

$$c= q+P/(I\{U+M-F-S'i[1+s]-T-D\}-Q)$$

$$= 1.0100+6,497/(099.00\{64,720$$
$$+41,548-29,750-48,851$$
$$*0.0100[1+0.0500]-3,385$$
$$-2,370\}-35,351)$$

$$= \underline{1.20} \text{ times}$$

Corporate IFRS-GAAP (B/S-I/S),ISBN-13: **978-1721066308**, ISBN-10: **1721066306**

Law-8196:

If both (**Q**= Quoted Longterm Liabilities= 35,351
millions USD), (**s**= [S/S']-1= Sales Growth= 5.00%),
(**l**= L/E= Leverage or Gearing Ratio= 99.00%), (**P**=
Procured Inventories= 6,497 millions USD), (**c**=
C/X= Current Ratio= 1.20 times), (**D**= Dividend
Paid= 2,370 millions USD), (**T**= Tax Paid= 3,385
millions USD), (**M**= Margin of Contribution= 41,548
millions USD), (**S'**= Sales of Past Year= 48,851
millions USD), (**i**= I/S= Interest Portion= 1.00%),
(**F**= Fixed Cost= 29,740 millions USD), and (**U**=
Utilized or Starting Capital= 64,720 millions USD),
then it's (**q**= Quick or Acid Test Ratio planned), is:

$$q= c\text{-}P/(l\ \{U+M\text{-}F\text{-}S'i[1+s]\text{-}T\text{-}D\}\text{-}Q)$$
$$= 1.2000\text{-}6,497/(0.9900\{64,720$$
$$+41,548\text{-}29,750\text{-}48,851$$
$${}^{*}0.0100[1+0.0500]\text{-}3,385$$
$$-2,370\}\text{-}35,351)$$
$$= \underline{1.01}\ \text{times}$$

Corporate IFRS-GAAP (B/S-I/S),ISBN-13: **978-1721066308**, ISBN-10: **1721066306**

Law-8197:

If both (**q**= [C-P]/X= Quick or Acid Test Ratio= 1.01 times), (**s**= [S/S']-1= Sales Growth= 5.00%), (**f**= L/E= Leverage or Gearing Ratio= 99.00%), (**p**= 360 P/V= Procured Inventory Days= 240 Days), (**V**= Variable Cost= 9,746 millions USD), (**c**= C/X= Current Ratio= 1.20 times), (**D**= Dividend Paid= 2,370 millions USD), (**T**= Tax Paid= 3,385 millions USD), (**M**= Margin of Contribution= 41,548 millions USD), (**S'**= Sales of Past Year= 48,851 millions USD), (**i**= I/S= Interest Portion= 1.00%), (**F**= Fixed Cost= 29,740 millions USD), and (**U**= Utilized or Starting Capital= 64,720 millions USD), then it's (**Q**= Quoted Longterm Liabilities planned), is:

$$Q= f\{U+M-F-S'i[1+s]-T-D\}-Vp/\{360[c-q]\}$$
$$= 0.9900\{64,720+41,548-29,750$$
$$-48,851*0.0100[1+0.0500]$$
$$-3,385-2,370\}-9,746*240$$
$$/\{360[1.2000-1.0100]\}$$
$$= 35,351 \text{ millions USD}$$

Corporate IFRS-GAAP (B/S-I/S),ISBN-13: **978-1721066308**, ISBN-10: **1721066306**

Law-8198:

If both (**q**= [C-P]/X= Quick or Acid Test Ratio= 1.01 times), (**s**= [S/S']-1= Sales Growth= 5.00%), (**Q**= Quoted Longterm Liabilities= 35,351 millions USD), (**p**= 360 P/V= Procured Inventory Days= 240 Days), (**V**= Variable Cost= 9,746 millions USD), (**c**= C/X= Current Ratio= 1.20 times), (**D**= Dividend Paid= 2,370 millions USD), (**T**= Tax Paid= 3,385 millions USD), (**M**= Margin of Contribution= 41,548 millions USD), (**S'**= Sales of Past Year= 48,851 millions USD), (**i**= I/S= Interest Portion= 1.00%), (**F**= Fixed Cost= 29,740 millions USD), and (**U**= Utilized or Starting Capital= 64,720 millions USD), then it's (**I** = Leverage or Gearing Ratio planned), is:

$$I = (Q+Vp/\{360[c-q]\})$$
$$/\{U+M-F-S'i[1+s]-T-D\}$$
$$- (35,351+9,746*240/\{360[1.2000$$
$$-1.0100]\})/\{64,720+41,548$$
$$-29,750-48,851*0.0100$$
$$[1+0.0500]-3,385-2,370\}$$
$$= 99.00\%$$

Corporate IFRS-GAAP (B/S-I/S),ISBN-13: **978-1721066308**, ISBN-10: **1721066306**

Law-8199:

If both (q= [C-P]/X= Quick or Acid Test Ratio= 1.01 times), (s= [S/S']-1= Sales Growth= 5.00%), (Q= Quoted Longterm Liabilities= 35,351 millions USD), (p= 360 P/V= Procured Inventory Days= 240 Days), (V= Variable Cost= 9,746 millions USD), (c= C/X= Current Ratio= 1.20 times), (D= Dividend Paid= 2,370 millions USD), (T= Tax Paid= 3,385 millions USD), (M= Margin of Contribution= 41,548 millions USD), (S'= Sales of Past Year= 48,851 millions USD), (i= I/S= Interest Portion= 1.00%), (F= Fixed Cost= 29,740 millions USD), and (I= L/E= Leverage or Gearing Ratio= 99.00%), then it's (U= Utilized or Starting Capital planned), is:

$$U = F + S'i[1+s] + T + D$$
$$+ (Q + Vp/\{360[c-q]\})/I - M$$
$$= 29,750 + 48,851*0.0100[1+0.0500]$$
$$+ 3,385 + 2,370 + (35,351$$
$$+ 9,746*240/\{360[1.2000$$
$$- 1.0100]\})/0.9900 - 41,548$$
$$= 64,720 \text{ millions USD}$$

Corporate IFRS-GAAP (B/S-I/S),ISBN-13: **978-1721066308**, ISBN-10: **1721066306**

Law-8200:

If both (q= [C-P]/X= Quick or Acid Test Ratio= 1.01 times), (s= [S/S']-1= Sales Growth= 5.00%), (Q= Quoted Longterm Liabilities= 35,351 millions USD), (p= 360 P/V= Procured Inventory Days= 240 Days), (V= Variable Cost= 9,746 millions USD), (c= C/X= Current Ratio= 1.20 times), (D= Dividend Paid= 2,370 millions USD), (T= Tax Paid= 3,385 millions USD), (U= Utilized or Starting Capital= 64,720 millions USD), (S'= Sales of Past Year= 48,851 millions USD), (i= I/S= Interest Portion= 1.00%), (F= Fixed Cost= 29,740 millions USD), and (l= L/E= Leverage or Gearing Ratio= 99.00%), then it's (M= Margin of Contribution planned), is:

$$M= F+S'i[1+s]+T+D$$
$$+(Q+Vp/\{360[c-q]\})/l-U$$

$$= 29,750+48,851*0.0100[1+0.0500]$$
$$+3,385+2,370+(35,351$$
$$+9,746*240/\{360[1.2000$$
$$-1.0100]\})/0.9900-64,720$$

$$= 41,548 \text{ millions USD}$$

Corporate IFRS-GAAP (B/S-I/S),ISBN-13: **978-1721066308**, ISBN-10: **1721066306**

Law-8201:

If both (**q**= [C-P]/X= Quick or Acid Test Ratio= 1.01 times), (**s**= [S/S']-1= Sales Growth= 5.00%), (**Q**= Quoted Longterm Liabilities= 35,351 millions USD), (**p**= 360 P/V= Procured Inventory Days= 240 Days), (**V**= Variable Cost= 9,746 millions USD), (**c**= C/X= Current Ratio= 1.20 times), (**D**= Dividend Paid= 2,370 millions USD), (**T**= Tax Paid= 3,385 millions USD), (**U**= Utilized or Starting Capital= 64,720 millions USD), (**S'**= Sales of Past Year= 48,851 millions USD), (**i**= I/S= Interest Portion= 1.00%), (**M**= Margin of Contribution= 41,548 millions USD), and (**I**= L/E= Leverage or Gearing Ratio= 99.00%), then it's (**F**= Fixed Cost planned), is:

$$F= U+M-S'i[1+s]-T-D$$
$$-(Q+Vp/\{360[c-q]\})/I$$
$$= 64,720+41,548-48,851*0.0100$$
$$[1+0.0500]-3,385-2,370$$
$$-(35,351+9,746*240/\{360$$
$$[1.2000-1.0100]\})/0.9900$$
$$= 29,750 \text{ millions USD}$$

Corporate IFRS-GAAP (B/S-I/S),ISBN-13: **978-1721066308**, ISBN-10: **1721066306**

Law-8202:

If both (**q**= [C-P]/X= Quick or Acid Test Ratio= 1.01 times), (**s**= [S/S']-1= Sales Growth= 5.00%), (**Q**= Quoted Longterm Liabilities= 35,351 millions USD), (**p**= 360 P/V= Procured Inventory Days= 240 Days), (**V**= Variable Cost= 9,746 millions USD), (**c**= C/X= Current Ratio= 1.20 times), (**D**= Dividend Paid= 2,370 millions USD), (**T**= Tax Paid= 3,385 millions USD), (**U**= Utilized or Starting Capital= 64,720 millions USD), (**F**= Fixed Cost= 29,740 millions USD), (**i**= I/S= Interest Portion= 1.00%), (**M**= Margin of Contribution= 41,548 millions USD), and (**/**= L/E= Leverage or Gearing Ratio= 99.00%), then it's (**S'**= Sales Past), must be:

$$S'= (U+M-F-T-D-(Q+Vp/\{360[c-q]\})//)$$
$$/\{i[1+s]\}$$
$$- (64,720+41,548-29,750-3,385-2,370$$
$$-(35,351+9,746*240/\{360$$
$$[1.2000-1.0100]\})/0.9900)$$
$$/\{0.0100[1+0.0500]\}$$
$$= 48,851 \text{ millions USD}$$

Corporate IFRS-GAAP (B/S-I/S),ISBN-13: **978-1721066308**, ISBN-10: **1721066306**

Law-8203:

If both (**q**= [C-P]/X= Quick or Acid Test Ratio= 1.01 times), (**s**= [S/S']-1= Sales Growth= 5.00%), (**Q**= Quoted Longterm Liabilities= 35,351 millions USD), (**p**= 360 P/V= Procured Inventory Days= 240 Days), (**V**= Variable Cost= 9,746 millions USD), (**c**= C/X= Current Ratio= 1.20 times), (**D**= Dividend Paid= 2,370 millions USD), (**T**= Tax Paid= 3,385 millions USD), (**U**= Utilized or Starting Capital= 64,720 millions USD), (**F**= Fixed Cost= 29,740 millions USD), (**S'**= Sales of Past Year= 48,851 millions USD), (**M**= Margin of Contribution= 41,548 millions USD), and (**l**= L/E= Leverage or Gearing Ratio= 99.00%), then it's (**i**= Interest Portion planned), is:

$$i= [U+M-F-T-D-(Q+Vp/\{360[c-q]\})/l]$$
$$/\{S'[1+s]\}$$
$$= [64,720+41,548-29,750-3,385$$
$$-2,370-(35,351+9,746*240$$
$$/\{360[1.2000-1.0100]\})$$
$$/0.9900]/\{48,851[1+0.0500]\}$$
$$= 1.00\%$$

Corporate IFRS-GAAP (B/S-I/S),ISBN-13: **978-1721066308**, ISBN-10: **1721066306**

Law-8204:

If both (**q**= [C-P]/X= Quick or Acid Test Ratio= 1.01 times), (**i**= I/S= Interest Portion= 1.00%), (**Q**= Quoted Longterm Liabilities= 35,351 millions USD), (**p**= 360 P/V= Procured Inventory Days= 240 Days), (**V**= Variable Cost= 9,746 millions USD), (**c**= C/X= Current Ratio= 1.20 times), (**D**= Dividend Paid= 2,370 millions USD), (**T**= Tax Paid= 3,385 millions USD), (**U**= Utilized or Starting Capital= 64,720 millions USD), (**F**= Fixed Cost= 29,740 millions USD), (**S'**= Sales of Past Year= 48,851 millions USD), (**M**= Margin of Contribution= 41,548 millions USD), and (**l**= L/E= Leverage or Gearing Ratio= 99.00%), then it's (**s**= Sales Growth planned), is:

$$s= (U+M-F-T-D-(Q+Vp/\{360[c-q]\})/l)/[S'i]-1$$
$$= (64,720+41,548-29,750-3,385$$
$$-2,370-(35,351+9,746*240$$
$$/\{360[1.2000-1.0100]\})$$
$$/0.9900)/[48,851*0.0100]-1$$
$$= 5.00\%$$

Corporate IFRS-GAAP (B/S-I/S),ISBN-13: **978-1721066308**, ISBN-10: **1721066306**

Law-8205:

If both (**q**= [C-P]/X= Quick or Acid Test Ratio= 1.01 times), (**i**= I/S= Interest Portion= 1.00%), (**Q**= Quoted Longterm Liabilities= 35,351 millions USD), (**p**= 360 P/V= Procured Inventory Days= 240 Days), (**V**= Variable Cost= 9,746 millions USD), (**c**= C/X= Current Ratio= 1.20 times), (**D**= Dividend Paid= 2,370 millions USD), (**s**= [S/S']-1= Sales Growth= 5.00%), (**U**= Utilized or Starting Capital= 64,720 millions USD), (**F**= Fixed Cost= 29,740 millions USD), (**S'**= Sales of Past Year= 48,851 millions USD), (**M**= Margin of Contribution= 41,548 millions USD), and (**l**= L/E= Leverage or Gearing Ratio= 99.00%), then it's (**T**= Tax Paid planned), is:

$$T= U+M-F-S'i[1+s]-D-(Q+Vp/\{360[c-q]\})/l$$
$$= (64,720+41,548-29,750-48,851$$
$$*0.0100[1+0.0500]-2,370$$
$$-(35,351+9,746*240/\{360$$
$$[1.2000-1.0100]\})/0.9900$$
$$= 3,385 \text{ millions USD}$$

Corporate IFRS-GAAP (B/S-I/S),ISBN-13: **978-1721066308**, ISBN-10: **1721066306**

<u>Law-8206</u>:

If both (**q**= [C-P]/X= Quick or Acid Test Ratio= <u>1.01</u> times), (**i**= I/S= Interest Portion= <u>1.00%</u>), (**Q**= Quoted Longterm Liabilities= <u>35,351</u> millions USD), (**p**= 360 P/V= Procured Inventory Days= <u>240</u> Days), (**V**= Variable Cost= <u>9,746</u> millions USD), (**c**= C/X= Current Ratio= <u>1.20</u> times), (**T**= Tax Paid= <u>3,385</u> millions USD), (**s**= [S/S']-1= Sales Growth= <u>5.00%</u>), (**U**= Utilized or Starting Capital= <u>64,720</u> millions USD), (**F**= Fixed Cost= <u>29,740</u> millions USD), (**S'**= Sales of Past Year= <u>48,851</u> millions USD), (**M**= Margin of Contribution= <u>41,548</u> millions USD), and (**I**= L/E= Leverage or Gearing Ratio= <u>99.00%</u>), then it's (**D**= Dividend Paid planned), is:

$$D= U+M-F-S'i[1+s]-T-(Q+Vp/\{360[c-q]\})/I$$
$$= (64,720+41,548-29,750-48,851$$
$$*0.0100[1+0.0500]-3,385$$
$$-(35,351+9,746*240/\{360$$
$$[1.2000-1.0100]\})/0.9900$$
$$= \underline{2,370} \text{ millions USD}$$

Corporate IFRS-GAAP (B/S-I/S),ISBN-13: **978-1721066308**, ISBN-10: **1721066306**

<u>Law-8207</u>:

If both (**q**= [C-P]/X= Quick or Acid Test Ratio= <u>1.01</u> times), (**i**= I/S= Interest Portion= <u>1.00%</u>), (**Q**= Quoted Longterm Liabilities= <u>35,351</u> millions USD), (**p**= 360 P/V= Procured Inventory Days= <u>240</u> Days), (**D**= Dividend Paid= <u>2,370</u> millions USD), (**c**= C/X= Current Ratio= <u>1.20</u> times), (**T**= Tax Paid= <u>3,385</u> millions USD), (**s**= [S/S']-1= Sales Growth= <u>5.00%</u>), (**U**= Utilized or Starting Capital= <u>64,720</u> millions USD), (**F**= Fixed Cost= <u>29,740</u> millions USD), (**S'**= Sales of Past Year= <u>48,851</u> millions USD), (**M**= Margin of Contribution= <u>41,548</u> millions USD), and (**I**= L/E= Leverage or Gearing Ratio= <u>99.00%</u>), then it's (**V**= Variable Cost planned), is:

$$V= [360[c-q](I \{U+M-F-S'i[1+s]$$
$$-T-D\}-Q)]/p$$
$$= [360[1.2000-1.0100](0.9900$$
$$\{64,720+41,548-29,750$$
$$-48,851*0.0100[1+0.0500]$$
$$-3,385-2,370\}-35,351)]/240$$
$$= \underline{9,746} \text{ millions USD}$$

Corporate IFRS-GAAP (B/S-I/S),ISBN-13: **978-1721066308**, ISBN-10: **1721066306**

Law-8208:

If both (**q**= [C-P]/X= Quick or Acid Test Ratio= 1.01 times), (**i**= I/S= Interest Portion= 1.00%), (**Q**= Quoted Longterm Liabilities= 35,351 millions USD), (**V**= Variable Cost= 9,746 millions USD), (**D**= Dividend Paid= 2,370 millions USD), (**c**= C/X= Current Ratio= 1.20 times), (**T**= Tax Paid= 3,385 millions USD), (**s**= [S/S']-1= Sales Growth= 5.00%), (**U**= Utilized or Starting Capital= 64,720 millions USD), (**F**= Fixed Cost= 29,740 millions USD), (**S'**= Sales of Past Year= 48,851 millions USD), (**M**= Margin of Contribution= 41,548 millions USD), and (**I**= L/E= Leverage or Gearing Ratio= 99.00%), then it's (**p**= Procured Inventory Days planned), is:

$$p= [360[c-q](I\{U+M-F-S'i[1+s]-T-D\}-Q)]/V$$
$$= [360[1.2000-1.0100](0.9900\{64,720$$
$$+41,548-29,750 +48,851$$
$$*0.0100[1+0.0500]-3,385$$
$$-2,370\}-35,351)]/9,746$$
$$= \underline{240} \text{ days}$$

Corporate IFRS-GAAP (B/S-I/S),ISBN-13: **978-1721066308**, ISBN-10: **1721066306**

Law-8209:

If both (q= [C-P]/X= Quick or Acid Test Ratio= 1.01 times), (i= I/S= Interest Portion= 1.00%), (Q= Quoted Longterm Liabilities= 35,351 millions USD), (V= Variable Cost= 9,746 millions USD), (D= Dividend Paid= 2,370 millions USD), (p= 360 P/V= Procured Inventory Days= 240 Days), (T= Tax Paid= 3,385 millions USD), (s= [S/S']-1= Sales Growth= 5.00%), (U= Utilized or Starting Capital= 64,720 millions USD), (F= Fixed Cost= 29,740 millions USD), (S'= Sales of Past Year= 48,851 millions USD), (M= Margin of Contribution= 41,548 millions USD), and (I= L/E= Leverage or Gearing Ratio= 99.00%), then it's (c= Current Ratio planned), is:

$$c= q+Vp/[360(I \{U+M-F-S'i[1+s]-T-D\}-Q)]$$
$$= 1.0100+9,746*240/[360(0.9900$$
$$\{64,720+41,548-29,750$$
$$-48,851*0.0100[1+0.0500]$$
$$-3,385-2,370\}-35,351)]$$
$$= 1.20 \text{ times}$$

Corporate IFRS-GAAP (B/S-I/S),ISBN-13: **978-1721066308**, ISBN-10: **1721066306**

Law-8210:

If both (**c**= C/X= Current Ratio= 1.20 times), (**i**= I/S= Interest Portion= 1.00%), (**Q**= Quoted Longterm Liabilities= 35,351 millions USD), (**V**= Variable Cost= 9,746 millions USD), (**D**= Dividend Paid= 2,370 millions USD), (**p**= 360 P/V= Procured Inventory Days= 240 Days), (**T**= Tax Paid= 3,385 millions USD), (**s**= [S/S']-1= Sales Growth= 5.00%), (**U**= Utilized or Starting Capital= 64,720 millions USD), (**F**= Fixed Cost= 29,740 millions USD), (**S'**= Sales of Past Year= 48,851 millions USD), (**M**= Margin of Contribution= 41,548 millions USD), and (**l**= L/E= Leverage or Gearing Ratio= 99.00%), then it's (**q**= Quick or Acid Test Ratio planned), is:

$$q = c\text{-}Vp/[360(l\,\{U+M\text{-}F\text{-}S'i[1+s]\text{-}T\text{-}D\}\text{-}Q)]$$

$$= 1.2000\text{-}9,746*240/[360(0.9900$$
$$\{64,720+41,548\text{-}29,750$$
$$48,851*0.0100[1+0.0500]$$
$$-3,385\text{-}2,370\}\text{-}35,351)]$$

$$= 1.01 \text{ times}$$

Corporate IFRS-GAAP (B/S-I/S),ISBN-13: **978-1721066308**, ISBN-10: **1721066306**

Law-8211:

If both (**c**= C/X= Current Ratio= 1.20 times), (**i**= I/S= Interest Portion= 1.00%), (**q**= [C-P]/X= Quick or Acid Test Ratio= 1.01 times), (**$**= Sales or Revenues= 51,294 millions USD), (**v**= V/S= Variable Portion= 19.00%), (**D**= Dividend Paid= 2,370 millions USD), (**p**= 360 P/V= Procured Inventory Days= 240 Days), (**T**= Tax Paid= 3,385 millions USD), (**s**= [S/S']-1= Sales Growth= 5.00%), (**U**= Utilized or Starting Capital= 64,720 millions USD), (**F**= Fixed Cost= 29,740 millions USD), (**$'**= Sales of Past Year= 48,851 millions USD), (**M**= Margin of Contribution= 41,548 millions USD), and (**$**= L/E= Leverage or Gearing Ratio= 99.00%), then it's (**Q**= Quoted Longterm Liabilities planned), is:

$$Q = I\{U+M-F-S'i[1+s]-T-D\}-Svp/\{360[c-q]\}$$
$$= 0.9900\{64,720+41,548-29,750$$
$$-48,851*0.0100[1+0.0500]$$
$$-3,385-2,370\}-51,294$$
$$*0.1900*240/\{360[1.2000$$
$$-1.0100]\}$$
$$= 35,351 \text{ millions USD}$$

Corporate IFRS-GAAP (B/S-I/S),ISBN-13: **978-1721066308**, ISBN-10: **1721066306**

<u>Law-8212</u>:

If both (**c**= C/X= Current Ratio= <u>1.20</u> times), (**i**= I/S= Interest Portion= <u>1.00%</u>), (**q**= [C-P]/X= Quick or Acid Test Ratio= <u>1.01</u> times), (**S**= Sales or Revenues= <u>51,294</u> millions USD), (**v**= V/S= Variable Portion= <u>19.00%</u>), (**D**= Dividend Paid= <u>2,370</u> millions USD), (**p**= 360 P/V= Procured Inventory Days= <u>240</u> Days), (**T**= Tax Paid= <u>3,385</u> millions USD), (**s**= [S/S']-1= Sales Growth= <u>5.00%</u>), (**U**= Utilized or Starting Capital= <u>64,720</u> millions USD), (**F**= Fixed Cost= <u>29,740</u> millions USD), (**S'**= Sales of Past Year= <u>48,851</u> millions USD), (**M**= Margin of Contribution= <u>41,548</u> millions USD), and (**Q**= Quoted Longterm Liabilities= <u>35,351</u> millions USD), then it's (**I** = Leverage or Gearing Ratio planned), is:

$$I = (Q+Svp/\{360[c-q]\})$$
$$/\{U+M-F-S'i[1+s]-T-D\}$$
$$- (35,351 + 51,294*0.1900*240/\{360$$
$$[1.2000-1.0100]\})/\{64,720$$
$$+41,548-29,750-48,851$$
$$*0.0100[1+0.0500]-3,385$$
$$-2,370\}$$
$$= \underline{99.00\%}$$

Corporate IFRS-GAAP (B/S-I/S),ISBN-13: **978-1721066308**, ISBN-10: **1721066306**

Law-8213:

If both (**c**= C/X= Current Ratio= 1.20 times), (**i**= I/S= Interest Portion= 1.00%), (**q**= [C-P]/X= Quick or Acid Test Ratio= 1.01 times), (**S**= Sales or Revenues= 51,294 millions USD), (**v**= V/S= Variable Portion= 19.00%), (**D**= Dividend Paid= 2,370 millions USD), (**p**= 360 P/V= Procured Inventory Days= 240 Days), (**T**= Tax Paid= 3,385 millions USD), (**s**= [S/S']-1= Sales Growth= 5.00%), (**Q**= Quoted Longterm Liabilities= 35,351 millions USD), (**F**= Fixed Cost= 29,740 millions USD), (**S'**= Sales of Past Year= 48,851 millions USD), (**M**= Margin of Contribution= 41,548 millions USD), and (**l**= L/E= Leverage or Gearing Ratio= 99.00%), then it's (**U**= Utilized or Starting Capital planned), is:

$$U= F+S'i[1+s]+T+D$$
$$+(Q+Svp/\{360[c-q]\})/l-M$$
$$= 29,750+48,851*0.0100[1+0.0500]$$
$$+3,385+2,370+(35,351$$
$$+51,294*0.1900*240/\{360$$
$$[1.2000-1.0100]\})/0.9900$$
$$-41,548$$
$$= \underline{64,720} \text{ millions USD}$$

Corporate IFRS-GAAP (B/S-I/S),ISBN-13: **978-1721066308**, ISBN-10: **1721066306**

Law-8214:

If both (**c**= C/X= Current Ratio= 1.20 times), (**i**= I/S=
Interest Portion= 1.00%), (**q**= [C-P]/X= Quick or
Acid Test Ratio= 1.01 times), (**S**= Sales or Revenues=
51,294 millions USD), (**v**= V/S= Variable Portion=
19.00%), (**D**= Dividend Paid= 2,370 millions USD),
(**p**= 360 P/V= Procured Inventory Days= 240 Days),
(**T**= Tax Paid= 3,385 millions USD), (**s**= [S/S']-1=
Sales Growth= 5.00%), (**Q**= Quoted Longterm
Liabilities= 35,351 millions USD), (**F**= Fixed Cost=
29,740 millions USD), (**S'**= Sales of Past Year=
48,851 millions USD), (**U**= Utilized or Starting
Capital= 64,720 millions USD), and (**I**= L/E=
Leverage or Gearing Ratio= 99.00%), then it's (**M**=
Margin of Contribution planned), is:

$$M= F+S'i[1+s]+T+D$$
$$+(Q+Svp/\{360[c-q]\})/I-U$$
$$= 29,750+48,851*0.0100[1+0.0500]$$
$$+3,385+2,370+(35,351$$
$$+51,294*0.1900*240/\{360$$
$$[1.2000-1.0100]\})/0.9900$$
$$-64,720$$
$$= 41,548 \text{ millions USD}$$

Corporate IFRS-GAAP (B/S-I/S),ISBN-13: **978-1721066308**, ISBN-10: **1721066306**

<u>Law-8215</u>:

If both (**c**= C/X= Current Ratio= <u>1.20</u> times), (**i**= I/S= Interest Portion= <u>1.00%</u>), (**q**= [C-P]/X= Quick or Acid Test Ratio= <u>1.01</u> times), (**S**= Sales or Revenues= <u>51,294</u> millions USD), (**v**= V/S= Variable Portion= <u>19.00%</u>), (**D**= Dividend Paid= <u>2,370</u> millions USD), (**p**= 360 P/V= Procured Inventory Days= <u>240</u> Days), (**T**= Tax Paid= <u>3,385</u> millions USD), (**s**= [S/S']-1= Sales Growth= <u>5.00%</u>), (**Q**= Quoted Longterm Liabilities= <u>35,351</u> millions USD), (**M**= Margin of Contribution= <u>41,548</u> millions USD), (**S'**= Sales of Past Year= <u>48,851</u> millions USD), (**U**= Utilized or Starting Capital= <u>64,720</u> millions USD), and (**I**= L/E= Leverage or Gearing Ratio= <u>99.00%</u>), then it's (**F**= Fixed Cost planned), is:

$$F= U+M-S'i[1+s]-T-D$$
$$-(Q+Svp/\{360[c-q]\})/I$$
$$= 64,720+41,548-48,851*0.0100$$
$$[1+0.0500]-3,385-2,370$$
$$-(35,351+51,294*0.1900$$
$$*240/\{360[1.2000$$
$$-1.0100]\})/0.9900$$
$$= \underline{29,750} \text{ millions USD}$$

Corporate IFRS-GAAP (B/S-I/S),ISBN-13: **978-1721066308**, ISBN-10: **1721066306**

Law-8216:

If both (**c**= C/X= Current Ratio= 1.20 times), (**i**= I/S= Interest Portion= 1.00%), (**q**= [C-P]/X= Quick or Acid Test Ratio= 1.01 times), (**S**= Sales or Revenues= 51,294 millions USD), (**v**= V/S= Variable Portion= 19.00%), (**D**= Dividend Paid= 2,370 millions USD), (**p**= 360 P/V= Procured Inventory Days= 240 Days), (**T**= Tax Paid= 3,385 millions USD), (**s**= [S/S']-1= Sales Growth= 5.00%), (**Q**= Quoted Longterm Liabilities= 35,351 millions USD), (**M**= Margin of Contribution= 41,548 millions USD), (**F**= Fixed Cost= 29,740 millions USD), (**U**= Utilized or Starting Capital= 64,720 millions USD), and (**I**= L/E= Leverage or Gearing Ratio= 99.00%), then it's (**S'**= Sales Past), must be:

$$S'= [U+M-F-T-D-(Q+Svp/\{360[c-q]\})/I]$$
$$/\{i[1+s]\}$$
$$- [64,720+41,548-29,750-3,385-2,370$$
$$-(35,351+51,294*0.1900*240$$
$$/\{360[1.2000-1.0100]\})$$
$$/0.9900]/\{0.0100[1+0.0500]\}$$
$$= 48,851 \text{ millions USD}$$

Corporate IFRS-GAAP (B/S-I/S),ISBN-13: **978-1721066308**, ISBN-10: **1721066306**

Law-8217:

If both (**c**= C/X= Current Ratio= 1.20 times), (**S'**= Sales of Past Year= 48,851 millions USD), (**q**= Quick or Acid Test Ratio= [C-P]/X= 1.01 times), (**S**= Sales or Revenues= 51,294 millions USD), (**v**= V/S= Variable Portion= 19.00%), (**D**= Dividend Paid= 2,370 millions USD), (**p**= 360 P/V= Procured Inventory Days= 240 Days), (**T**= Tax Paid= 3,385 millions USD), (**s**= [S/S']-1= Sales Growth= 5.00%), (**Q**= Quoted Longterm Liabilities= 35,351 millions USD), (**M**= Margin of Contribution= 41,548 millions USD), (**F**= Fixed Cost= 29,740 millions USD), (**U**= Utilized or Starting Capital= 64,720 millions USD), and (**I**= L/E= Leverage or Gearing Ratio= 99.00%), then it's (**i**= Interest Portion planned), is:

$$i= [U+M-F-T-D-(Q+Svp/\{360[c-q]\})/I]/\{S'[1+s]\}$$

$$= [64,720+41,548-29,750-3,385$$
$$-2,370-(35,351+51,294$$
$$*0.1900*240/\{360[1.2000$$
$$-1.0100]\})/0.9900]/\{48,851$$
$$[1+0.0500]\}$$

$$= 1.00\%$$

Corporate IFRS-GAAP (B/S-I/S),ISBN-13: **978-1721066308**, ISBN-10: **1721066306**

<u>Law-8218</u>:

If both (**c**= C/X= Current Ratio= <u>1.20</u> times), (**S'**= Sales of Past Year= <u>48,851</u> millions USD), (**q**= Quick or Acid Test Ratio= [C-P]/X= <u>1.01</u> times), (**S**= Sales or Revenues= <u>51,294</u> millions USD), (**v**= V/S= Variable Portion= <u>19.00%</u>), (**D**= Dividend Paid= <u>2,370</u> millions USD), (**p**= 360 P/V= Procured Inventory Days= <u>240</u> Days), (**T**= Tax Paid= <u>3,385</u> millions USD), (**i**= I/S= Interest Portion= <u>1.00%</u>), (**Q**= Quoted Longterm Liabilities= <u>35,351</u> millions USD), (**M**= Margin of Contribution= <u>41,548</u> millions USD), (**F**= Fixed Cost= <u>29,740</u> millions USD), (**U**= Utilized or Starting Capital= <u>64,720</u> millions USD), and (**I**= L/E= Leverage or Gearing Ratio= <u>99.00%</u>), then it's (**s**= Sales Growth planned), is:

$$s= [U+M-F-T-D-(Q+Svp/\{360[c-q]\})/I]$$
$$/[S'i]-1$$
$$= [64,720+41,548-29,750-3,385$$
$$-2,370-(35,351+51,294$$
$$*0.1900*240/\{360[1.2000$$
$$-1.0100]\})/0.9900]$$
$$/[48,851*0.0100]-1$$
$$= \underline{5.00\%}$$

Corporate IFRS-GAAP (B/S-I/S),ISBN-13: **978-1721066308**, ISBN-10: **1721066306**

Law-8219:

If both (**c**= C/X= Current Ratio= 1.20 times), (**S'**= Sales of Past Year= 48,851 millions USD), (**q**= Quick or Acid Test Ratio= [C-P]/X= 1.01 times), (**S**= Sales or Revenues= 51,294 millions USD), (**v**= V/S= Variable Portion= 19.00%), (**D**= Dividend Paid= 2,370 millions USD), (**p**= 360 P/V= Procured Inventory Days= 240 Days), (**s**= [S/S']-1= Sales Growth= 5.00%), (**i**= I/S= Interest Portion= 1.00%), (**Q**= Quoted Longterm Liabilities= 35,351 millions USD), (**M**= Margin of Contribution= 41,548 millions USD), (**F**= Fixed Cost= 29,740 millions USD), (**U**= Utilized or Starting Capital= 64,720 millions USD), and (**I**= L/E= Leverage or Gearing Ratio= 99.00%), then it's (**T**= Tax Paid planned), is:

$$T= U+M-F-S'i[1+s]-D-(Q+Svp/\{360[c-q]\})/I$$
$$= 64,720+41,548-29,750-48,851$$
$$*0.0100[1+0.0500]-2,370$$
$$-(35,351+51,294*0.1900$$
$$*240/\{360[1.2000$$
$$-1.0100]\})/0.9900$$
$$= 3,385 \text{ millions USD}$$

Corporate IFRS-GAAP (B/S-I/S),ISBN-13: **978-1721066308**, ISBN-10: **1721066306**

<u>Law-8220</u>:

If both (**c**= C/X= Current Ratio= <u>1.20</u> times), (**S'**= Sales of Past Year= <u>48,851</u> millions USD), (**q**= Quick or Acid Test Ratio= [C-P]/X= <u>1.01</u> times), (**S**= Sales or Revenues= <u>51,294</u> millions USD), (**v**= V/S= Variable Portion= <u>19.00%</u>), (**T**= Tax Paid= <u>3,385</u> millions USD), (**p**= 360 P/V= Procured Inventory Days= <u>240</u> Days), (**s**= [S/S']-1= Sales Growth= <u>5.00%</u>), (**i**= I/S= Interest Portion= <u>1.00%</u>), (**Q**= Quoted Longterm Liabilities= <u>35,351</u> millions USD), (**M**= Margin of Contribution= <u>41,548</u> millions USD), (**F**= Fixed Cost= <u>29,740</u> millions USD), (**U**= Utilized or Starting Capital= <u>64,720</u> millions USD), and (**I**= L/E= Leverage or Gearing Ratio= <u>99.00%</u>), then it's (**D**= Dividend Paid planned), is:

$$D= U+M-F-S'i[1+s]-T-(Q+Svp/\{360[c-q]\})/I$$

$$= 64,720+41,548-29,750-48,851$$
$$*0.0100[1+0.0500]-3,385$$
$$-(35,351+51,294*0.1900$$
$$*240/\{360[1.2000$$
$$-1.0100]\})/0.9900$$

$$= \underline{2,370} \text{ millions USD}$$

Corporate IFRS-GAAP (B/S-I/S),ISBN-13: **978-1721066308**, ISBN-10: **1721066306**

Law-8221:

If both (**c**= C/X= Current Ratio= 1.20 times), (**S'**= Sales of Past Year= 48,851 millions USD), (**q**= Quick or Acid Test Ratio= [C-P]/X= 1.01 times), (**D**= Dividend Paid= 2,370 millions USD), (**v**= V/S= Variable Portion= 19.00%), (**T**= Tax Paid= 3,385 millions USD), (**p**= 360 P/V= Procured Inventory Days= 240 Days), (**s**= [S/S']-1= Sales Growth= 5.00%), (**i**= I/S= Interest Portion= 1.00%), (**Q**= Quoted Longterm Liabilities= 35,351 millions USD), (**M**= Margin of Contribution= 41,548 millions USD), (**F**= Fixed Cost= 29,740 millions USD), (**U**= Utilized or Starting Capital= 64,720 millions USD), and (**I**= L/E= Leverage or Gearing Ratio= 99.00%), then it's (**S**= Sales or Revenues planned), is:

$$S= 360[\textbf{c-q}](\textbf{I} \{\textbf{U+M-F-S'i}[1+\textbf{s}]\textbf{-T-D}\}\textbf{-Q})$$
$$/[\textbf{vp}]$$
$$= 360[1.2000-1.0100](0.9900$$
$$\{64,720+41,548-29,750$$
$$-48,851*0.0100[1+0.0500]$$
$$-3,385-2,370\}-35,351)$$
$$/[0.1900*240]$$
$$= 51,294 \text{ millions USD}$$

Corporate IFRS-GAAP (B/S-I/S),ISBN-13: **978-1721066308**, ISBN-10: **1721066306**

Law-8222:

If both (**c**= C/X= Current Ratio= 1.20 times), (**$'**= Sales of Past Year= 48,851 millions USD), (**q**= Quick or Acid Test Ratio= [C-P]/X= 1.01 times), (**D**= Dividend Paid= 2,370 millions USD), (**$**= Sales or Revenues= 51,294 millions USD), (**T**= Tax Paid= 3,385 millions USD), (**p**= 360 P/V= Procured Inventory Days= 240 Days), (**s**= [S/S']-1= Sales Growth= 5.00%), (**i**= I/S= Interest Portion= 1.00%), (**Q**= Quoted Longterm Liabilities= 35,351 millions USD), (**M**= Margin of Contribution= 41,548 millions USD), (**F**= Fixed Cost= 29,740 millions USD), (**U**= Utilized or Starting Capital= 64,720 millions USD), and (**$**= L/E= Leverage or Gearing Ratio= 99.00%), then it's (**v**= Variable Portion planned), is:

$$\textbf{v}= 360[\textbf{c-q}](\textbf{$}\{\textbf{U+M-F-$'i}[1+\textbf{s}]\textbf{-T-D}\}\textbf{-Q})/[\textbf{$p}]$$

$$= 360[1.2000-1.0100](0.9900\{64,720 +41,548-29,750-48,851 *0.0100[1+0.0500]-3,385 -2,370\}-35,351) /[51,294*240]$$

$$= 19.00\%$$

Corporate IFRS-GAAP (B/S-I/S),ISBN-13: **978-1721066308**, ISBN-10: **1721066306**

Law-8223:

If both (**c**= C/X= Current Ratio= 1.20 times), (**$'**= Sales of Past Year= 48,851 millions USD), (**q**= Quick or Acid Test Ratio= [C-P]/X= 1.01 times), (**D**= Dividend Paid= 2,370 millions USD), (**$**= Sales or Revenues= 51,294 millions USD), (**T**= Tax Paid= 3,385 millions USD), (**v**= V/S= Variable Portion= 19.00%), (**s**= [S/S']-1= Sales Growth= 5.00%), (**i**= I/S= Interest Portion= 1.00%), (**Q**= Quoted Longterm Liabilities= 35,351 millions USD), (**M**= Margin of Contribution= 41,548 millions USD), (**F**= Fixed Cost= 29,740 millions USD), (**U**= Utilized or Starting Capital= 64,720 millions USD), and (**I**= L/E= Leverage or Gearing Ratio= 99.00%), then it's (**p**= Procured Inventory Days planned), is:

$$p= 360[c\text{-}q](I\{U+M\text{-}F\text{-}\$'i[1+s]\text{-}T\text{-}D\}\text{-}Q)/[\$v]$$

$$= 360[1.2000\text{-}1.0100](0.9900$$
$$\{64,720+41,548\text{-}29,750$$
$$\text{-}48,851*0.0100[1+0.0500]$$
$$\text{-}3,385\text{-}2,370\}\text{-}35,351)$$
$$/[51,294*0.1900]$$

$$= \underline{240} \text{ days}$$

Corporate IFRS-GAAP (B/S-I/S),ISBN-13: **978-1721066308**, ISBN-10: **1721066306**

Law-8224:

If both (**p**= 360 P/V= Procured Inventory Days= 240
Days), (**$'**= Sales of Past Year= 48,851 millions
USD), (**q**= [C-P]/X= Quick or Acid Test Ratio= 1.01
times), (**D**= Dividend Paid= 2,370 millions USD), (**$**=
Sales or Revenues= 51,294 millions USD), (**T**= Tax
Paid= 3,385 millions USD), (**v**= V/S= Variable
Portion= 19.00%), (**s**= [S/S']-1= Sales Growth=
5.00%), (**i**= I/S= Interest Portion= 1.00%), (**Q**=
Quoted Longterm Liabilities= 35,351 millions USD),
(**M**= Margin of Contribution= 41,548 millions USD),
(**F**= Fixed Cost= 29,740 millions USD), (**U**= Utilized
or Starting Capital= 64,720 millions USD), and (**l**=
L/E= Leverage or Gearing Ratio= 99.00%), then it's
(**c**= Current Ratio planned), is:

$$c= q+\text{Svp}/[360(l\ \{U+M-F-\text{S'}i[1+s]-T-D\}-Q)]$$
$$= 1.0100+51,294*0.1900*240$$
$$/[360(0.9900\{64,720$$
$$+41,548-29,750-48,851$$
$$*0.0100[1+0.0500]-3,385$$
$$-2,370\}-35,351)]$$
$$= 1.20 \text{ times}$$

Corporate IFRS-GAAP (B/S-I/S),ISBN-13: **978-1721066308**, ISBN-10: **1721066306**

<u>Law-8225</u>:

If both (**p**= 360 P/V= Procured Inventory Days= <u>240</u>
Days), (**S'**= Sales of Past Year= <u>48,851</u> millions
USD), (**c**= C/X= Current Ratio= <u>1.20</u> times), (**D**=
Dividend Paid= <u>2,370</u> millions USD), (**S**= Sales or
Revenues= <u>51,294</u> millions USD), (**T**= Tax Paid=
<u>3,385</u> millions USD), (**v**= V/S= Variable Portion=
<u>19.00%</u>), (**s**= [S/S']-1= Sales Growth= <u>5.00%</u>), (**i**=
I/S= Interest Portion= <u>1.00%</u>), (**Q**= Quoted Longterm
Liabilities= <u>35,351</u> millions USD), (**M**= Margin of
Contribution= <u>41,548</u> millions USD), (**F**= Fixed
Cost= <u>29,740</u> millions USD), (**U**= Utilized or Starting
Capital= <u>64,720</u> millions USD), and (**I**= L/E=
Leverage or Gearing Ratio= <u>99.00%</u>), then it's (**q**=
Quick or Acid Test Ratio planned), is:

$$q= c\text{-}Svp/[360(I \{U+M\text{-}F\text{-}S'i[1+s]\text{-}T\text{-}D\}\text{-}Q)]$$
$$= 1.2000\text{-}51,294*0.1900*240/[360$$
$$(0.9900\{64,720+41,548$$
$$-29,750\text{-}48,851*0.0100$$
$$[1+0.0500]\text{-}3,385\text{-}2,370\}$$
$$-35,351)]$$
$$= \underline{1.01} \text{ times}$$

Corporate IFRS-GAAP (B/S-I/S),ISBN-13: **978-1721066308**, ISBN-10: **1721066306**

<u>Law-8226</u>:

If both (**p**= 360 P/V= Procured Inventory Days= <u>240</u> Days), (**S'**= Sales of Past Year= <u>48,851</u> millions USD), (**c**= C/X= Current Ratio= <u>1.20</u> times), (**D**= Dividend Paid= <u>2,370</u> millions USD), (**T**= Tax Paid= <u>3,385</u> millions USD), (**v**= V/S= Variable Portion= <u>19.00%</u>), (**s**= [S/S']-1= Sales Growth= <u>5.00%</u>), (**i**= I/S= Interest Portion= <u>1.00%</u>), (**q**= [C-P]/X= Quick or Acid Test Ratio= <u>1.01</u> times), (**M**= Margin of Contribution= <u>41,548</u> millions USD), (**F**= Fixed Cost= <u>29,740</u> millions USD), (**U**= Utilized or Starting Capital= <u>64,720</u> millions USD), and (**I**= L/E= Leverage or Gearing Ratio= <u>99.00%</u>), then it's (**Q**= Quoted Longterm Liabilities planned), is:

$$Q = I\{U+M-F-S'i[1+s]-T-D\}$$
$$-S'vp[1+s]/\{360[c-q]\}$$
$$= 0.9900\{64,720+41,548-29,750$$
$$-48,851*0.0100[1+0.0500]$$
$$-3,385-2,370\}-48,851$$
$$*0.1900*240[1+0.0500]$$
$$/\{360[1.2000-1.0100]\}$$
$$= \underline{35,351} \text{ millions USD}$$

Corporate IFRS-GAAP (B/S-I/S),ISBN-13: **978-1721066308**, ISBN-10: **1721066306**

<u>Law-8227</u>:

If both (**p**= 360 P/V= Procured Inventory Days= <u>240</u> Days), (**$'**= Sales of Past Year= <u>48,851</u> millions USD), (**c**= C/X= Current Ratio= <u>1.20</u> times), (**D**= Dividend Paid= <u>2,370</u> millions USD), (**T**= Tax Paid= <u>3,385</u> millions USD), (**v**= V/S= Variable Portion= <u>19.00%</u>), (**s**= [S/S']-1= Sales Growth= <u>5.00%</u>), (**i**= I/S= Interest Portion= <u>1.00%</u>), (**q**= [C-P]/X= Quick or Acid Test Ratio= <u>1.01</u> times), (**M**= Margin of Contribution= <u>41,548</u> millions USD), (**F**= Fixed Cost= <u>29,740</u> millions USD), (**U**= Utilized or Starting Capital= <u>64,720</u> millions USD), and (**Q**= Quoted Longterm Liabilities= <u>35,351</u> millions USD), then it's (**I** = Leverage or Gearing Ratio planned), is:

$$I = (\mathbf{Q+S'vp}[1+\mathbf{s}]/\{360[\mathbf{c\text{-}q}]\})$$
$$/\{\mathbf{U+M\text{-}F\text{-}S'i}[1+\mathbf{s}]\text{-}\mathbf{T\text{-}D}\}$$
$$= (35,351+48,851*0.1900*240$$
$$[1+0.0500]/\{360[1.2000$$
$$-1.0100]\})/\{64,720+41,548$$
$$-29,750-48,851*0.0100$$
$$[1+0.0500]-3,385-2,370\}$$
$$= \underline{99.00\%}$$

Corporate IFRS-GAAP (B/S-I/S),ISBN-13: **978-1721066308**, ISBN-10: **1721066306**

<u>Law-8228</u>:

If both (**p**= 360 P/V= Procured Inventory Days= <u>240</u> Days), (**S'**= Sales of Past Year= <u>48,851</u> millions USD), (**c**= C/X= Current Ratio= <u>1.20</u> times), (**D**= Dividend Paid= <u>2,370</u> millions USD), (**T**= Tax Paid= <u>3,385</u> millions USD), (**v**= V/S= Variable Portion= <u>19.00%</u>), (**s**= [S/S']-1= Sales Growth= <u>5.00%</u>), (**i**= I/S= Interest Portion= <u>1.00%</u>), (**q**= [C-P]/X= Quick or Acid Test Ratio= <u>1.01</u> times), (**M**= Margin of Contribution= <u>41,548</u> millions USD), (**F**= Fixed Cost= <u>29,740</u> millions USD), (**I**= L/E= Leverage or Gearing Ratio= <u>99.00%</u>), and (**Q**= Quoted Longterm Liabilities= <u>35,351</u> millions USD), then it's (**U**= Utilized or Starting Capital planned), is:

$$U= F+S'i[1+s]+T+D+(Q+S'vp[1+s]$$
$$/\{360[c-q]\})/I -M$$
$$= 29,750+48,851*0.0100[1+0.0500]$$
$$+3,385 \mid 2,370 \mid (35,351$$
$$+48,851*0.1900*240$$
$$[1+0.0500]//\{360[1.2000$$
$$-1.0100]\})/0.9900-41,548$$
$$= \underline{64,720} \text{ millions USD}$$

Corporate IFRS-GAAP (B/S-I/S),ISBN-13: **978-1721066308**, ISBN-10: **1721066306**

Law-8229:

If both (**p**= 360 P/V= Procured Inventory Days= 240 Days), (**S'**= Sales of Past Year= 48,851 millions USD), (**c**= C/X= Current Ratio= 1.20 times), (**D**= Dividend Paid= 2,370 millions USD), (**T**= Tax Paid= 3,385 millions USD), (**v**= V/S= Variable Portion= 19.00%), (**s**= [S/S']-1= Sales Growth= 5.00%), (**i**= I/S= Interest Portion= 1.00%), (**q**= [C-P]/X= Quick or Acid Test Ratio= 1.01 times), (**U**= Utilized or Starting Capital= 64,720 millions USD), (**F**= Fixed Cost= 29,740 millions USD), (**I**= L/E= Leverage or Gearing Ratio= 99.00%), and (**Q**= Quoted Longterm Liabilities= 35,351 millions USD), then it's (**M**= Margin of Contribution planned), is:

$$M= F+S'i[1+s]+T+D+(Q+S'vp[1+s]$$
$$/\{360[c-q]\})/I-U$$

$$= 29,750+48,851*0.0100[1+0.0500]$$
$$+3,385+2,370+(35,351$$
$$+48,851*0.1900*240$$
$$[1+0.0500]//\{360[1.2000$$
$$-1.0100]\})/0.9900-64,720$$

$$= 41,548 \text{ millions USD}$$

Corporate IFRS-GAAP (B/S-I/S),ISBN-13: **978-1721066308**, ISBN-10: **1721066306**

Law-8230:

If both (**c**= C/X= Current Ratio= <u>1.20</u> times), (**i**= I/S= Interest Portion= <u>1.00%</u>), (**q**= [C-P]/X= Quick or Acid Test Ratio= <u>1.01</u> times), (**v**= V/S= Variable Portion= <u>19.00%</u>), (**D**= Dividend Paid= <u>2,370</u> millions USD), (**p**= 360 P/V= Procured Inventory Days= <u>240</u> Days), (**T**= Tax Paid= <u>3,385</u> millions USD), (**s**= [S/S']-1= Sales Growth= <u>5.00%</u>), (**Q**= Quoted Longterm Liabilities= <u>35,351</u> millions USD), (**M**= Margin of Contribution= <u>41,548</u> millions USD), (**S'**= Sales of Past Year= <u>48,851</u> millions USD), (**U**= Utilized or Starting Capital= <u>64,720</u> millions USD), and (**F**= L/E= Leverage or Gearing Ratio= <u>99.00%</u>), then it's (**F**= Fixed Cost planned), is:

$$F= U+M-S'i[1+s]-T-D-(Q+S'vp[1+s]$$
$$/\{360[c-q]\})/F$$
$$= 64{,}720+41{,}548-48{,}851*0.0100$$
$$[1+0.0500]-3{,}385-2{,}370$$
$$-(35{,}351+48{,}851*0.1900$$
$$*240[1+s]/\{360[1.2000$$
$$-1.0100]\})/0.9900$$
$$= \underline{29{,}750} \text{ millions USD}$$

Corporate IFRS-GAAP (B/S-I/S),ISBN-13: **978-1721066308**, ISBN-10: **1721066306**

<u>Law-8231</u>:

If both (**c**= C/X= Current Ratio= <u>1.20</u> times), (**i**= I/S= Interest Portion= <u>1.00%</u>), (**q**= [C-P]/X= Quick or Acid Test Ratio= <u>1.01</u> times), (**v**= V/S= Variable Portion= <u>19.00%</u>), (**D**= Dividend Paid= <u>2,370</u> millions USD), (**p**= 360 P/V= Procured Inventory Days= <u>240</u> Days), (**T**= Tax Paid= <u>3,385</u> millions USD), (**s**= [S/S']-1= Sales Growth= <u>5.00%</u>), (**Q**= Quoted Longterm Liabilities= <u>35,351</u> millions USD), (**M**= Margin of Contribution= <u>41,548</u> millions USD), (**F**= Fixed Cost= <u>29,740</u> millions USD), (**U**= Utilized or Starting Capital= <u>64,720</u> millions USD), and (**l**= L/E= Leverage or Gearing Ratio= <u>99.00%</u>), then it's (**S'**= Sales Past), must be:

$$S' = [U+M-F-T-D-(Q+S'vp[1+s]$$
$$/\{360[c-q]\})/l\ ]/\{i[1+s]\}$$
$$= [64,720+41,548-29,750-3,385$$
$$-2,370-(35,351+48,851$$
$$*0.1900*240[1+0.0500]$$
$$/\{360[1.2000-1.0100]\})$$
$$/0.9900]/\{0.0100$$
$$[1+0.0500]\}$$
$$= \underline{48,851}\ \text{millions USD}$$

Corporate IFRS-GAAP (B/S-I/S),ISBN-13: **978-1721066308**, ISBN-10: **1721066306**

Law-8232:

If both (c= C/X= Current Ratio= 1.20 times), (S'= Sales of Past Year= 48,851 millions USD), (q= Quick or Acid Test Ratio= [C-P]/X= 1.01 times), (v= V/S= Variable Portion= 19.00%), (D= Dividend Paid= 2,370 millions USD), (p= 360 P/V= Procured Inventory Days= 240 Days), (T= Tax Paid= 3,385 millions USD), (s= [S/S']-1= Sales Growth= 5.00%), (Q= Quoted Longterm Liabilities= 35,351 millions USD), (M= Margin of Contribution= 41,548 millions USD), (F= Fixed Cost= 29,740 millions USD), (U= Utilized or Starting Capital= 64,720 millions USD), and (I= L/E= Leverage or Gearing Ratio= 99.00%), then it's (i= Interest Portion planned), is:

$$i= [U+M-F-T-D-(Q+S'vp[1+s]$$
$$/\{360[c-q]\})/I\,]/\{S'[1+s]\}$$
$$= [64,720+41,548-29,750-3,385$$
$$-2,370-(35,351+48,851$$
$$*0.1900*240[1+0.0500]$$
$$/\{360[1.2000-1.0100]\})$$
$$/0.9900]/\{48,851$$
$$[1+0.0500]\}$$
$$= 1.00\%$$

Corporate IFRS-GAAP (B/S-I/S),ISBN-13: **978-1721066308**, ISBN-10: **1721066306**

<u>Law-8233</u>:

If both (c= C/X= Current Ratio= 1.20 times), (S'= Sales of Past Year= 48,851 millions USD), (q= Quick or Acid Test Ratio= [C-P]/X= 1.01 times), (v= V/S= Variable Portion= 19.00%), (D= Dividend Paid= 2,370 millions USD), (p= 360 P/V= Procured Inventory Days= 240 Days), (T= Tax Paid= 3,385 millions USD), (i= I/S= Interest Portion= 1.00%), (Q= Quoted Longterm Liabilities= 35,351 millions USD), (M= Margin of Contribution= 41,548 millions USD), (F= Fixed Cost= 29,740 millions USD), (U= Utilized or Starting Capital= 64,720 millions USD), and (I= L/E= Leverage or Gearing Ratio= 99.00%), then it's (s= Sales Growth planned), is:

$$s= [U+M-F-T-D-(Q+S'vp[1+s]$$
$$/\{360[c-q]\})/I]/[S'i]-1$$
$$= [64,720+41,548-29,750-3,385$$
$$-2,370-(35,351+48,851$$
$$*0.1900*240[1+0.0500]$$
$$/\{360[1.2000-1.0100]\})$$
$$/0.9900]/[48,851*0.0100]-1$$
$$= 5.00\%$$

Corporate IFRS-GAAP (B/S-I/S),ISBN-13: **978-1721066308**, ISBN-10: **1721066306**

Law-8234:

If both (c= C/X= Current Ratio= 1.20 times), (S'= Sales of Past Year= 48,851 millions USD), (q= Quick or Acid Test Ratio= [C-P]/X= 1.01 times), (v= V/S= Variable Portion= 19.00%), (D= Dividend Paid= 2,370 millions USD), (p= 360 P/V= Procured Inventory Days= 240 Days), (s= [S/S']-1= Sales Growth= 5.00%), (i= I/S= Interest Portion= 1.00%), (Q= Quoted Longterm Liabilities= 35,351 millions USD), (M= Margin of Contribution= 41,548 millions USD), (F= Fixed Cost= 29,740 millions USD), (U= Utilized or Starting Capital= 64,720 millions USD), and (I= L/E= Leverage or Gearing Ratio= 99.00%), then it's (T= Tax Paid planned), is:

$$T= U+M-F-S'i[1+s]-D-(Q+S'vp[1+s]/\{360[c-q]\})/I$$

$$= 64,720+41,548-29,750-48,851$$
$$*0.0100[1+0.0500]-2,370$$
$$-(35,351+48,851*0.1900$$
$$*240[1+0.0500]/\{360$$
$$[1.2000-1.0100]\})/0.9900$$

$$= 3,385 \text{ millions USD}$$

Corporate IFRS-GAAP (B/S-I/S),ISBN-13: **978-1721066308**, ISBN-10: **1721066306**

Law-8235:

If both (**c**= C/X= Current Ratio= 1.20 times), (**S'**= Sales of Past Year= 48,851 millions USD), (**q**= Quick or Acid Test Ratio= [C-P]/X= 1.01 times), (**v**= V/S= Variable Portion= 19.00%), (**T**= Tax Paid= 3,385 millions USD), (**p**= 360 P/V= Procured Inventory Days= 240 Days), (**s**= [S/S']-1= Sales Growth= 5.00%), (**i**= I/S= Interest Portion= 1.00%), (**Q**= Quoted Longterm Liabilities= 35,351 millions USD), (**M**= Margin of Contribution= 41,548 millions USD), (**F**= Fixed Cost= 29,740 millions USD), (**U**= Utilized or Starting Capital= 64,720 millions USD), and (**I**= L/E= Leverage or Gearing Ratio= 99.00%), then it's (**D**= Dividend Paid planned), is:

$$D= U+M-F-S'i[1+s]-T-(Q+S'vp[1+s]$$
$$/\{360[c-q]\})/I$$
$$= 64,720+41,548-29,750-48,851$$
$$*0.0100[1+0.0500]-3,385$$
$$-(35,351+48,851*0.1900$$
$$*240[1+0.0500]/\{360$$
$$[1.2000-1.0100]\})/0.9900$$
$$= 2,370 \text{ millions USD}$$

Corporate IFRS-GAAP (B/S-I/S),ISBN-13: **978-1721066308**, ISBN-10: **1721066306**

Law-8236:

If both (**c**= C/X= Current Ratio= <u>1.20</u> times), (**$'**= Sales of Past Year= <u>48,851</u> millions USD), (**q**= Quick or Acid Test Ratio= [C-P]/X= <u>1.01</u> times), (**v**= V/S= Variable Portion= <u>19.00%</u>), (**T**= Tax Paid= <u>3,385</u> millions USD), (**p**= 360 P/V= Procured Inventory Days= <u>240</u> Days), (**s**= [S/S']-1= Sales Growth= <u>5.00%</u>), (**i**= I/S= Interest Portion= <u>1.00%</u>), (**Q**= Quoted Longterm Liabilities= <u>35,351</u> millions USD), (**M**= Margin of Contribution= <u>41,548</u> millions USD), (**F**= Fixed Cost= <u>29,740</u> millions USD), (**U**= Utilized or Starting Capital= <u>64,720</u> millions USD), and (**/**= L/E= Leverage or Gearing Ratio= <u>99.00%</u>), then it's (**D**= Dividend Paid planned), is:

$$D= U+M-F-\$'i[1+s]-T-(Q+\$'vp[1+s]$$
$$/\{360[c-q]\})/I$$
$$= 64,720+41,548-29,750-48,851$$
$$*0.0100[1+0.0500]-3,385$$
$$-(35,351+48,851*0.1900$$
$$*240[1+0.0500]/\{360$$
$$[1.2000-1.0100]\})/0.9900$$
$$= \underline{2,370} \text{ millions USD}$$

Corporate IFRS-GAAP (B/S-I/S),ISBN-13: **978-1721066308**, ISBN-10: **1721066306**

Law-8237:

If both (**c**= C/X= Current Ratio= 1.20 times), (**S'**= Sales of Past Year= 48,851 millions USD), (**q**= Quick or Acid Test Ratio= [C-P]/X= 1.01 times), (**D**= Dividend Paid= 2,370 millions USD), (**T**= Tax Paid= 3,385 millions USD), (**p**= 360 P/V= Procured Inventory Days= 240 Days), (**s**= [S/S']-1= Sales Growth= 5.00%), (**i**= I/S= Interest Portion= 1.00%), (**Q**= Quoted Longterm Liabilities= 35,351 millions USD), (**M**= Margin of Contribution= 41,548 millions USD), (**F**= Fixed Cost= 29,740 millions USD), (**U**= Utilized or Starting Capital= 64,720 millions USD), and (**I**= L/E= Leverage or Gearing Ratio= 99.00%), then it's (**v**= Variable Portion planned), is:

$$\mathbf{v}= [360[\mathbf{c}\text{-}\mathbf{q}](\mathbf{I}\{\mathbf{U}+\mathbf{M}\text{-}\mathbf{F}\text{-}\mathbf{S'}\mathbf{i}[1+\mathbf{s}]\text{-}\mathbf{T}\text{-}\mathbf{D}\}\text{-}\mathbf{Q})]$$
$$/\{\mathbf{S'}\mathbf{p}[1+\mathbf{s}]\}$$
$$= 360[1.2000\text{-}1.0100](0.9900$$
$$\{64,720+41,548\text{-}29,750$$
$$-48,851*0.0100[1+0.0500]$$
$$-3,385\text{-}2,370\}\text{-}35,351)]$$
$$/\{48,851*240[1+0.0500]\}$$
$$= \underline{19.00\%}$$

Corporate IFRS-GAAP (B/S-I/S),ISBN-13: **978-1721066308**, ISBN-10: **1721066306**

Law-8238:

If both (c= C/X= Current Ratio= 1.20 times), (S'= Sales of Past Year= 48,851 millions USD), (q= Quick or Acid Test Ratio= [C-P]/X= 1.01 times), (D= Dividend Paid= 2,370 millions USD), (T= Tax Paid= 3,385 millions USD), (v= V/S= Variable Portion= 19.00%), (s= [S/S']-1= Sales Growth= 5.00%), (i= I/S= Interest Portion= 1.00%), (Q= Quoted Longterm Liabilities= 35,351 millions USD), (M= Margin of Contribution= 41,548 millions USD), (F= Fixed Cost= 29,740 millions USD), (U= Utilized or Starting Capital= 64,720 millions USD), and (I= L/E= Leverage or Gearing Ratio= 99.00%), then it's (p= Procured Inventory Days planned), is:

$$p = [360[c\text{-}q](I \{U+M-F-S'i[1+s]-T-D\}-Q)]$$
$$/\{S'v[1+s]\}$$

$$= 360[1.2000\text{-}1.0100](0.9900$$
$$\{64,720+41,548\ 29,750$$
$$-48,851*0.0100[1+0.0500]$$
$$-3,385\text{-}2,370\}-35,351)]$$
$$/\{48,851*0.1900$$
$$[1+0.0500]\}$$

$$= \underline{240} \text{ days}$$

Corporate IFRS-GAAP (B/S-I/S),ISBN-13: **978-1721066308**, ISBN-10: **1721066306**

Law-8239:

If both (**p**= 360 P/V= Procured Inventory Days= 240 Days), (**$'**= Sales of Past Year= 48,851 millions USD), (**q**= [C-P]/X= Quick or Acid Test Ratio= 1.01 times), (**D**= Dividend Paid= 2,370 millions USD), (**T**= Tax Paid= 3,385 millions USD), (**v**= V/S= Variable Portion= 19.00%), (**s**= [S/S']-1= Sales Growth= 5.00%), (**i**= I/S= Interest Portion= 1.00%), (**Q**= Quoted Longterm Liabilities= 35,351 millions USD), (**M**= Margin of Contribution= 41,548 millions USD), (**F**= Fixed Cost= 29,740 millions USD), (**U**= Utilized or Starting Capital= 64,720 millions USD), and (**I**= L/E= Leverage or Gearing Ratio= 99.00%), then it's (**c**= Current Ratio planned), is:

$$c= q+\$'vp[1+s]/[360[c-q](I\{U+M-F$$
$$-\$'i[1+s]-T-D\}-Q)]$$
$$= 1.0100+48,851*0.1900*240$$
$$[1+0.0500]/[360[1.2000$$
$$-1.0100](0.9900\{64,720$$
$$+41,548-29,750-48,851$$
$$*0.0100[1+0.0500]-3,385$$
$$-2,370\}-35,351)]$$
$$= 1.20 \text{ times}$$

Corporate IFRS-GAAP (B/S-I/S),ISBN-13: **978-1721066308**, ISBN-10: **1721066306**

Law-8240:

If both (**p**= 360 P/V= Procured Inventory Days= 240 Days), (**$'**= Sales of Past Year= 48,851 millions USD), (**c**= C/X= Current Ratio= 1.20 times), (**D**= Dividend Paid= 2,370 millions USD), (**T**= Tax Paid= 3,385 millions USD), (**v**= V/S= Variable Portion= 19.00%), (**s**= [S/S']-1= Sales Growth= 5.00%), (**i**= I/S= Interest Portion= 1.00%), (**Q**= Quoted Longterm Liabilities= 35,351 millions USD), (**M**= Margin of Contribution= 41,548 millions USD), (**F**= Fixed Cost= 29,740 millions USD), (**U**= Utilized or Starting Capital= 64,720 millions USD), and (**I**= L/E= Leverage or Gearing Ratio= 99.00%), then it's (**q**= Quick or Acid Test Ratio planned), is:

$$q= c\text{-}\textbf{\$'vp}[1+s]/[360[\textbf{c-q}](\textbf{\textit{I}} \{\textbf{U+M-F}$$
$$-\textbf{\$'i}[1+s]\text{-}\textbf{T-D}\}\text{-}\textbf{Q})]$$
$$= 1.2000\text{-}48,851*0.1900*240$$
$$[1\mid0.0500]/[360[1.2000$$
$$-1.0100](0.9900\{64,720$$
$$+41,548\text{-}29,750\text{-}48,851$$
$$*0.0100[1+0.0500]\text{-}3,385$$
$$-2,370\}\text{-}35,351)]$$
$$= 1.01 \text{ times}$$

Corporate IFRS-GAAP (B/S-I/S),ISBN-13: **978-1721066308**, ISBN-10: **1721066306**

Law-8241:

If both ($\pmb{S'}$= Sales of Past Year= 48,851 millions USD), ($\pmb{d}$= D/A= Dividend Portion or Payout= 30.00%), ($\pmb{T}$= Tax Paid= 3,385 millions USD), ($\pmb{s}$= [S/S']-1= Sales Growth= 5.00%), ($\pmb{i}$= I/S= Interest Portion= 1.00%), ($\pmb{X}$= Xpress or Current Debt= 34,196 millions USD), ($\pmb{M}$= Margin of Contribution= 41,548 millions USD), ($\pmb{F}$= Fixed Cost= 29,740 millions USD), ($\pmb{U}$= Utilized or Starting Capital= 64,720 millions USD), and ($\pmb{l}$= L/E= Leverage or Gearing Ratio= 99.00%), then it's ($\pmb{Q}$= Quoted Longterm Liabilities planned), is:

$$Q = l\,(U + \{M - F - S'i[1+s] - T\}[1-d]) - X$$
$$= 0.9900(64,720 + \{41,548 - 29,750$$
$$-48,851*0.0100[1+0.0500]$$
$$-3,385\}[1-0.3000]) - 34,196$$
$$= 35,351 \text{ millions USD}$$

Corporate IFRS-GAAP (B/S-I/S),ISBN-13: **978-1721066308**, ISBN-10: **1721066306**

Law-8242:

If both (**$'**= Sales of Past Year= <u>48,851</u> millions USD), (**d**= D/A= Dividend Portion or Payout= <u>30.00%</u>), (**T**= Tax Paid= <u>3,385</u> millions USD), (**s**= [S/S']-1= Sales Growth= <u>5.00%</u>), (**i**= I/S= Interest Portion= <u>1.00%</u>), (**X**= Xpress or Current Debt= <u>34,196</u> millions USD), (**M**= Margin of Contribution= <u>41,548</u> millions USD), (**F**= Fixed Cost= <u>29,740</u> millions USD), (**U**= Utilized or Starting Capital= <u>64,720</u> millions USD), and (**Q**= Quoted Longterm Liabilities= <u>35,351</u> millions USD), then it's (**I** = Leverage or Gearing Ratio planned), is:

$$I = [Q+X]/(U+\{M\text{-}F\text{-}\$'i[1+s]\text{-}T\}[1\text{-}d])$$
$$= [35,351+34,196]/(64,720$$
$$+\{41,548\text{-}29,750\text{-}48,851$$
$$*0.0100[1+0.0500]\text{-}3,385\}$$
$$[1\text{-}0.3000])$$
$$- 99.00\%$$

Corporate IFRS-GAAP (B/S-I/S),ISBN-13: **978-1721066308**, ISBN-10: **1721066306**

Law-8243:

If both (**$'**= Sales of Past Year= 48,851 millions
USD), (**d**= D/A= Dividend Portion or Payout=
30.00%), (**T**= Tax Paid= 3,385 millions USD), (**s**=
[S/S']-1= Sales Growth= 5.00%), (**i**= I/S= Interest
Portion= 1.00%), (**X**= Xpress or Current Debt=
34,196 millions USD), (**M**= Margin of Contribution=
41,548 millions USD), (**F**= Fixed Cost= 29,740
millions USD), (**I**= L/E= Leverage or Gearing Ratio=
99.00%), and (**Q**= Quoted Longterm Liabilities=
35,351 millions USD), then it's (**U**=Utilized or
Starting Capital planned), is:

$$U= [Q+X]/I - \{M-F-S'i[1+s]-T\}[1-d]$$
$$= [35,351+34,196]/0.9900-\{41,548$$
$$-29,750-48,851*0.0100$$
$$[1+0.0500]-3,385\}$$
$$[1-0.3000])$$
$$= 64,720 \text{ millions USD}$$

Corporate IFRS-GAAP (B/S-I/S),ISBN-13: **978-1721066308**, ISBN-10: **1721066306**

Law-8244:

If both (I= L/E= Leverage or Gearing Ratio= 99.00%), (S'= Sales of Past Year= 48,851 millions USD), (d= D/A= Dividend Portion or Payout= 30.00%), (T= Tax Paid= 3,385 millions USD), (s= [S/S']-1= Sales Growth= 5.00%), (U= Utilized or Starting Capital= 64,720 millions USD), (i= I/S= Interest Portion= 1.00%), (X= Xpress or Current Debt= 34,196 millions USD), (F= Fixed Cost= 29,740 millions USD), and (Q= Quoted Longterm Liabilities= 35,351 millions USD), then it's (M=Margin of Contribution planned), is:

$$M = F + S'i[1+s] + T + \{[Q+X]/I - U\}/[1-d]$$
$$= 29,750 + 48,851*0.0100$$
$$[1+0.0500] + 3,385$$
$$+ \{[35,351+34,196]$$
$$/0.9900 - 64,720\}$$
$$/[1-0.3000]$$
$$= 41,548 \text{ millions USD}$$

Corporate IFRS-GAAP (B/S-I/S),ISBN-13: **978-1721066308**, ISBN-10: **1721066306**

Law-8245:

If both (**$'**= Sales of Past Year= 48,851 millions USD), (**d**= D/A= Dividend Portion or Payout= 30.00%), (**T**= Tax Paid= 3,385 millions USD), (**s**= [S/S']-1= Sales Growth= 5.00%), (**U**= Utilized or Starting Capital= 64,720 millions USD), (**i**= I/S= Interest Portion= 1.00%), (**X**= Xpress or Current Debt= 34,196 millions USD), (**M**= Margin of Contribution= 41,548 millions USD), (**I**= L/E= Leverage or Gearing Ratio= 99.00%), and (**Q**= Quoted Longterm Liabilities= 35,351 millions USD), then it's (**F**=Fixed Cost planned), is:

$$F = M\text{-}S'i[1+s]\text{-}T\text{-}\{[Q+X]/I\text{-}U\}/[1\text{-}d]$$
$$= 41,548\text{-}48,851*0.0100[1+0.0500]$$
$$-3,385\text{-}\{[35,351+34,196]$$
$$/0.9900\text{-}64,720\}/[1\text{-}0.3000]$$
$$= 29,750 \text{ millions USD}$$

Corporate IFRS-GAAP (B/S-I/S),ISBN-13: **978-1721066308**, ISBN-10: **1721066306**

Law-8246:

If both (**F**= Fixed Cost= 29,740 millions USD), (**d**= D/A= Dividend Portion or Payout= 30.00%), (**T**= Tax Paid= 3,385 millions USD), (**s**= [S/S']-1= Sales Growth= 5.00%), (**U**= Utilized or Starting Capital= 64,720 millions USD), (**i**= I/S= Interest Portion= 1.00%), (**X**= Xpress or Current Debt= 34,196 millions USD), (**M**= Margin of Contribution= 41,548 millions USD), (**l**= L/E= Leverage or Gearing Ratio= 99.00%), and (**Q**= Quoted Longterm Liabilities= 35,351 millions USD), then it's (**S'**=Sales Past), must be:

$$\textbf{S'}= (\textbf{M-F-T-}\{[\textbf{Q+X}]/\textbf{l -U}\}/[1-\textbf{d}])/\{\textbf{i}[1+\textbf{s}]\}$$
$$= (41,548-29,750-3,385-\{[35,351$$
$$+34,196]/0.9900-64,720\}$$
$$/[1-0.3000])/\{0.0100$$
$$[1+0.0500]\}$$
$$= 48,851 \text{ millions USD}$$

Corporate IFRS-GAAP (B/S-I/S),ISBN-13: **978-1721066308**, ISBN-10: **1721066306**

Law-8247:

If both (**F**= Fixed Cost= 29,740 millions USD), (**d**= D/A= Dividend Portion or Payout= 30.00%), (**T**= Tax Paid= 3,385 millions USD), (**s**= [S/S']-1= Sales Growth= 5.00%), (**U**= Utilized or Starting Capital= 64,720 millions USD), (**S'**= Sales of Past Year= 48,851 millions USD), (**X**= Xpress or Current Debt= 34,196 millions USD), (**M**= Margin of Contribution= 41,548 millions USD), (**I**= L/E= Leverage or Gearing Ratio= 99.00%), and (**Q**= Quoted Longterm Liabilities= 35,351 millions USD), then it's (**i**=Interest Portion planned), is:

$$i= (M\text{-}F\text{-}T\text{-}\{[Q+X]/I\text{-}U\}/[1\text{-}d])/\{S'[1+s]\}$$
$$= (41{,}548\text{-}29{,}750\text{-}3{,}385\text{-}\{[35{,}351$$
$$+34{,}196]/0.9900\text{-}64{,}720\}$$
$$/[1\text{-}0.3000])/\{48{,}851$$
$$[1+0.0500]\}$$
$$= \underline{1.00\%}$$

Corporate IFRS-GAAP (B/S-I/S),ISBN-13: **978-1721066308**, ISBN-10: **1721066306**

Law-8248:

If both (**F**= Fixed Cost= 29,740 millions USD), (**d**= D/A= Dividend Portion or Payout= 30.00%), (**T**= Tax Paid= 3,385 millions USD), (**i**= I/S= Interest Portion= 1.00%), (**U**= Utilized or Starting Capital= 64,720 millions USD), (**S'**= Sales of Past Year= 48,851 millions USD), (**X**= Xpress or Current Debt= 34,196 millions USD), (**M**= Margin of Contribution= 41,548 millions USD), (**I**= L/E= Leverage or Gearing Ratio= 99.00%), and (**Q**= Quoted Longterm Liabilities= 35,351 millions USD), then it's (**s**=Sales Growth planned), is:

$$s= (M\text{-}F\text{-}T\text{-}\{[Q+X]/I\text{-}U\}/[1\text{-}d])/[S\text{'}i]\text{-}1$$
$$= (41,548\text{-}29,750\text{-}3,385\text{-}\{[35,351 +34,196]/0.9900\text{-}64,720\} /[1\text{-}0.3000])/[48,851 *0.0100]\text{-}1$$
$$- \underline{5.00\%}$$

Corporate IFRS-GAAP (B/S-I/S),ISBN-13: **978-1721066308**, ISBN-10: **1721066306**

Law-8249:

If both (**F**= Fixed Cost= 29,740 millions USD), (**d**= D/A= Dividend Portion or Payout= 30.00%), (**s**= [S/S']-1= Sales Growth= 5.00%), (**i**= I/S= Interest Portion= 1.00%), (**U**= Utilized or Starting Capital= 64,720 millions USD), (**S'**= Sales of Past Year= 48,851 millions USD), (**X**= Xpress or Current Debt= 34,196 millions USD), (**M**= Margin of Contribution= 41,548 millions USD), (**I**= L/E= Leverage or Gearing Ratio= 99.00%), and (**Q**= Quoted Longterm Liabilities= 35,351 millions USD), then it's (**T**=Tax Paid planned), is:

$$T= M-F-S'i[1+s]-\{[Q+X]/I-U\}/[1-d]$$
$$= 41,548-29,750-48,851*0.0100$$
$$[1+0.0500]-\{[35,351$$
$$+34,196]/0.9900-64,720\}$$
$$/[1-0.3000]$$
$$= 3,385 \text{ millions USD}$$

Corporate IFRS-GAAP (B/S-I/S),ISBN-13: **978-1721066308**, ISBN-10: **1721066306**

Law-8250:

If both (**F**= Fixed Cost= 29,740 millions USD), (**T**= Tax Paid= 3,385 millions USD), (**s**= [S/S']-1= Sales Growth= 5.00%), (**i**= I/S= Interest Portion= 1.00%), (**U**= Utilized or Starting Capital= 64,720 millions USD), (**S'**= Sales of Past Year= 48,851 millions USD), (**X**= Xpress or Current Debt= 34,196 millions USD), (**M**= Margin of Contribution= 41,548 millions USD), (**I**= L/E= Leverage or Gearing Ratio= 99.00%), and (**Q**= Quoted Longterm Liabilities= 35,351 millions USD), then it's (**d**=Dividend Portion or Payout planned), is:

$$d= 1-\{[\mathbf{Q}+\mathbf{X}]/\mathbf{I}-\mathbf{U}\}/\{\mathbf{M}-\mathbf{F}-\mathbf{S'i}[1+\mathbf{s}]-\mathbf{T}\}$$
$$= 1-\{[35,351+34,196]/0.9900$$
$$-64,720\}/\{41,548-29,750$$
$$-48,851*0.0100$$
$$[1+0.0500]-3,385\}$$
$$= 30.00\%$$

Corporate IFRS-GAAP (B/S-I/S),ISBN-13: **978-1721066308**, ISBN-10: **1721066306**

<u>Law-8251</u>:

If both (**F**= Fixed Cost= <u>29,740</u> millions USD), (**T**= Tax Paid= <u>3,385</u> millions USD), (**s**= [S/S']-1= Sales Growth= <u>5.00%</u>), (**i**= I/S= Interest Portion= <u>1.00%</u>), (**U**= Utilized or Starting Capital= <u>64,720</u> millions USD), (**S'**= Sales of Past Year= <u>48,851</u> millions USD), (**d**= D/A= Dividend Portion or Payout= <u>30.00%</u>), (**M**= Margin of Contribution= <u>41,548</u> millions USD), (**l**= L/E= Leverage or Gearing Ratio= <u>99.00%</u>), and (**Q**= Quoted Longterm Liabilities= <u>35,351</u> millions USD), then it's (**X**=Xpress or Current Debt planned), is :

$$X= I(U+\{M-F-S'i[1+s]-T\}[1-d])-Q$$
$$= 0.9900(64,720+\{41,548-29,750$$
$$-48,851*0.0100[1+0.0500]$$
$$-3,385\}[1-0.3000])-35,351$$
$$= \underline{34,196} \text{ millions USD}$$

Law-8252:

If both (**q**= [C-P]/X= Quick or Acid Test Ratio= 1.01 times), (**P**= Procured Inventories= 6,497 millions USD), (**F**= Fixed Cost= 29,740 millions USD), (**T**= Tax Paid= 3,385 millions USD), (**s**= [S/S']-1= Sales Growth= 5.00%), (**i**= I/S= Interest Portion= 1.00%), (**U**= Utilized or Starting Capital= 64,720 millions USD), (**S'**= Sales of Past Year= 48,851 millions USD), (**d**= D/A= Dividend Portion or Payout= 30.00%), (**M**= Margin of Contribution= 41,548 millions USD), (**l**= L/E= Leverage or Gearing Ratio= 99.00%), and (**c**= C/X= Current Ratio= 1.20 times), then it's (**Q**= Quoted Longterm Liabilities planned), is:

$$Q= I(U+\{M-F-S'i[1+s]-T\}[1-d])-P/[c-q]$$
$$= 0.9900(64,720+\{41,548-29,750$$
$$-48,851*0.0100[1+0.0500]$$
$$-3,385\}[1\ 0.3000])\ 6,497$$
$$/[1.2000-1.0100]$$
$$= \underline{35,351} \text{ millions USD}$$

Corporate IFRS-GAAP (B/S-I/S),ISBN-13: **978-1721066308**, ISBN-10: **1721066306**

Law-8253:

If both (**q**= [C-P]/X= Quick or Acid Test Ratio= 1.01
times), (**P**= Procured Inventories= 6,497 millions
USD), (**F**= Fixed Cost= 29,740 millions USD), (**T**=
Tax Paid= 3,385 millions USD), (**s**= [S/S']-1= Sales
Growth= 5.00%), (**i**= I/S= Interest Portion= 1.00%),
(**U**= Utilized or Starting Capital= 64,720 millions
USD), (**S'**= Sales of Past Year= 48,851 millions
USD), (**d**= D/A= Dividend Portion or Payout=
30.00%), (**M**= Margin of Contribution= 41,548
millions USD), (**Q**= Quoted Longterm Liabilities=
35,351 millions USD), and (**c**= C/X= Current Ratio=
1.20 times), then it's (**I** = Leverage or gearing Ratio
planned), is:

$$I = \{Q+P/[c-q]\}/(U+\{M-F-S'i[1+s]-T\}[1-d])$$
$$= \{35,351+6,497/[1.2000-1.0100]\}$$
$$/(64,720+\{41,548-29,750$$
$$-48,851*0.0100[1+0.0500]$$
$$-3,385\}[1-0.3000])$$
$$= 99.00\%$$

Corporate IFRS-GAAP (B/S-I/S),ISBN-13: **978-1721066308**, ISBN-10: **1721066306**

Law-8254:

If both ($\mathbf{q}$= [C-P]/X= Quick or Acid Test Ratio= 1.01 times), ($\mathbf{P}$= Procured Inventories= 6,497 millions USD), ($\mathbf{F}$= Fixed Cost= 29,740 millions USD), ($\mathbf{T}$= Tax Paid= 3,385 millions USD), ($\mathbf{s}$= [S/S']-1= Sales Growth= 5.00%), ($\mathbf{i}$= I/S= Interest Portion= 1.00%), ($\mathbf{l}$= L/E= Leverage or Gearing Ratio= 99.00%), ($\mathbf{S'}$= Sales of Past Year= 48,851 millions USD), ($\mathbf{d}$= D/A= Dividend Portion or Payout= 30.00%), ($\mathbf{M}$= Margin of Contribution= 41,548 millions USD), ($\mathbf{Q}$= Quoted Longterm Liabilities= 35,351 millions USD), and ($\mathbf{c}$= C/X= Current Ratio= 1.20 times), then it's ($\mathbf{U}$= Utilized or Starting Capital planned), is:

$$\mathbf{U} = \{\mathbf{Q} + \mathbf{P}/[\mathbf{c}-\mathbf{q}]\}/\mathbf{l} - \{\mathbf{M}-\mathbf{F}-\mathbf{S'i}[1+\mathbf{s}]-\mathbf{T}\}[1-\mathbf{d}]$$

$$= \{35,351+6,497/[1.2000-1.0100]\}$$
$$/0.9900-\{41,548-29,750$$
$$-48,851*0.0100[1+0.0500]$$
$$-3,385\}[1-0.3000]$$

$$= \underline{64,720} \text{ millions USD}$$

Corporate IFRS-GAAP (B/S-I/S),ISBN-13: **978-1721066308**, ISBN-10: **1721066306**

Law-8255:

If both (**q**= [C-P]/X= Quick or Acid Test Ratio= 1.01
times), (**P**= Procured Inventories= 6,497 millions
USD), (**F**= Fixed Cost= 29,740 millions USD), (**T**=
Tax Paid= 3,385 millions USD), (**s**= [S/S']-1= Sales
Growth= 5.00%), (**i**= I/S= Interest Portion= 1.00%),
(**l**= L/E= Leverage or Gearing Ratio= 99.00%), (**S'**=
Sales of Past Year= 48,851 millions USD), (**d**= D/A=
Dividend Portion or Payout= 30.00%), (**U**= Utilized
or Starting Capital= 64,720 millions USD), (**Q**=
Quoted Longterm Liabilities= 35,351 millions USD),
and (**c**= C/X= Current Ratio= 1.20 times), then it's
(**M**= Margin of Contribution planned), is:

$$M = F + S'i[1+s] + T + (\{Q+P/[c-q]\}/l - U)/[1-d]$$
$$= 29,750 + 48,851*0.0100[1+0.0500]$$
$$+3,385 + (\{35,351+6,497$$
$$/[1.2000-1.0100]\}/0.9900$$
$$-64,720) /[1-0.3000]$$
$$= 41,547 \text{ millions USD}$$

Corporate IFRS-GAAP (B/S-I/S),ISBN-13: **978-1721066308**, ISBN-10: **1721066306**

Law-8256:

If both (**q**= [C-P]/X= Quick or Acid Test Ratio= 1.01 times), (**P**= Procured Inventories= 6,497 millions USD), (**M**= Margin of Contribution= 41,548 millions USD), (**T**= Tax Paid= 3,385 millions USD), (**s**= [S/S']-1= Sales Growth= 5.00%), (**i**= I/S= Interest Portion= 1.00%), (**l**= L/E= Leverage or Gearing Ratio= 99.00%), (**S'**= Sales of Past Year= 48,851 millions USD), (**d**= D/A= Dividend Portion or Payout= 30.00%), (**U**= Utilized or Starting Capital= 64,720 millions USD), (**Q**= Quoted Longterm Liabilities= 35,351 millions USD), and (**c**= C/X= Current Ratio= 1.20 times), then it's (**F**= Fixed Cost planned), is:

$$\mathbf{F}= \mathbf{M}\text{-}\mathbf{S'i}[1+\mathbf{s}]\text{-}\mathbf{T}\text{-}(\{\mathbf{Q}+\mathbf{P}/[\mathbf{c}\text{-}\mathbf{q}]\}/\mathbf{l}\text{-}\mathbf{U})/[1\text{-}\mathbf{d}]$$

$$= 41,548\text{-}48,851*0.0100[1+0.0500]$$
$$-3,385\text{-}(\{35,351\text{-}6,497$$
$$/[1.2000\text{-}1.0100]\}/0.9900$$
$$-64,720) /[1\text{-}0.3000]$$

$$= \underline{29,750} \text{ millions USD}$$

Corporate IFRS-GAAP (B/S-I/S),ISBN-13: **978-1721066308**, ISBN-10: **1721066306**

Law-8257:

If both (**q**= [C-P]/X= Quick or Acid Test Ratio= 1.01 times), (**P**= Procured Inventories= 6,497 millions USD), (**M**= Margin of Contribution= 41,548 millions USD), (**T**= Tax Paid= 3,385 millions USD), (**s**= [S/S']-1= Sales Growth= 5.00%), (**i**= I/S= Interest Portion= 1.00%), (**I**= L/E= Leverage or Gearing Ratio= 99.00%), (**F**= Fixed Cost= 29,740 millions USD), (**d**= D/A= Dividend Portion or Payout= 30.00%), (**U**= Utilized or Starting Capital= 64,720 millions USD), (**Q**= Quoted Longterm Liabilities= 35,351 millions USD), and (**c**= C/X= Current Ratio= 1.20 times), then it's (**S'**= Sales Past), must be:

$$S' = [M\text{-}F\text{-}T\text{-}(\{Q+P/[c\text{-}q]\}/I\text{-}U)/[1\text{-}d]]$$
$$/\{i[1+s]\}$$
$$= [41,548\text{-}29,750\text{-}3,385\text{-}(\{35,351$$
$$-6,497/[1.2000\text{-}1.0100]\}$$
$$/0.9900\text{-}64,720)/[1\text{-}0.3000]]$$
$$/\{0.0100[1+0.0500]\}$$
$$= \underline{48,851} \text{ millions USD}$$

Corporate IFRS-GAAP (B/S-I/S),ISBN-13: **978-1721066308**, ISBN-10: **1721066306**

Law-8258:

If both (**q**= [C-P]/X= Quick or Acid Test Ratio= 1.01 times), (**P**= Procured Inventories= 6,497 millions USD), (**M**= Margin of Contribution= 41,548 millions USD), (**T**= Tax Paid= 3,385 millions USD), (**s**= [S/S']-1= Sales Growth= 5.00%), (**S'**= Sales of Past Year= 48,851 millions USD), (**l**= L/E= Leverage or Gearing Ratio= 99.00%), (**F**= Fixed Cost= 29,740 millions USD), (**d**= D/A= Dividend Portion or Payout= 30.00%), (**U**= Utilized or Starting Capital= 64,720 millions USD), (**Q**= Quoted Longterm Liabilities= 35,351 millions USD), and (**c**= C/X= Current Ratio= 1.20 times), then it's (**i**= Interest Portion planned), is:

$$i= [M-F-T-(\{Q+P/[c-q]\}/l-U)/[1-d]]$$
$$/\{S'[1+s]\}$$
$$= [41,548-29,750-3,385-(\{35,351$$
$$6,497/[1.2000-1.0100]\}$$
$$/0.9900-64,720)$$
$$/[1-0.3000]]$$
$$/\{48,851[1+0.0500]\}$$
$$= 1.00\%$$

Corporate IFRS-GAAP (B/S-I/S),ISBN-13: **978-1721066308**, ISBN-10: **1721066306**

Law-8259:

If both (**q**= [C-P]/X= Quick or Acid Test Ratio= 1.01 times), (**P**= Procured Inventories= 6,497 millions USD), (**M**= Margin of Contribution= 41,548 millions USD), (**T**= Tax Paid= 3,385 millions USD), (**i**= I/S= Interest Portion= 1.00%), (**$'**= Sales of Past Year= 48,851 millions USD), (**l**= L/E= Leverage or Gearing Ratio= 99.00%), (**F**= Fixed Cost= 29,740 millions USD), (**d**= D/A= Dividend Portion or Payout= 30.00%), (**U**= Utilized or Starting Capital= 64,720 millions USD), (**Q**= Quoted Longterm Liabilities= 35,351 millions USD), and (**c**= C/X= Current Ratio= 1.20 times), then it's (**$**= Sales Growth planned), is:

$$\mathbf{s}= [\mathbf{M}\text{-}\mathbf{F}\text{-}\mathbf{T}\text{-}(\{\mathbf{Q}+\mathbf{P}/[\mathbf{c}\text{-}\mathbf{q}]\}/\mathbf{l}\text{-}\mathbf{U})/[1\text{-}\mathbf{d}]]$$
$$/[\mathbf{\$'i}]\text{-}1$$
$$= [41,548\text{-}29,750\text{-}3,385\text{-}(\{35,351$$
$$\text{-}6,497/[1.2000\text{-}1.0100]\}$$
$$/0.9900\text{-}64,720)/[1\text{-}0.3000]]$$
$$/[48,851*0.0100]\text{-}1$$
$$= 5.00\%$$

Corporate IFRS-GAAP (B/S-I/S),ISBN-13: **978-1721066308**, ISBN-10: **1721066306**

Law-8260:

If both (**q**= [C-P]/X= Quick or Acid Test Ratio= 1.01 times), (**P**= Procured Inventories= 6,497 millions USD), (**M**= Margin of Contribution= 41,548 millions USD), (**s**= [S/S']-1= Sales Growth= 5.00%), (**i**= I/S= Interest Portion= 1.00%), (**S'**= Sales of Past Year= 48,851 millions USD), (**I**= L/E= Leverage or Gearing Ratio= 99.00%), (**F**= Fixed Cost= 29,740 millions USD), (**d**= D/A= Dividend Portion or Payout= 30.00%), (**U**= Utilized or Starting Capital= 64,720 millions USD), (**Q**= Quoted Longterm Liabilities= 35,351 millions USD), and (**c**= C/X= Current Ratio= 1.20 times), then it's (**T**= Tax Paid planned), is:

$$\mathbf{T= M\text{-}F\text{-}S'i}[1+\mathbf{s}]\text{-}(\{\mathbf{Q}+\mathbf{P}/[\mathbf{c}\text{-}\mathbf{q}]\}/\mathbf{I}\text{-}\mathbf{U})/[1\text{-}\mathbf{d}]$$

$$= 41{,}548\text{-}29{,}750\text{-}48{,}851^*0.0100$$
$$[1+0.0500]\text{-}(\{35{,}351$$
$$-6{,}497/[1.2000\text{-}1.0100]\}$$
$$/0.9900\text{-}64{,}720)/[1\ 0.3000]$$

$$= 3{,}385 \text{ millions USD}$$

Corporate IFRS-GAAP (B/S-I/S),ISBN-13: **978-1721066308**, ISBN-10: **1721066306**

Law-8261:

If both (**q**= [C-P]/X= Quick or Acid Test Ratio= 1.01 times), (**P**= Procured Inventories= 6,497 millions USD), (**M**= Margin of Contribution= 41,548 millions USD), (**s**= [S/S']-1= Sales Growth= 5.00%), (**i**= I/S= Interest Portion= 1.00%), (**S'**= Sales of Past Year= 48,851 millions USD), (**I**= L/E= Leverage or Gearing Ratio= 99.00%), (**F**= Fixed Cost= 29,740 millions USD), (**T**= Tax Paid= 3,385 millions USD), (**U**= Utilized or Starting Capital= 64,720 millions USD), (**Q**= Quoted Longterm Liabilities= 35,351 millions USD), and (**c**= C/X= Current Ratio= 1.20 times), then it's (**d**= Dividend Portion or Payout planned), is:

$$d= 1-(\{Q+P/[c-q]\}/I -U)/\{M-F-S'i[1+s]-T\}$$
$$= 1-(\{35,351-6,497/[1.2000-1.0100]\}$$
$$/0.9900-64,720)/\{41,548$$
$$-29,750-48,851*0.0100$$
$$[1+0.0500]-3,385\}$$
$$= 30.00\%$$

Corporate IFRS-GAAP (B/S-I/S),ISBN-13: **978-1721066308**, ISBN-10: **1721066306**

Law-8262:

If both (**q**= [C-P]/X= Quick or Acid Test Ratio= 1.01 times), (**d**= D/A= Dividend Portion or Payout= 30.00%), (**M**= Margin of Contribution= 41,548 millions USD), (**s**= [S/S']-1= Sales Growth= 5.00%), (**i**= I/S= Interest Portion= 1.00%), (**S'**= Sales of Past Year= 48,851 millions USD), (**l**= L/E= Leverage or Gearing Ratio= 99.00%), (**F**= Fixed Cost= 29,740 millions USD), (**T**= Tax Paid= 3,385 millions USD), (**U**= Utilized or Starting Capital= 64,720 millions USD), (**Q**= Quoted Longterm Liabilities= 35,351 millions USD), and (**c**= C/X= Current Ratio= 1.20 times), then it's (**P**= Procured Inventories planned), is:

$$P= [c-q][l(U+\{M-F-S'i[1+s]-T\}[1-d])-Q]$$
$$= [1.2000-1.0100][0.9900(64,720$$
$$+\{41,548-29,750-48,851$$
$$*0.0100[1+0.0500]-3,385\}$$
$$[1-0.3000])-35,351]$$
$$= 6,497 \text{ millions USD}$$

Corporate IFRS-GAAP (B/S-I/S),ISBN-13: **978-1721066308**, ISBN-10: **1721066306**

Law-8263:

If both (**q**= [C-P]/X= Quick or Acid Test Ratio= 1.01 times), (**d**= D/A= Dividend Portion or Payout= 30.00%), (**M**= Margin of Contribution= 41,548 millions USD), (**s**= [S/S']-1= Sales Growth= 5.00%), (**i**= I/S= Interest Portion= 1.00%), (**S'**= Sales of Past Year= 48,851 millions USD), (**l**= L/E= Leverage or Gearing Ratio= 99.00%), (**F**= Fixed Cost= 29,740 millions USD), (**T**= Tax Paid= 3,385 millions USD), (**U**= Utilized or Starting Capital= 64,720 millions USD), (**Q**= Quoted Longterm Liabilities= 35,351 millions USD), and (**P**= Procured Inventories= 6,497 millions USD), then it's (**c**= Current Ratio planned), is:

$$c= q+P/[l\,(U+\{M\text{-}F\text{-}S'i[1+s]\text{-}T\}[1\text{-}d])\text{-}Q]$$
$$= 1.0100+6,497/[0.9900(64,720$$
$$+\{41,548\text{-}29,750\text{-}48,851$$
$$*0.0100[1+0.0500]\text{-}3,385\}$$
$$[1\text{-}0.3000])\text{-}35,351]$$
$$= 1.20 \text{ times}$$

Corporate IFRS-GAAP (B/S-I/S),ISBN-13: **978-1721066308**, ISBN-10: **1721066306**

Law-8264:

If both (**c**= C/X= Current Ratio= 1.20 times), (**d**= D/A= Dividend Portion or Payout= 30.00%), (**M**= Margin of Contribution= 41,548 millions USD), (**s**= [S/S']-1= Sales Growth= 5.00%), (**i**= I/S= Interest Portion= 1.00%), (**$'**= Sales of Past Year= 48,851 millions USD), (**l**= L/E= Leverage or Gearing Ratio= 99.00%), (**F**= Fixed Cost= 29,740 millions USD), (**T**= Tax Paid= 3,385 millions USD), (**U**= Utilized or Starting Capital= 64,720 millions USD), (**Q**= Quoted Longterm Liabilities= 35,351 millions USD), and (**P**= Procured Inventories= 6,497 millions USD), then it's (**q**= Quick or Acid Test Ratio planned), is:

$$q= c\text{-}P/[l\,(U+\{M\text{-}F\text{-}\$'i[1+s]\text{-}T\}[1\text{-}d])\text{-}Q]$$
$$= 1.2000\text{-}6,497/[0.9900(64,720$$
$$+\{41,548\text{-}29,750\text{-}48,851$$
$$*0.0100[1+0.0500]\text{-}3,385\}$$
$$[1\ 0.3000])\text{-}35,351]$$
$$= 1.01 \text{ times}$$

Corporate IFRS-GAAP (B/S-I/S),ISBN-13: **978-1721066308**, ISBN-10: **1721066306**

<u>Law-8265</u>:

If both (**c**= C/X= Current Ratio= <u>1.20</u> times), (**d**= D/A= Dividend Portion or Payout= <u>30.00%</u>), (**M**= Margin of Contribution= <u>41,548</u> millions USD), (**s**= [S/S']-1= Sales Growth= <u>5.00%</u>), (**i**= I/S= Interest Portion= <u>1.00%</u>), (**S'**= Sales of Past Year= <u>48,851</u> millions USD), (**l**= L/E= Leverage or Gearing Ratio= <u>99.00%</u>), (**F**= Fixed Cost= <u>29,740</u> millions USD), (**T**= Tax Paid= <u>3,385</u> millions USD), (**U**= Utilized or Starting Capital= <u>64,720</u> millions USD), (**V**= Variable Cost= <u>9,746</u> millions USD), (**p**= 360 P/V= Procured Inventory Days= <u>240</u> Days), and (**q**= Quick or Acid Test Ratio= [C-P]/X= <u>1.01</u> times), then it's (**Q**= Quoted Longterm Liabilities planned), is:

$$Q = l\,(U+\{M\text{-}F\text{-}S'i[1+s]\text{-}T\}[1\text{-}d])$$
$$-Vp/\{360[c\text{-}q]\}$$
$$= 0.9900(64,720+\{41,548\text{-}29,750$$
$$-48,851*0.0100[1+0.0500]$$
$$-3,385\}[1\text{-}0.3000])\text{-}9,746$$
$$*240/\{360[1.2000\text{-}1.0100]\}$$
$$= \underline{35,351} \text{ millions USD}$$

Corporate IFRS-GAAP (B/S-I/S),ISBN-13: **978-1721066308**, ISBN-10: **1721066306**

Law-8266:

If both (**c**= C/X= Current Ratio= 1.20 times), (**d**= D/A= Dividend Portion or Payout= 30.00%), (**M**= Margin of Contribution= 41,548 millions USD), (**s**= [S/S']-1= Sales Growth= 5.00%), (**i**= I/S= Interest Portion= 1.00%), (**S'**= Sales of Past Year= 48,851 millions USD), (**Q**= Quoted Longterm Liabilities= 35,351 millions USD), (**F**= Fixed Cost= 29,740 millions USD), (**T**= Tax Paid= 3,385 millions USD), (**U**= Utilized or Starting Capital= 64,720 millions USD), (**V**= Variable Cost= 9,746 millions USD), (**p**= 360 P/V= Procured Inventory Days= 240 Days), and (**q**= [C-P]/X= Quick or Acid Test Ratio= 1.01 times), then it's (**I** = Leverage or Gearing Ratio planned), is:

$$I = (Q+Vp/\{360[c-q]\})/(U+\{M-F$$
$$-S'i[1+s]-T\}[1-d])$$
$$= (35,351+9,746*240/\{360[1.2000$$
$$-1.0100]\})/(64,720+\{41,548$$
$$-29,750-48,851*0.0100$$
$$[1+0.0500]-3,385\}[1-0.3000])$$
$$= \underline{99.00\%}$$

Corporate IFRS-GAAP (B/S-I/S),ISBN-13: **978-1721066308**, ISBN-10: **1721066306**

Law-8267:

If both (**c**= C/X= Current Ratio= 1.20 times), (**d**= D/A= Dividend Portion or Payout= 30.00%), (**M**= Margin of Contribution= 41,548 millions USD), (**s**= [S/S']-1= Sales Growth= 5.00%), (**i**= I/S= Interest Portion= 1.00%), (**S'**= Sales of Past Year= 48,851 millions USD), (**l**= L/E= Leverage or Gearing Ratio= 99.00%), (**F**= Fixed Cost= 29,740 millions USD), (**T**= Tax Paid= 3,385 millions USD), (**Q**= Quoted Longterm Liabilities= 35,351 millions USD), (**V**= Variable Cost= 9,746 millions USD), (**p**= 360P/V= Procured Inventory Days= 240 Days), and (**q**= Quick or Acid Test Ratio= [C-P]/X= 1.01 times), then it's (**U**= Utilized or Starting Capital planned), is:

$$U= (Q+Vp/\{360[c\text{-}q]\})/l$$
$$-\{M\text{-}F\text{-}S'i[1+s]\text{-}T\}[1\text{-}d]$$
$$= (35,351+9,746*240/\{360[1.2000$$
$$-1.0100]\})/0.9900-\{41,548$$
$$-29,750-48,851*0.0100$$
$$[1+0.0500]-3,385\}[1-0.3000])$$
$$= \underline{64,720} \text{ millions USD}$$

Corporate IFRS-GAAP (B/S-I/S),ISBN-13: **978-1721066308**, ISBN-10: **1721066306**

Law-8268:

If both (**c**= C/X= Current Ratio= 1.20 times), (**d**= D/A= Dividend Portion or Payout= 30.00%), (**U**= Utilized or Starting Capital= 64,720 millions USD), (**s**= [S/S']-1= Sales Growth= 5.00%), (**i**= I/S= Interest Portion= 1.00%), (**S'**= Sales of Past Year= 48,851 millions USD), (**l**= L/E= Leverage or Gearing Ratio= 99.00%), (**F**= Fixed Cost= 29,740 millions USD), (**T**= Tax Paid= 3,385 millions USD), (**Q**= Quoted Longterm Liabilities= 35,351 millions USD), (**V**= Variable Cost= 9,746 millions USD), (**p**= 360 P/V= Procured Inventory Days= 240 Days), and (**q**= Quick or Acid Test Ratio= [C-P]/X= 1.01 times), then it's (**M**= Margin of Contribution planned), is:

$$M= F+S'i[1+s]+T+[(Q+Vp/\{360[c-q]\})/l-U]/[1-d]$$

$$= 29,750+48,851*0.0100[1+0.0500]$$
$$+3,385+[(35,351+9,746$$
$$*240/\{360[1.2000-1.0100]\})$$
$$/0.9900-64,720]/[1-0.3000])$$

$$= 41,548 \text{ millions USD}$$

273

Corporate IFRS-GAAP (B/S-I/S),ISBN-13: **978-1721066308**, ISBN-10: **1721066306**

Law-8269:

If both (**c**= C/X= Current Ratio= 1.20 times), (**d**= D/A= Dividend Portion or Payout= 30.00%), (**U**= Utilized or Starting Capital= 64,720 millions USD), (**s**= [S/S']-1= Sales Growth= 5.00%), (**i**= I/S= Interest Portion= 1.00%), (**S'**= Sales of Past Year= 48,851 millions USD), (**l**= L/E= Leverage or Gearing Ratio= 99.00%), (**M**= Margin of Contribution= 41,548 millions USD), (**T**= Tax Paid= 3,385 millions USD), (**Q**= Quoted Longterm Liabilities= 35,351 millions USD), (**V**= Variable Cost= 9,746 millions USD), (**p**= 360 P/V= Procured Inventory Days= 240 Days), and (**q**= [C-P]/X= Quick or Acid Test Ratio= 1.01 times), then it's (**F**= Fixed Cost planned), is:

$$\mathbf{F}= \mathbf{M}\text{-}\mathbf{S'i}[1+\mathbf{s}]\text{-}\mathbf{T}\text{-}[(\mathbf{Q}+\mathbf{Vp}/\{360[\mathbf{c}\text{-}\mathbf{q}]\})/\mathbf{l}\text{-}\mathbf{U}]$$
$$/[1\text{-}\mathbf{d}]$$

$$= 41,548\text{-}48,851*0.0100[1+0.0500]$$
$$-3,385\text{-}[(35,351+9,746$$
$$*240/\{360[1.2000\text{-}1.0100]\})$$
$$/0.9900\text{-}64,720]/[1\text{-}0.3000]$$

$$= 29,750 \text{ millions USD}$$

Corporate IFRS-GAAP (B/S-I/S),ISBN-13: **978-1721066308**, ISBN-10: **1721066306**

<u>Law-8270</u>:

If both (**c**= C/X= Current Ratio= <u>1.20</u> times), (**d**= D/A= Dividend Portion or Payout= <u>30.00%</u>), (**U**= Utilized or Starting Capital= <u>64,720</u> millions USD), (**s**= [S/S']-1= Sales Growth= <u>5.00%</u>), (**i**= I/S= Interest Portion= <u>1.00%</u>), (**F**= Fixed Cost= <u>29,740</u> millions USD), (**l**= L/E= Leverage or Gearing Ratio= <u>99.00%</u>), (**M**= Margin of Contribution= <u>41,548</u> millions USD), (**T**= Tax Paid= <u>3,385</u> millions USD), (**Q**= Quoted Longterm Liabilities= <u>35,351</u> millions USD), (**V**= Variable Cost= <u>9,746</u> millions USD), (**p**= 360 P/V= Procured Inventory Days= <u>240</u> Days), and (**q**= Quick or Acid Test Ratio= [C-P]/X= <u>1.01</u> times), then it's (**S'**= Sales Past), must be:

$$\textbf{S'} = \{\textbf{M-F-T-}[(\textbf{Q+Vp}/\{360[\textbf{c-q}]\})/\textbf{l -U}]/[1-\textbf{d}]\}$$
$$/\{\textbf{i}[1+\textbf{s}]\}$$

$$= \{41,548-29,750-3,385-[(35,351$$
$$+9,746*240/\{360[1.2000$$
$$-1.0100]\})/0.9900-64,720]$$
$$/[1-0.3000]\}/\{0.0100$$
$$[1+0.0500]\}$$

$$= \underline{48,851} \text{ millions USD}$$

Corporate IFRS-GAAP (B/S-I/S),ISBN-13: **978-1721066308**, ISBN-10: **1721066306**

Law-8271:

If both (**c**= C/X= Current Ratio= 1.20 times), (**d**= D/A= Dividend Portion or Payout= 30.00%), (**U**= Utilized or Starting Capital= 64,720 millions USD), (**s**= [S/S']-1= Sales Growth= 5.00%), (**S'**= Sales of Past Year= 48,851 millions USD), (**F**= Fixed Cost= 29,740 millions USD), (**l**= L/E= Leverage or Gearing Ratio= 99.00%), (**M**= Margin of Contribution= 41,548 millions USD), (**T**= Tax Paid= 3,385 millions USD), (**Q**= Quoted Longterm Liabilities= 35,351 millions USD), (**V**= Variable Cost= 9,746 millions USD), (**p**= 360 P/V= Procured Inventory Days= 240 Days), and (**q**= [C-P]/X= Quick or Acid Test Ratio= 1.01 times), then it's (**i**= Interest Portion planned), is:

$$i= \{M\text{-}F\text{-}T\text{-}[(Q+Vp/\{360[c\text{-}q]\})/l\text{-}U]/[1\text{-}d]\}/\{S'[1+s]\}$$

$$= \{41,548\text{-}29,750\text{-}3,385\text{-}[(35,351+9,746*240/\{360[1.2000-1.0100]\})/0.9900\text{-}64,720]/[1\text{-}0.3000]\}/\{48,851[1+0.0500]\}$$

$$= 1.00\%$$

Corporate IFRS-GAAP (B/S-I/S),ISBN-13: **978-1721066308**, ISBN-10: **1721066306**

Law-8272:

If both (**c**= C/X= Current Ratio= 1.20 times), (**d**= D/A= Dividend Portion or Payout= 30.00%), (**U**= Utilized or Starting Capital= 64,720 millions USD), (**i**= I/S= Interest Portion= 1.00%), (**$'**= Sales of Past Year= 48,851 millions USD), (**F**= Fixed Cost= 29,740 millions USD), (**I**= L/E= Leverage or Gearing Ratio= 99.00%), (**M**= Margin of Contribution= 41,548 millions USD), (**T**= Tax Paid= 3,385 millions USD), (**Q**= Quoted Longterm Liabilities= 35,351 millions USD), (**V**= Variable Cost= 9,746 millions USD), (**p**= 360P/V= Procured Inventory Days= 240 Days), and (**q**= [C-P]/X= Quick or Acid Test Ratio= 1.01 times), then it's (**s**= Sales Growth planned), is:

$$s = \{M\text{-}F\text{-}T\text{-}[(Q+Vp/\{360[c\text{-}q]\})/I\text{-}U]$$
$$/[1\text{-}d]\}/[\$'i]\text{-}1$$
$$= \{41,548\text{-}29,750\text{-}3,385\text{-}[(35,351$$
$$+9,746*240/\{360[1.2000$$
$$-1.0100]\})/0.9900\text{-}64,720]$$
$$/[1\text{-}0.3000]\}$$
$$/[48,851*0.0100]\text{-}1$$
$$= \underline{5.00\%}$$

Corporate IFRS-GAAP (B/S-I/S),ISBN-13: **978-1721066308**, ISBN-10: **1721066306**

<u>Law-8273</u>:

If both (**c**= C/X= Current Ratio= <u>1.20</u> times), (**d**= D/A= Dividend Portion or Payout= <u>30.00%</u>), (**U**= Utilized or Starting Capital= <u>64,720</u> millions USD), (**i**= I/S= Interest Portion= <u>1.00%</u>), (**$'**= Sales of Past Year= <u>48,851</u> millions USD), (**F**= Fixed Cost= <u>29,740</u> millions USD), (**⌐**= L/E= Leverage or Gearing Ratio= <u>99.00%</u>), (**M**= Margin of Contribution= <u>41,548</u> millions USD), (**s**= [S/S']-1= Sales Growth= <u>5.00%</u>), (**Q**= Quoted Longterm Liabilities= <u>35,351</u> millions USD), (**V**= Variable Cost= <u>9,746</u> millions USD), (**p**= 360 P/V= Procured Inventory Days= <u>240</u> Days), and (**q**= [C-P]/X= Quick or Acid Test Ratio= <u>1.01</u> times), then it's (**T**= Tax Paid planned), is:

$$T= M\text{-}F\text{-}\$'i[1+s]\text{-}[(Q+Vp/\{360[c\text{-}q]\})/\mathit{I}\text{-}U]$$
$$/[1\text{-}d]$$
$$= 41,548\text{-}29,750\text{-}48,851*0.0100$$
$$[1+0.0500]\text{-}[(35,351$$
$$+9,746*240/\{360[1.2000$$
$$-1.0100]\})/0.9900\text{-}64,720]$$
$$/[1\text{-}0.3000]$$
$$= \underline{3,385} \text{ millions USD}$$

Corporate IFRS-GAAP (B/S-I/S),ISBN-13: **978-1721066308**, ISBN-10: **1721066306**

Law-8274:

If both (**c**= C/X= Current Ratio= <u>1.20</u> times), (**T**= Tax Paid= <u>3,385</u> millions USD), (**U**= Utilized or Starting Capital= <u>64,720</u> millions USD), (**i**= I/S= Interest Portion= <u>1.00%</u>), (**S'**= Sales of Past Year= <u>48,851</u> millions USD), (**F**= Fixed Cost= <u>29,740</u> millions USD), (**l**= L/E= Leverage or Gearing Ratio= <u>99.00%</u>), (**M**= Margin of Contribution= <u>41,548</u> millions USD), (**s**= [S/S']-1= Sales Growth= <u>5.00%</u>), (**Q**= Quoted Longterm Liabilities= <u>35,351</u> millions USD), (**V**= Variable Cost= <u>9,746</u> millions USD), (**p**= 360 P/V= Procured Inventory Days= <u>240</u> Days), and (**q**= Quick or Acid Test Ratio= [C-P]/X= <u>1.01</u> times), then it's (**d**= Dividend Portion or payout planned), is:

$$d= 1-[(\mathbf{Q}+\mathbf{Vp}/\{360[\mathbf{c}-\mathbf{q}]\})/\mathbf{l}-\mathbf{U}]$$
$$/\{\mathbf{M}-\mathbf{F}-\mathbf{S'i}[1+\mathbf{s}]-\mathbf{T}\}$$
$$= 1-[(35,351+9,746*240/\{360$$
$$[1.2000-1.0100]\})/0.9900$$
$$-64,720]/\{41,548-29,750$$
$$-48,851*0.0100[1+0.0500]$$
$$-3,385\}$$
$$= \underline{30.00\%}$$

Corporate IFRS-GAAP (B/S-I/S),ISBN-13: **978-1721066308**, ISBN-10: **1721066306**

Law-8275:

If both (**c**= C/X= Current Ratio= 1.20 times), (**T**= Tax Paid= 3,385 millions USD), (**U**= Utilized or Starting Capital= 64,720 millions USD), (**i**= I/S= Interest Portion= 1.00%), (**S'**= Sales of Past Year= 48,851 millions USD), (**F**= Fixed Cost= 29,740 millions USD), (**l**= L/E= Leverage or Gearing Ratio= 99.00%), (**M**= Margin of Contribution= 41,548 millions USD), (**s**= [S/S']-1= Sales Growth= 5.00%), (**Q**= Quoted Longterm Liabilities= 35,351 millions USD), (**d**= D/A= Dividend Portion or Payout= 30.00%), (**p**= 360 P/V= Procured Inventory Days= 240 Days), and (**q**= [C-P]/X= Quick or Acid Test Ratio= 1.01 times), then it's (**V**= Variable Cost planned), is:

$$V= 360[\mathbf{c}-\mathbf{q}][\mathbf{l}\,(\mathbf{U}+\{\mathbf{M}-\mathbf{F}-\mathbf{S'i}[1+\mathbf{s}]-\mathbf{T}\}$$
$$[1-\mathbf{d}])-\mathbf{Q}]/\mathbf{p}$$
$$= \{360[1.2000-1.0100][0.9900$$
$$(64,720+\{41,548-29,750$$
$$-48,851*0.0100[1+0.0500]$$
$$-3,385\}[1-0.3000])$$
$$-35,351]\}/240$$
$$= \underline{9,746}\text{ millions USD}$$

Corporate IFRS-GAAP (B/S-I/S),ISBN-13: **978-1721066308**, ISBN-10: **1721066306**

Law-8276:

If both (**c**= C/X= Current Ratio= 1.20 times), (**T**= Tax Paid= 3,385 millions USD), (**U**= Utilized or Starting Capital= 64,720 millions USD), (**i**= I/S= Interest Portion= 1.00%), (**$'**= Sales of Past Year= 48,851 millions USD), (**F**= Fixed Cost= 29,740 millions USD), (**l**= L/E= Leverage or Gearing Ratio= 99.00%), (**M**= Margin of Contribution= 41,548 millions USD), (**s**= [S/S']-1= Sales Growth= 5.00%), (**Q**= Quoted Longterm Liabilities= 35,351 millions USD), (**d**= D/A= Dividend Portion or Payout= 30.00%), (**V**= Variable Cost= 9,746 millions USD), and (**q**= Quick or Acid Test Ratio= [C-P]/X= 1.01 times), then it's (**p**= Procured Inventory Days planned), is:

$$\mathbf{p}= 360[\mathbf{c}\text{-}\mathbf{q}][\mathbf{l}\,(\mathbf{U}+\{\mathbf{M}\text{-}\mathbf{F}\text{-}\mathbf{\$'i}[1+\mathbf{s}]\text{-}\mathbf{T}\}$$
$$[1\text{-}d])\text{-}\mathbf{Q}]/\mathbf{V}$$

$$= 360[1.2000\text{-}1.0100][0.9900$$
$$(64,720+\{41,548\text{-}29,750$$
$$-48,851*0.0100[1+0.0500]$$
$$-3,385\}[1\text{-}0.3000])$$
$$-35,351]/9,746$$

$$= \underline{240}\ \text{days}$$

Corporate IFRS-GAAP (B/S-I/S),ISBN-13: **978-1721066308**, ISBN-10: **1721066306**

Law-8277:

If both (**p**= 360 P/V= Procured Inventory Days= <u>240</u>
Days), (**T**= Tax Paid= <u>3,385</u> millions USD), (**U**=
Utilized or Starting Capital= <u>64,720</u> millions USD),
(**i**= I/S= Interest Portion= <u>1.00%</u>), (**S'**= Sales of Past
Year= <u>48,851</u> millions USD), (**F**= Fixed Cost=
<u>29,740</u> millions USD), (**l**= L/E= Leverage or Gearing
Ratio= <u>99.00%</u>), (**M**= Margin of Contribution=
<u>41,548</u> millions USD), (**s**= [S/S']-1= Sales Growth=
<u>5.00%</u>), (**Q**= Quoted Longterm Liabilities= <u>35,351</u>
millions USD), (**d**= D/A= Dividend Portion or
Payout= <u>30.00%</u>), (**V**= Variable Cost= <u>9,746</u> millions
USD), and (**q**= [C-P]/X= Quick or Acid Test Ratio=
<u>1.01</u> times), then it's (**c**= Current Ratio planned), is:

$$c= q+Vp/\{360[l\,(U+\{M\text{-}F\text{-}S'i[1+s]\text{-}T\}$$
$$[1\text{-}d])\text{-}Q]\}$$
$$= 1.0100+9,746*240/\{360[0.9900$$
$$(64,720+\{41,548\text{-}29,750$$
$$-48,851*0.0100[1+0.0500]$$
$$-3,385\}[1\text{-}0.3000])\text{-}35,351]\}$$

$$= \underline{1.20}\text{ times}$$

Corporate IFRS-GAAP (B/S-I/S),ISBN-13: **978-1721066308**, ISBN-10: **1721066306**

Law-8278:

If both (**p**= 360 P/V= Procured Inventory Days= 240 Days), (**c**= C/X= Current Ratio= 1.20 times), (**T**= Tax Paid= 3,385 millions USD), (**U**= Utilized or Starting Capital= 64,720 millions USD), (**i**= I/S= Interest Portion= 1.00%), (**S'**= Sales of Past Year= 48,851 millions USD), (**F**= Fixed Cost= 29,740 millions USD), (**l**= L/E= Leverage or Gearing Ratio= 99.00%), (**M**= Margin of Contribution= 41,548 millions USD), (**s**= [S/S']-1= Sales Growth= 5.00%), (**Q**= Quoted Longterm Liabilities= 35,351 millions USD), (**d**= D/A= Dividend Portion or Payout= 30.00%), and (**V**= Variable Cost= 9,746 millions USD), then it's (**q**= Quick or Acid Test Ratio planned), is:

$$q= c\text{-}Vp/\{360[l\,(U+\{M\text{-}F\text{-}S'i[1+s]\text{-}T\}$$
$$[1\text{-}d])\text{-}Q]\}$$
$$= 1.2000\text{-}9,746*240/\{360[0.9900$$
$$(64,720+\{41,548\text{-}29,750$$
$$-48,851*0.0100[1+0.0500]$$
$$-3,385\}[1\text{-}0.3000])\text{-}35,351]\}$$
$$= 1.01 \text{ times}$$

Corporate IFRS-GAAP (B/S-I/S),ISBN-13: **978-1721066308**, ISBN-10: **1721066306**

<u>Law-8279</u>:

If both (**p**= 360 P/V= Procured Inventory Days= <u>240</u> Days), (**c**= C/X= Current Ratio= <u>1.20</u> times), (**T**= Tax Paid= <u>3,385</u> millions USD), (**U**= Utilized or Starting Capital= <u>64,720</u> millions USD), (**i**= I/S= Interest Portion= <u>1.00%</u>), (**$'**= Sales of Past Year= <u>48,851</u> millions USD), (**F**= Fixed Cost= <u>29,740</u> millions USD), (**l**= L/E= Leverage or Gearing Ratio= <u>99.00%</u>), (**M**= Margin of Contribution= <u>41,548</u> millions USD), (**s**= [S/S']-1= Sales Growth= <u>5.00%</u>), (**q**= [C-P]/X= Quick or Acid Test Ratio= <u>1.01</u> times), (**$**= Sales or Revenues= <u>51,294</u> millions USD), (**d**= D/A= Dividend Portion or Payout= <u>30.00%</u>), and (**v**= V/S= Variable Portion= <u>19.00%</u>), then it's (**Q**= Quoted Longterm Liabilities planned), is:

$$Q= \textbf{\textit{l}}(U+\{M-F-\$'i[1+s]-T\}[1-d])$$
$$-\$vp/\{360[c-q]\}$$
$$= 0.9900(64,720+\{41,548-29,750$$
$$-48,851*0.0100[1+0.0500]$$
$$-3,385\}[1-0.3000])-51,294$$
$$*0.1900*240/\{360[1.2000$$
$$-1.0100]\}$$
$$= \underline{35,351} \text{ millions USD}$$

Corporate IFRS-GAAP (B/S-I/S),ISBN-13: **978-1721066308**, ISBN-10: **1721066306**

Law-8280:

If both (**p**= 360 P/V= Procured Inventory Days= 240 Days), (**c**= C/X= Current Ratio= 1.20 times), (**T**= Tax Paid= 3,385 millions USD), (**U**= Utilized or Starting Capital= 64,720 millions USD), (**i**= I/S= Interest Portion= 1.00%), (**S'**= Sales of Past Year= 48,851 millions USD), (**F**= Fixed Cost= 29,740 millions USD), (**Q**= Quoted Longterm Liabilities= 35,351 millions USD), (**M**= Margin of Contribution= 41,548 millions USD), (**s**= [S/S']-1= Sales Growth= 5.00%), (**q**= [C-P]/X= Quick or Acid Test Ratio= 1.01 times), (**$**= Sales or Revenues= 51,294 millions USD), (**d**= D/A= Dividend Portion or Payout= 30.00%), and (**v**= V/S= Variable Portion= 19.00%), then it's (**l** = Leverage or Gearing Ratio planned), is:

$$l = (Q + \$vp / \{360[c-q]\})$$
$$/(U + \{M - F - \$'i[1+s] - T\}[1-d])$$
$$= (35,351 + 51,294 * 0.1900 * 240$$
$$/\{360[1.2000 - 1.0100]\})$$
$$/(64,720 + \{41,548 - 29,750$$
$$-48,851 * 0.0100[1 + 0.0500]$$
$$-3,385\}[1 - 0.3000])$$
$$= 99.00\%$$

Corporate IFRS-GAAP (B/S-I/S),ISBN-13: **978-1721066308**, ISBN-10: **1721066306**

<u>Law-8281</u>:

If both (**p**= 360 P/V= Procured Inventory Days= <u>240</u> Days), (**c**= C/X= Current Ratio= <u>1.20</u> times), (**T**= Tax Paid= <u>3,385</u> millions USD), (**I**= L/E= Leverage or Gearing Ratio= <u>99.00%</u>), (**i**= I/S= Interest Portion= <u>1.00%</u>), (**S'**= Sales of Past Year= <u>48,851</u> millions USD), (**F**= Fixed Cost= <u>29,740</u> millions USD), (**Q**= Quoted Longterm Liabilities= <u>35,351</u> millions USD), (**M**= Margin of Contribution= <u>41,548</u> millions USD), (**s**= [S/S']-1= Sales Growth= <u>5.00%</u>), (**q**= [C-P]/X= Quick or Acid Test Ratio= <u>1.01</u> times), (**S**= Sales or Revenues= <u>51,294</u> millions USD), (**d**= D/A= Dividend Portion or Payout= <u>30.00%</u>), and (**v**= V/S= Variable Portion= <u>19.00%</u>), then it's (**U**= Utilized or Starting Capital planned), is:

$$U= (Q+Svp/\{360[c-q]\})/I$$
$$-\{M-F-S'i[1+s]-T\}[1-d]$$
$$= (35,351+51,294*0.1900*240$$
$$/\{360[1.2000-1.0100]\})$$
$$/0.9900-\{41,548-29,750$$
$$-48,851*0.0100[1+0.0500]$$
$$-3,385\}[1-0.3000]$$
$$= \underline{64,720} \text{ millions USD}$$

Corporate IFRS-GAAP (B/S-I/S),ISBN-13: **978-1721066308**, ISBN-10: **1721066306**

Law-8282:

If both (**p**= 360 P/V= Procured Inventory Days= <u>240</u> Days), (**c**= C/X= Current Ratio= <u>1.20</u> times), (**T**= Tax Paid= <u>3,385</u> millions USD), (**l**= L/E= Leverage or Gearing Ratio= <u>99.00%</u>), (**i**= I/S= Interest Portion= <u>1.00%</u>), (**$'**= Sales of Past Year= <u>48,851</u> millions USD), (**F**= Fixed Cost= <u>29,740</u> millions USD), (**Q**= Quoted Longterm Liabilities= <u>35,351</u> millions USD), (**U**= Utilized or Starting Capital= <u>64,720</u> millions USD), (**s**= [S/S']-1= Sales Growth= <u>5.00%</u>), (**q**= Quick or Acid Test Ratio= [C-P]/X= <u>1.01</u> times), (**$**= Sales or Revenues= <u>51,294</u> millions USD), (**d**= D/A= Dividend Portion or Payout= <u>30.00%</u>), and (**v**= V/S= Variable Portion= <u>19.00%</u>), then it's (**M**= Margin of Contribution planned), is:

$$M = F + S'i[1+s] + T + [(Q + Svp$$
$$/\{360[c-q]\})/l - U]/[1-d]$$
$$= 29,750 + 48,851 * 0.0100[1+0.0500]$$
$$+3,385 + [(35,351 + 51,294$$
$$*0.1900 * 240/\{360[1.2000$$
$$-1.0100]\})/0.9900 - 64,720]$$
$$/[1-0.3000]$$
$$= \underline{41,548} \text{ millions USD}$$

Corporate IFRS-GAAP (B/S-I/S),ISBN-13: **978-1721066308**, ISBN-10: **1721066306**

Law-8283:

If both (**p**= 360 P/V= Procured Inventory Days= 240 Days), (**c**= C/X= Current Ratio= 1.20 times), (**T**= Tax Paid= 3,385 millions USD), (**f**= L/E= Leverage or Gearing Ratio= 99.00%), (**i**= I/S= Interest Portion= 1.00%), (**S'**= Sales of Past Year= 48,851 millions USD), (**M**= Margin of Contribution= 41,548 millions USD), (**Q**= Quoted Longterm Liabilities= 35,351 millions USD), (**U**= Utilized or Starting Capital= 64,720 millions USD), (**s**= [S/S']-1= Sales Growth= 5.00%), (**q**= [C-P]/X= Quick or Acid Test Ratio= 1.01 times), (**S**= Sales or Revenues= 51,294 millions USD), (**d**= D/A= Dividend Portion or Payout= 30.00%), and (**v**= V/S= Variable Portion= 19.00%), then it's (**F**= Fixed Cost planned), is:

$$F= M\text{-}S'i[1+s]\text{-}T\text{-}[(Q+Svp/\{360[c\text{-}q]\})/f\text{-}U]$$
$$/[1\text{-}d]$$
$$= 41{,}548\text{-}48{,}851*0.0100[1+0.0500]$$
$$-3{,}385\text{-}[(35{,}351+51{,}294$$
$$*0.1900*240/\{360[1.2000$$
$$-1.0100]\})/0.9900\text{-}64{,}720]$$
$$/[1\text{-}0.3000]$$
$$= \underline{29{,}750} \text{ millions USD}$$

Corporate IFRS-GAAP (B/S-I/S),ISBN-13: **978-1721066308**, ISBN-10: **1721066306**

Law-8284:

If both (**p**= 360 P/V= Procured Inventory Days= <u>240</u> Days), (**c**= C/X= Current Ratio= <u>1.20</u> times), (**T**= Tax Paid= <u>3,385</u> millions USD), (**F**= L/E= Leverage or Gearing Ratio= <u>99.00%</u>), (**i**= I/S= Interest Portion= <u>1.00%</u>), (**F**= Fixed Cost= <u>29,740</u> millions USD), (**M**= Margin of Contribution= <u>41,548</u> millions USD), (**Q**= Quoted Longterm Liabilities= <u>35,351</u> millions USD), (**U**= Utilized or Starting Capital= <u>64,720</u> millions USD), (**s**= [S/S']-1= Sales Growth= <u>5.00%</u>), (**q**= Quick or Acid Test Ratio= [C-P]/X= <u>1.01</u> times), (**S**= Sales or Revenues= <u>51,294</u> millions USD), (**d**= D/A= Dividend Portion or Payout= <u>30.00%</u>), and (**v**= V/S= Variable Portion= <u>19.00%</u>), then it's (**S'**= Sales Past), must be:

$$S'= \{M\text{-}F\text{-}T\text{-}[(Q+Svp/\{360[c\text{-}q]\})/F\text{-}U]$$
$$/[1\text{-}d]\}/\{i[1+s]\}$$
$$= \{41,548\text{-}29,750\text{-}3,385\text{-}[(35,351$$
$$+51,294*0.1900*240/\{360$$
$$[1.2000\text{-}1.0100]\})/0.9900$$
$$-64,720]/[1\text{-}0.3000]\}$$
$$/\{0.0100[1+0.0500]\}$$

= <u>48,851</u> millions USD

Corporate IFRS-GAAP (B/S-I/S),ISBN-13: **978-1721066308**, ISBN-10: **1721066306**

Law-8285:

If both (**p**= 360 P/V= Procured Inventory Days= 240 Days), (**c**= C/X= Current Ratio= 1.20 times), (**T**= Tax Paid= 3,385 millions USD), (**F**= L/E= Leverage or Gearing Ratio= 99.00%), (**$'**= Sales of Past Year= 48,851 millions USD), (**F**= Fixed Cost= 29,740 millions USD), (**M**= Margin of Contribution= 41,548 millions USD), (**Q**= Quoted Longterm Liabilities= 35,351 millions USD), (**U**= Utilized or Starting Capital= 64,720 millions USD), (**s**= [S/S']-1= Sales Growth= 5.00%), (**q**= [C-P]/X= Quick or Acid Test Ratio= 1.01 times), (**$**= Sales or Revenues= 51,294 millions USD), (**d**= D/A= Dividend Portion or Payout= 30.00%), and (**v**= V/S= Variable Portion= 19.00%), then it's (**i**= Interest Portion planned), is :

$$\mathbf{i} = \{M\text{-}F\text{-}T\text{-}[(Q+\mathbf{S}\mathbf{v}\mathbf{p}/\{360[\mathbf{c}\text{-}\mathbf{q}]\})/\mathbf{F}\text{-}U]$$
$$/[1\text{-}\mathbf{d}]\}/\{\mathbf{\$'}[1+\mathbf{s}]\}$$

$$= \{41{,}548\text{-}29{,}750\text{-}3{,}385\text{-}[(35{,}351$$
$$+51{,}294*0.1900*240/\{360$$
$$[1.2000\text{-}1.0100]\})/0.9900$$
$$-64{,}720]/[1\text{-}0.3000]\}$$
$$/\{48{,}851[1+0.0500]\}$$

$$= \underline{1.00\%}$$

Corporate IFRS-GAAP (B/S-I/S),ISBN-13: **978-1721066308**, ISBN-10: **1721066306**

Law-8286:

If both (**p**= 360 P/V= Procured Inventory Days= 240 Days), (**c**= C/X= Current Ratio= 1.20 times), (**T**= Tax Paid= 3,385 millions USD), (**l**= L/E= Leverage or Gearing Ratio= 99.00%), (**S'**= Sales of Past Year= 48,851 millions USD), (**F**= Fixed Cost= 29,740 millions USD), (**M**= Margin of Contribution= 41,548 millions USD), (**Q**= Quoted Longterm Liabilities= 35,351 millions USD), (**U**= Utilized or Starting Capital= 64,720 millions USD), (**i**= I/S= Interest Portion= 1.00%), (**q**= [C-P]/X= Quick or Acid Test Ratio= 1.01 times), (**S**= Sales or Revenues= 51,294 millions USD), (**d**= D/A= Dividend Portion or Payout= 30.00%), and (**v**= V/S= Variable Portion= 19.00%), then it's (**s**= Sales Growth planned), is:

$$s= \{M\text{-}F\text{-}T\text{-}[(Q+Svp/\{360[c\text{-}q]\})/l\text{-}U]$$
$$/[1\text{-}d]\}/[S'i]\text{-}1$$
$$= \{41,548\text{-}29,750\text{-}3,385\text{-}[(35,351$$
$$+51,294*0.1900*240/\{360$$
$$[1.2000\text{-}1.0100]\})/0.9900$$
$$\text{-}64,720]/[1\text{-}0.3000]\}$$
$$/[48,851*0.0100]\text{-}1$$
$$= 5.00\%$$

Corporate IFRS-GAAP (B/S-I/S),ISBN-13: **978-1721066308**, ISBN-10: **1721066306**

Law-8287:

If both (**p**= 360 P/V= Procured Inventory Days= 240 Days), (**c**= C/X= Current Ratio= 1.20 times), (**s**= [S/S']-1= Sales Growth= 5.00%), (**f**= L/E= Leverage or Gearing Ratio= 99.00%), (**S'**= Sales of Past Year= 48,851 millions USD), (**F**= Fixed Cost= 29,740 millions USD), (**M**= Margin of Contribution= 41,548 millions USD), (**Q**= Quoted Longterm Liabilities= 35,351 millions USD), (**U**= Utilized or Starting Capital= 64,720 millions USD), (**i**= I/S= Interest Portion= 1.00%), (**q**= [C-P]/X= Quick or Acid Test Ratio= 1.01 times), (**S**= Sales or Revenues= 51,294 millions USD), (**d**= D/A= Dividend Portion or Payout= 30.00%), and (**v**= V/S= Variable Portion= 19.00%), then it's (**T**= Tax Paid planned), is:

$$T= M\text{-}F\text{-}S'i[1+s]\text{-}[(Q+Svp$$
$$/\{360[c\text{-}q]\})/f\text{-}U]/[1\text{-}d]$$
$$= 41{,}548\text{-}29{,}750\text{-}48{,}851*0.0100$$
$$[1+0.0500]\text{-}[(35{,}351+51{,}294$$
$$*0.1900*240/\{360[1.2000$$
$$-1.0100]\})/0.9900\text{-}64{,}720]$$
$$/[1\text{-}0.3000]$$
$$= 3{,}385 \text{ millions USD}$$

Corporate IFRS-GAAP (B/S-I/S),ISBN-13: **978-1721066308**, ISBN-10: **1721066306**

Law-8288:

If both (**p**= 360 P/V= Procured Inventory Days= 240
Days), (**c**= C/X= Current Ratio= 1.20 times), (**s**=
[S/S']-1= Sales Growth= 5.00%), (**f**= L/E= Leverage
or Gearing Ratio= 99.00%), (**S'**= Sales of Past Year=
48,851 millions USD), (**F**= Fixed Cost= 29,740
millions USD), (**M**= Margin of Contribution= 41,548
millions USD), (**Q**= Quoted Longterm Liabilities=
35,351 millions USD), (**U**= Utilized or Starting
Capital= 64,720 millions USD), (**i**= I/S= Interest
Portion= 1.00%), (**q**= [C-P]/X= Quick or Acid Test
Ratio= 1.01 times), (**S**= Sales or Revenues= 51,294
millions USD), (**T**= Tax Paid= 3,385 millions USD),
and (**v**= V/S= Variable Portion= 19.00%), then it's
(**d**= Dividend Portion or Payout planned), is:

$$d= 1-[(Q+Svp/\{360[c-q]\})/f-U]$$
$$/\{M-F-S'i[1+s]-T\}$$
$$= 1-[(35,351+51,294*0.1900*240$$
$$/\{360[1.2000-1.0100]\})$$
$$/0.9900-64,720]/\{41,548$$
$$-29,750-48,851*0.0100$$
$$[1+0.0500]-3,385\}$$
$$= 30.00\%$$

Corporate IFRS-GAAP (B/S-I/S),ISBN-13: **978-1721066308**, ISBN-10: **1721066306**

Law-8289:

If both (**p**= 360 P/V= Procured Inventory Days= 240 Days), (**c**= C/X= Current Ratio= 1.20 times), (**s**= [S/S']-1= Sales Growth= 5.00%), (**l**= L/E= Leverage or Gearing Ratio= 99.00%), (**S'**= Sales of Past Year= 48,851 millions USD), (**F**= Fixed Cost= 29,740 millions USD), (**M**= Margin of Contribution= 41,548 millions USD), (**Q**= Quoted Longterm Liabilities= 35,351 millions USD), (**U**= Utilized or Starting Capital= 64,720 millions USD), (**i**= I/S= Interest Portion= 1.00%), (**q**= [C-P]/X= Quick or Acid Test Ratio= 1.01 times), (**d**= D/A=Dividend Portion or Payout= 30.00%), (**T**= Tax Paid= 3,385 millions USD), and (**v**= V/S= Variable Portion= 19.00%), then it's (**S**= Sales or Revenues planned), is:

$$S= 360[\mathbf{c\text{-}q}][\mathbf{l}\,(\mathbf{U} + \{\mathbf{M\text{-}F\text{-}S'i}[1 + \mathbf{s}] - \mathbf{T}\}[1 - \mathbf{d}]) - \mathbf{Q}]/[\mathbf{vp}]$$

$$= 360[1.2000 - 1.0100][0.9900$$
$$(64{,}720 + \{41{,}548 - 29{,}750$$
$$-48{,}851*0.0100[1 + 0.0500]$$
$$-3{,}385\}[1 - 0.3000]) - 35{,}351]$$
$$/[0.1900*240]$$

$$= 51{,}294 \text{ millions USD}$$

Corporate IFRS-GAAP (B/S-I/S),ISBN-13: **978-1721066308**, ISBN-10: **1721066306**

Law-8290:

If both (**p**= 360 P/V= Procured Inventory Days= 240
Days), (**c**= C/X= Current Ratio= 1.20 times), (**s**=
[S/S']-1= Sales Growth= 5.00%), (**l**= L/E= Leverage
or Gearing Ratio= 99.00%), (**S'**= Sales of Past Year=
48,851 millions USD), (**F**= Fixed Cost= 29,740
millions USD), (**M**= Margin of Contribution= 41,548
millions USD), (**Q**= Quoted Longterm Liabilities=
35,351 millions USD), (**U**= Utilized or Starting
Capital= 64,720 millions USD), (**i**= I/S= Interest
Portion= 1.00%), (**q**= [C-P]/X= Quick or Acid Test
Ratio= 1.01 times), (**d**= D/A= Dividend Portion or
Payout= 30.00%), (**T**= Tax Paid= 3,385 millions
USD), and (**S**= Sales or Revenues= 51,294 millions
USD), then it's (**v**= Variable Portion planned), is :

$$v= 360[c\text{-}q][l\,(U+\{M\text{-}F\text{-}S'i[1+s]\text{-}T\}$$
$$[1\text{-}d])\text{-}Q]/[Sp]$$

$$= 360[1.2000\text{-}1.0100][0.9900$$
$$(64{,}720+\{41{,}548\text{-}29{,}750$$
$$-48{,}851*0.0100[1+0.0500]$$
$$-3{,}385\}[1\text{-}0.3000])\text{-}35{,}351]$$
$$/[51{,}294*240]$$

$$= \underline{19.00\%}$$

Corporate IFRS-GAAP (B/S-I/S),ISBN-13: **978-1721066308**, ISBN-10: **1721066306**

<u>Law-8291</u>:

If both ($\mathbf{v}$= V/S= Variable Portion= <u>19.00%</u>), ($\mathbf{c}$= C/X= Current Ratio= <u>1.20</u> times), ($\mathbf{s}$= [S/S']-1= Sales Growth= <u>5.00%</u>), ($\mathbf{l}$= L/E= Leverage or Gearing Ratio= <u>99.00%</u>), ($\mathbf{S'}$= Sales of Past Year= <u>48,851</u> millions USD), ($\mathbf{F}$= Fixed Cost= <u>29,740</u> millions USD), ($\mathbf{M}$= Margin of Contribution= <u>41,548</u> millions USD), ($\mathbf{Q}$= Quoted Longterm Liabilities= <u>35,351</u> millions USD), ($\mathbf{U}$= Utilized or Starting Capital= <u>64,720</u> millions USD), ($\mathbf{i}$= I/S= Interest Portion= <u>1.00%</u>), ($\mathbf{q}$= [C-P]/X= Quick or Acid Test Ratio= <u>1.01</u> times), ($\mathbf{d}$= D/A= Dividend Portion or Payout= <u>30.00%</u>), ($\mathbf{T}$= Tax Paid= <u>3,385</u> millions USD), and ($\mathbf{S}$= Sales or Revenues= <u>51,294</u> millions USD), then it's ($\mathbf{p}$= Procured Inventory Days planned), is:

$$\mathbf{p}= 360[\mathbf{c}\text{-}\mathbf{q}][\mathbf{l}\,(\mathbf{U}+\{\mathbf{M}\text{-}\mathbf{F}\text{-}\mathbf{S'i}[1+\mathbf{s}]\text{-}\mathbf{T}\}[1\text{-}\mathbf{d}])\text{-}\mathbf{Q}]/[\mathbf{Sv}]$$

$$= 360[1.2000\text{-}1.0100][0.9900$$
$$(64,720+\{41,548\text{-}29,750$$
$$-48,851*0.0100[1+0.0500]$$
$$-3,385\}[1\text{-}0.3000])\text{-}35,351]$$
$$/[51,294*0.1900]$$

$$= \underline{19.00\%}$$

Corporate IFRS-GAAP (B/S-I/S),ISBN-13: **978-1721066308**, ISBN-10: **1721066306**

Law-8292:

If both (v= V/S= Variable Portion= 19.00%), (p= 360 P/V= Procured Inventory Days= 240 Days), (s= [S/S']-1= Sales Growth= 5.00%), (l= L/E= Leverage or Gearing Ratio= 99.00%), (S'= Sales of Past Year= 48,851 millions USD), (F= Fixed Cost= 29,740 millions USD), (M= Margin of Contribution= 41,548 millions USD), (Q= Quoted Longterm Liabilities= 35,351 millions USD), (U= Utilized or Starting Capital= 64,720 millions USD), (i= I/S= Interest Portion= 1.00%), (q= [C-P]/X= Quick or Acid Test Ratio= 1.01 times), (d= D/A= Dividend Portion or Payout= 30.00%), (T= Tax Paid= 3,385 millions USD), and (S= Sales or Revenues= 51,294 millions USD), then it's (c= Current Ratio planned), is:

$$c= q+Svp/\{360[l\,(U+\{M-F-S'i[1+s]-T\}[1-d])-Q]\}$$

$$= 1.0100+51,294*0.1900*240$$
$$/\{360[0.9900(64,720$$
$$+\{41,548-29,750-48,851$$
$$*0.0100[1+0.0500]-3,385\}$$
$$[1-0.3000])-35,351]\}$$

$$= 1.20 \text{ times}$$

Corporate IFRS-GAAP (B/S-I/S),ISBN-13: **978-1721066308**, ISBN-10: **1721066306**

Law-8293:

If both (v= V/S= Variable Portion= 19.00%), (p= 360 P/V= Procured Inventory Days= 240 Days), (s= [S/S']-1= Sales Growth= 5.00%), (l= L/E= Leverage or Gearing Ratio= 99.00%), (s'= Sales of Past Year= 48,851 millions USD), (F= Fixed Cost= 29,740 millions USD), (M= Margin of Contribution= 41,548 millions USD), (Q= Quoted Longterm Liabilities= 35,351 millions USD), (U= Utilized or Starting Capital= 64,720 millions USD), (i= I/S= Interest Portion= 1.00%), (c= C/X= Current Ratio= 1.20 times), (d= D/A= Dividend Portion or Payout= 30.00%), (T= Tax Paid= 3,385 millions USD), and (s= Sales or Revenues= 51,294 millions USD), then it's (q= Quick or Acid Test Ratio planned), is:

$$q= c\text{-}svp/\{360[l\,(U+\{M\text{-}F\text{-}s'i[1+s]\text{-}T\}[1\text{-}d])\text{-}Q]\}$$

$$= 1.2000\text{-}51,294*0.1900*240/\{360[0.9900(64,720+\{41,548-29,750\text{-}48,851*0.0100[1+0.0500]\text{-}3,385\}[1\text{-}0.3000])\text{-}35,351]\}$$

$$= 1.01 \text{ times}$$

Corporate IFRS-GAAP (B/S-I/S),ISBN-13: **978-1721066308**, ISBN-10: **1721066306**

Law-8294:

If both (**v**= V/S= Variable Portion= 19.00%), (**p**= 360 P/V= Procured Inventory Days= 240 Days), (**s**= [S/S']-1= Sales Growth= 5.00%), (**l**= L/E= Leverage or Gearing Ratio= 99.00%), (**F**= Fixed Cost= 29,740 millions USD), (**M**= Margin of Contribution= 41,548 millions USD), (**q**= [C-P]/X= Quick or Acid Test Ratio= 1.01 times), (**U**= Utilized or Starting Capital= 64,720 millions USD), (**i**= I/S= Interest Portion= 1.00%), (**c**= C/X= Current Ratio= 1.20 times), (**d**= D/A= Dividend Portion or Payout= 30.00%), (**T**= Tax Paid= 3,385 millions USD), and (**S'**= Sales of Past Year= 48,851 millions USD), then it's (**Q**= Quoted Longterm Liabilities planned), is:

$$Q= \textbf{\textit{l}}\,(\textbf{U}+\{\textbf{M-F-S'i}[1+\textbf{s}]\textbf{-T}\}[1\textbf{-d}])$$
$$-\textbf{S'vp}[1+\textbf{s}]/\{360[\textbf{c-q}]\}$$
$$= 0.9900(64,720+\{41,548-29,750$$
$$-48,851*0.0100[1+0.0500]$$
$$-3,385\}[1-0.3000])-48,851$$
$$*0.1900*240[1+0.0500]$$
$$/\{360[1.2000-1.0100]\}$$
$$= 35,351 \text{ millions USD}$$

Corporate IFRS-GAAP (B/S-I/S),ISBN-13: **978-1721066308**, ISBN-10: **1721066306**

Law-8295:

If both ($\boldsymbol{v}$= V/S= Variable Portion= 19.00%), ($\boldsymbol{p}$= 360 P/V= Procured Inventory Days= 240 Days), ($\boldsymbol{s}$= [S/S']-1= Sales Growth= 5.00%), ($\boldsymbol{F}$= Fixed Cost= 29,740 millions USD), ($\boldsymbol{M}$= Margin of Contribution= 41,548 millions USD), ($\boldsymbol{Q}$= Quoted Longterm Liabilities= 35,351 millions USD), ($\boldsymbol{q}$= [C-P]/X= Quick or Acid Test Ratio= 1.01 times), ($\boldsymbol{U}$= Utilized or Starting Capital= 64,720 millions USD), ($\boldsymbol{i}$= I/S= Interest Portion= 1.00%), ($\boldsymbol{c}$= C/X= Current Ratio= 1.20 times), ($\boldsymbol{d}$= D/A= Dividend Portion or Payout= 30.00%), ($\boldsymbol{T}$= Tax Paid= 3,385 millions USD), and ($\boldsymbol{S'}$= Sales of Past Year= 48,851 millions USD), then it's ($\boldsymbol{I}$= Leverage or Gearing Ratio planned), is:

$$\boldsymbol{I} = (\boldsymbol{Q}+\boldsymbol{S'vp}[1+\boldsymbol{s}]/\{360[\boldsymbol{c\text{-}q}]\})$$
$$/(\boldsymbol{U}+\{\boldsymbol{M\text{-}F\text{-}S'i}[1+\boldsymbol{s}]\text{-}\boldsymbol{T}\}[1\text{-}\boldsymbol{d}])$$
$$= (35,351+48,851*0.1900*240$$
$$[1+0.0500]/\{360[1.2000$$
$$-1.0100]\})/(64,720+\{41,548$$
$$-29,750-48,851*0.0100$$
$$[1+0.0500]-3,385\}[1-0.3000]$$
$$= 99.00\%$$

Corporate IFRS-GAAP (B/S-I/S),ISBN-13: **978-1721066308**, ISBN-10: **1721066306**

Law-8296:

If both (v= V/S= Variable Portion= 19.00%), (p= 360 P/V= Procured Inventory Days= 240 Days), (s= [S/S']-1= Sales Growth= 5.00%), (F= Fixed Cost= 29,740 millions USD), (M= Margin of Contribution= 41,548 millions USD), (Q= Quoted Longterm Liabilities= 35,351 millions USD), (q= [C-P]/X= Quick or Acid Test Ratio= 1.01 times), (I= L/E= Leverage or Gearing Ratio= 99.00%), (i= I/S= Interest Portion= 1.00%), (c= C/X= Current Ratio= 1.20 times), (d= D/A= Dividend Portion or Payout= 30.00%), (T= Tax Paid= 3,385 millions USD), and (S'= Sales of Past Year= 48,851 millions USD), then it's (U= Utilized or Starting Capital planned), is:

$$U= (Q+S'vp[1+s]/\{360[c-q]\})/I$$
$$-\{M-F-S'i[1+s]-T\}[1-d]$$
$$= (35,351+48,851*0.1900*240$$
$$[1+0.0500]/\{360[1.2000$$
$$-1.0100]\})/0.9900-\{41,548$$
$$-29,750-48,851*0.0100$$
$$[1+0.0500]-3,385\}$$
$$[1-0.3000]$$
$$= 64,720 \text{ millions USD}$$

Corporate IFRS-GAAP (B/S-I/S),ISBN-13: **978-1721066308**, ISBN-10: **1721066306**

<u>Law-8297</u>:

If both (**υ**= V/S= Variable Portion= <u>19.00%</u>), (**p**= 360 P/V= Procured Inventory Days= <u>240</u> Days), (**s**= [S/S']-1= Sales Growth= <u>5.00%</u>), (**F**= Fixed Cost= <u>29,740</u> millions USD), (**U**= Utilized or Starting Capital= <u>64,720</u> millions USD), (**Q**= Quoted Longterm Liabilities= <u>35,351</u> millions USD), (**q**= Quick or Acid Test Ratio= [C-P]/X= <u>1.01</u> times), (**ℓ**= L/E= Leverage or Gearing Ratio= <u>99.00%</u>), (**i**= I/S= Interest Portion= <u>1.00%</u>), (**c**= C/X= Current Ratio= <u>1.20</u> times), (**d**= D/A= Dividend Portion or Payout= <u>30.00%</u>), (**T**= Tax Paid= <u>3,385</u> millions USD), and (**S'**= Sales of Past Year= <u>48,851</u> millions USD), then it's (**M**= Margin of Contribution planned), is:

$$M = F+S'i[1+s]+T+[(Q+S'υp[1+s]$$
$$/\{360[c-q]\})/\ell-U]/[1-d]$$
$$= 29,750+48,851*0.0100[1+0.0500]$$
$$+3,385+[(35,351+48,851$$
$$*0.1900*240[1+0.0500]/\{360$$
$$[1.2000-1.0100]\})/0.9900$$
$$-64,720]/[1-0.3000]$$
$$= \underline{41,548} \text{ millions USD}$$

Corporate IFRS-GAAP (B/S-I/S),ISBN-13: **978-1721066308**, ISBN-10: **1721066306**

Law-8298:

If both (**v**= V/S= Variable Portion= 19.00%), (**p**= 360 P/V= Procured Inventory Days= 240 Days), (**s**= [S/S']-1= Sales Growth= 5.00%), (**M**= Margin of Contribution= 41,548 millions USD), (**U**= Utilized or Starting Capital= 64,720 millions USD), (**Q**= Quoted Longterm Liabilities= 35,351 millions USD), (**q**= Quick or Acid Test Ratio= [C-P]/X= 1.01 times), (**l**= L/E= Leverage or Gearing Ratio= 99.00%), (**i**= I/S= Interest Portion= 1.00%), (**c**= C/X= Current Ratio= 1.20 times), (**d**= D/A= Dividend Portion or Payout= 30.00%), (**T**= Tax Paid= 3,385 millions USD), and (**S'**= Sales of Past Year= 48,851 millions USD), then it's (**F**= Fixed Cost planned), is:

$$F= M\text{-}S'i[1+s]\text{-}T\text{-}[(Q+S'vp[1+s]$$
$$/\{360[c\text{-}q]\})/l\text{-}U]/[1\text{-}d]$$
$$= 41,548\text{-}48,851*0.0100[1+0.0500]$$
$$\text{-}3,385\text{-}[(35,351+48,851$$
$$*0.1900*240[1+0.0500]$$
$$/\{360[1.2000\text{-}1.0100]\})$$
$$/0.9900\text{-}64,720]/[1\text{-}0.3000]$$
$$= 29,750 \text{ millions USD}$$

Corporate IFRS-GAAP (B/S-I/S),ISBN-13: **978-1721066308**, ISBN-10: **1721066306**

Law-8299:

If both (v= V/S= Variable Portion= 19.00%), (p= 360 P/V= Procured Inventory Days= 240 Days), (s= [S/S']-1= Sales Growth= 5.00%), (M= Margin of Contribution= 41,548 millions USD), (U= Utilized or Starting Capital= 64,720 millions USD), (Q= Quoted Longterm Liabilities= 35,351 millions USD), (q= Quick or Acid Test Ratio= [C-P]/X= 1.01 times), (l= L/E= Leverage or Gearing Ratio= 99.00%), (i= I/S= Interest Portion= 1.00%), (c= C/X= Current Ratio= 1.20 times), (d= D/A= Dividend Portion or Payout= 30.00%), (T= Tax Paid= 3,385 millions USD), and (F= Fixed Cost= 29,740 millions USD), then it's (S'= Sales Past), must be:

$$S'= \{M\text{-}F\text{-}T\text{-}[(Q+S'vp[1+s]/\{360[c\text{-}q]\})/l$$
$$-U]/[1\text{-}d]\}/\{i[1+s]\}$$
$$= \{41,548\text{-}29,750\text{-}3,385\text{-}[(35,351$$
$$+48,851*0.1900*240$$
$$[1+0.0500]/\{360[1.2000$$
$$-1.0100]\})/0.9900\text{-}64,720]$$
$$/[1\text{-}0.3000]\}/\{0.0100$$
$$[1+0.0500]\}$$
$$= 48,851 \text{ millions USD}$$

Corporate IFRS-GAAP (B/S-I/S),ISBN-13: **978-1721066308**, ISBN-10: **1721066306**

Law-8300:

If both (**v**= V/S= Variable Portion= 19.00%), (**p**= 360 P/V= Procured Inventory Days= 240 Days), (**s**= [S/S']-1= Sales Growth= 5.00%), (**M**= Margin of Contribution= 41,548 millions USD), (**U**= Utilized or Starting Capital= 64,720 millions USD), (**Q**= Quoted Longterm Liabilities= 35,351 millions USD), (**q**= Quick or Acid Test Ratio= [C-P]/X= 1.01 times), (**l**= L/E= Leverage or Gearing Ratio= 99.00%), (**S'**= Sales of Past Year= 48,851 millions USD), (**c**= C/X= Current Ratio= 1.20 times), (**d**= D/A= Dividend Portion or Payout= 30.00%), (**T**= Tax Paid= 3,385 millions USD), and (**F**= Fixed Cost= 29,740 millions USD), then it's (**i**= Interest Portion planned), is:

$$i= \{M\text{-}F\text{-}T\text{-}[(Q+S'vp[1+s]/\{360[c\text{-}q]\})/l\text{-}U]$$
$$/[1\text{-}d]\}/\{S'[1+s]\}$$

$$= \{41,548\text{-}29,750\text{-}3,385\text{-}[(35,351$$
$$+48,851*0.1900*240$$
$$[1+0.0500]/\{360[1.2000$$
$$-1.0100]\})/0.9900\text{-}64,720]$$
$$/[1\text{-}0.3000]\}/\{48,851$$
$$[1+0.0500]\}$$

$$= 1.00\%$$

Corporate IFRS-GAAP (B/S-I/S),ISBN-13: **978-1721066308**, ISBN-10: **1721066306**

Law-8301:

If both (**v**= V/S= Variable Portion= 19.00%), (**p**= 360 P/V= Procured Inventory Days= 240 Days), (**i**= I/S= Interest Portion= 1.00%), (**M**= Margin of Contribution= 41,548 millions USD), (**U**= Utilized or Starting Capital= 64,720 millions USD), (**Q**= Quoted Longterm Liabilities= 35,351 millions USD), (**q**= Quick or Acid Test Ratio= [C-P]/X= 1.01 times), (**l**= L/E= Leverage or Gearing Ratio= 99.00%), (**S'**= Sales of Past Year= 48,851 millions USD), (**c**= C/X= Current Ratio= 1.20 times), (**d**= D/A= Dividend Portion or Payout= 30.00%), (**T**= Tax Paid= 3,385 millions USD), and (**F**= Fixed Cost= 29,740 millions USD), then it's (**s**= Sales Growth planned), is:

$$s= \{M\text{-}F\text{-}T\text{-}[(Q+S'vp[1+s]/\{360[c\text{-}q]\})/l\text{-}U]$$
$$/[1\text{-}d]\}/[S'i]\text{-}1$$

$$= \{41,548\text{-}29,750\text{-}3,385\text{-}[(35,351$$
$$+48,851*0.1900*240$$
$$[1+0.0500]/\{360[1.2000$$
$$-1.0100]\})/0.9900\text{-}64,720]$$
$$/[1\text{-}0.3000]\}$$
$$/[48,851*0.0100]\text{-}1$$

$$= 5.00\%$$

Corporate IFRS-GAAP (B/S-I/S),ISBN-13: **978-1721066308**, ISBN-10: **1721066306**

<u>Law-8302</u>:

If both (**v**= V/S= Variable Portion= <u>19.00%</u>), (**p**= 360
P/V= Procured Inventory Days= <u>240</u> Days), (**i**= I/S=
Interest Portion= <u>1.00%</u>), (**M**= Margin of
Contribution= <u>41,548</u> millions USD), (**U**= Utilized or
Starting Capital= <u>64,720</u> millions USD), (**Q**= Quoted
Longterm Liabilities= <u>35,351</u> millions USD), (**q**=
Quick or Acid Test Ratio= [C-P]/X= <u>1.01</u> times), (**l**=
L/E= Leverage or Gearing Ratio= <u>99.00%</u>), (**S'**=
Sales of Past Year= <u>48,851</u> millions USD), (**c**= C/X=
Current Ratio= <u>1.20</u> times), (**d**= D/A= Dividend
Portion or Payout= <u>30.00%</u>), (**s**= [S/S']-1= Sales
Growth= <u>5.00%</u>), and (**F**= Fixed Cost= <u>29,740</u>
millions USD), then it's (**T**= Tax Paid planned), is:

$$T= M\text{-}F\text{-}S'i[1+s]\text{-}[(Q+S'vp[1+s]$$
$$/\{360[c\text{-}q]\})/l\text{-}U]/[1\text{-}d]$$
$$= \{41,548\text{-}29,750\text{-}48,851*0.0100$$
$$[1+0.0500]\text{-}[(35,351+48,851$$
$$*0.1900*240[1+0.0500]/\{360$$
$$[1.2000\text{-}1.0100]\})/0.9900$$
$$-64,720]/[1\text{-}0.3000]$$
$$= \underline{3,385} \text{ millions USD}$$

Corporate IFRS-GAAP (B/S-I/S),ISBN-13: **978-1721066308**, ISBN-10: **1721066306**

<u>Law-8303</u>:

If both (**v**= V/S= Variable Portion= <u>19.00%</u>), (**p**= 360 P/V= Procured Inventory Days= <u>240</u> Days), (**i**= I/S= Interest Portion= <u>1.00%</u>), (**M**= Margin of Contribution= <u>41,548</u> millions USD), (**U**= Utilized or Starting Capital= <u>64,720</u> millions USD), (**Q**= Quoted Longterm Liabilities= <u>35,351</u> millions USD), (**q**= Quick or Acid Test Ratio= [C-P]/X= <u>1.01</u> times), (**l**= L/E= Leverage or Gearing Ratio= <u>99.00%</u>), (**S'**= Sales of Past Year= <u>48,851</u> millions USD), (**c**= C/X= Current Ratio= <u>1.20</u> times), (**T**= Tax Paid= <u>3,385</u> millions USD), (**s**= [S/S']-1= Sales Growth= <u>5.00%</u>), and (**F**= Fixed Cost= <u>29,740</u> millions USD), then it's (**d**= Dividend Portion or Payout planned), is:

$$d= 1-[(Q+S'vp[1+s]/\{360[c-q]\})/l -U]$$
$$/\{M-F-S'i[1+s]-T\}$$

$$= 1-[(35,351+48,851*0.1900+240$$
$$[1+0.0500]/\{360[1.2000$$
$$-1.0100]\})/0.9900-64,720]$$
$$/\{41,548-29,750-48,851$$
$$*0.0100[1+0.0500]-3,385\}$$

$$= \underline{30.00\%}$$

Corporate IFRS-GAAP (B/S-I/S),ISBN-13: **978-1721066308**, ISBN-10: **1721066306**

Law-8304:

If both (**d**= D/A= Dividend Portion or Payout= 30.00%), (**p**= 360 P/V= Procured Inventory Days= 240 Days), (**i**= I/S= Interest Portion= 1.00%), (**M**= Margin of Contribution= 41,548 millions USD), (**U**= Utilized or Starting Capital= 64,720 millions USD), (**Q**= Quoted Longterm Liabilities= 35,351 millions USD), (**q**= [C-P]/X= Quick or Acid Test Ratio= 1.01 times), (**l**= L/E= Leverage or Gearing Ratio= 99.00%), (**$'**= Sales of Past Year= 48,851 millions USD), (**c**= C/X= Current Ratio= 1.20 times), (**T**= Tax Paid= 3,385 millions USD), (**s**= [S/S']-1= Sales Growth= 5.00%), and (**F**= Fixed Cost= 29,740 millions USD), then it's (**v**= Variable Portion planned), is:

$$v= 360[c\text{-}q][l\text{-}(U+\{M\text{-}F\text{-}\$'i[1+s]\text{-}T\}$$
$$[1\text{-}d])\text{-}Q]/\{\$'p[1+s]\}$$
$$= 360[1.2000\text{-}1.0100][0.9900$$
$$(64,720+\{41,548\text{-}29,750$$
$$-48,851*0.0100[1+0.0500]$$
$$-3,385\}[1\text{-}0.3000])\text{-}35,351]$$
$$/\{48,851*240[1+0.0500]\}$$
$$= 19.00\%$$

Corporate IFRS-GAAP (B/S-I/S),ISBN-13: **978-1721066308**, ISBN-10: **1721066306**

<u>Law-8306</u>:

If both (**d**= D/A= Dividend Portion or Payout= 30.00%), (**v**= V/S= Variable Portion= 19.00%), (**i**= I/S= Interest Portion= 1.00%), (**M**= Margin of Contribution= 41,548 millions USD), (**U**= Utilized or Starting Capital= 64,720 millions USD), (**Q**= Quoted Longterm Liabilities= 35,351 millions USD), (**q**= Quick or Acid Test Ratio= [C-P]/X= 1.01 times), (**l**= L/E= Leverage or Gearing Ratio= 99.00%), (**S'**= Sales of Past Year= 48,851 millions USD), (**p**= 360 P/V= Procured Inventory Days= 240 Days), (**T**= Tax Paid= 3,385 millions USD), (**s**= [S/S']-1= Sales Growth= 5.00%), and (**F**= Fixed Cost= 29,740 millions USD), then it's (**c**= Current Ratio planned), is:

$$c= q+S'vp[1+s]/\{360[/(U+\{M\text{-}F\text{-}S'i[1+s]\text{-}T\}[1\text{-}d])\text{-}Q]\}$$

$$= 1.0100+48,851*0.1900*240$$
$$[1+0.0500]/\{360[0.9900$$
$$(64,720+\{41,548\text{-}29,750$$
$$-48,851*0.0100[1+0.0500]$$
$$-3,385\}[1\text{-}0.3000])\text{-}35,351]\}$$

$$= 1.20 \text{ times}$$

Corporate IFRS-GAAP (B/S-I/S),ISBN-13: **978-1721066308**, ISBN-10: **1721066306**

Law-8307:

If both (**d**= D/A= Dividend Portion or Payout= 30.00%), (**v**= V/S= Variable Portion= 19.00%), (**i**= I/S= Interest Portion= 1.00%), (**M**= Margin of Contribution= 41,548 millions USD), (**U**= Utilized or Starting Capital= 64,720 millions USD), (**Q**= Quoted Longterm Liabilities= 35,351 millions USD), (**c**= C/X= Current Ratio= 1.20 times), (**l**= L/E= Leverage or Gearing Ratio= 99.00%), (**S'**= Sales of Past Year= 48,851 millions USD), (**p**= 360 P/V= Procured Inventory Days= 240 Days), (**T**= Tax Paid= 3,385 millions USD), (**s**= [S/S']-1= Sales Growth= 5.00%), and (**F**= Fixed Cost= 29,740 millions USD), then it's (**q**= Quick or Acid Test Ratio planned), is:

$$q= c\text{-}S'vp[1+s]/\{360[l\,(U+\{M\text{-}F\text{-}S'i[1+s]\text{-}T\}$$
$$[1\text{-}d])\text{-}Q]\}$$

$$= 1.2000\text{-}48,851*0.1900*240$$
$$[1+0.0500]/\{360[0.9900$$
$$(64,720+\{41,548\text{-}29,750$$
$$-48,851*0.0100[1+0.0500]$$
$$-3,385\}[1\text{-}0.3000])\text{-}35,351]\}$$

$$= 1.01 \text{ times}$$

Corporate IFRS-GAAP (B/S-I/S),ISBN-13: **978-1721066308**, ISBN-10: **1721066306**

Law-8308:

If both (**d**= D/A= Dividend Portion or Payout= 30.00%), (**X**= Xpress or Current Debt= 34,196 millions USD), (**T**= Tax Paid= 3,385 millions USD), (**A**= After Tax Income= 7,899 millions USD), (**i**= I/S= Interest Portion= 1.00%), (**M**= Margin of Contribution= 41,548 millions USD), (**U**= Utilized or Starting Capital= 64,720 millions USD), (**l**= L/E= Leverage or Gearing Ratio= 99.00%), (**S'**= Sales of Past Year= 48,851 millions USD), (**s**= [S/S']-1= Sales Growth= 5.00%), and (**F**= Fixed Cost= 29,740 millions USD), then it's (**Q**= Quoted Longterm Liabilities planned), is:

$$Q= l\{U+M-F-S'i[1+s]-T-Ad\}-X$$
$$= 0.9900\{64,720+41,548-29,750$$
$$-48,851*0.0100[1+0.0500]$$
$$-3,385-7,899*0.3000\}$$
$$-34,196$$
$$= 35,351 \text{ millions USD}$$

Corporate IFRS-GAAP (B/S-I/S),ISBN-13: **978-1721066308**, ISBN-10: **1721066306**

Law-8309:

If both (**d**= D/A= Dividend Portion or Payout= 30.00%), (**Q**= Quoted Longterm Liabilities= 35,351 millions USD), (**X**= Xpress or Current Debt= 34,196 millions USD), (**T**= Tax Paid= 3,385 millions USD), (**A**= After Tax Income= 7,899 millions USD), (**i**= I/S= Interest Portion= 1.00%), (**M**= Margin of Contribution= 41,548 millions USD), (**U**= Utilized or Starting Capital= 64,720 millions USD), (**$'**= Sales of Past Year= 48,851 millions USD), (**s**= [S/S']-1= Sales Growth= 5.00%), and (**F**= Fixed Cost= 29,740 millions USD), then it's (**I** = Leverage or Gearing Ratio planned), is:

$$I = [Q+X]/\{U+M-F-S'i[1+s]-T-Ad\}$$
$$= [35,351+34,196]/\{64,720+41,548$$
$$-29,750-48,851*0.0100$$
$$[1+0.0500]-3,385-7,899$$
$$*0.3000\}$$
$$= \underline{99.00\%}$$

Corporate IFRS-GAAP (B/S-I/S),ISBN-13: **978-1721066308**, ISBN-10: **1721066306**

Law-8310:

If both (**d**= D/A= Dividend Portion or Payout= 30.00%), (**X**= Xpress or Current Debt= 34,196 millions USD), (**T**= Tax Paid= 3,385 millions USD), (**A**= After Tax Income= 7,899 millions USD), (**i**= I/S= Interest Portion= 1.00%), (**M**= Margin of Contribution= 41,548 millions USD), (**Q**= Quoted Longterm Liabilities= 35,351 millions USD), (**l**= L/E= Leverage or Gearing Ratio= 99.00%), (**S'**= Sales of Past Year= 48,851 millions USD), (**s**= [S/S']-1= Sales Growth= 5.00%), and (**F**= Fixed Cost= 29,740 millions USD), then it's (**U**= Utilized or Starting Capital planned), is:

$$U= F+S'i[1+s]+T+Ad+[Q+X]/l-M$$
$$= 29,750+48,851*0.0100[1+0.0500]$$
$$+3,385+7,899*0.3000$$
$$+[35,351+34,196]/0.9900$$
$$-41,548$$
$$= \underline{64,720} \text{ millions USD}$$

Corporate IFRS-GAAP (B/S-I/S),ISBN-13: **978-1721066308**, ISBN-10: **1721066306**

Law-8311:

If both (**d**= D/A= Dividend Portion or Payout= 30.00%), (**X**= Xpress or Current Debt= 34,196 millions USD), (**T**= Tax Paid= 3,385 millions USD), (**A**= After Tax Income= 7,899 millions USD), (**i**= I/S= Interest Portion= 1.00%), (**U**= Utilized or Starting Capital= 64,720 millions USD), (**Q**= Quoted Longterm Liabilities= 35,351 millions USD), (**l**= L/E= Leverage or Gearing Ratio= 99.00%), (**S'**= Sales of Past Year= 48,851 millions USD), (**s**= [S/S']-1= Sales Growth= 5.00%), and (**F**= Fixed Cost= 29,740 millions USD), then it's (**M**= Margin of Contribution planned), is:

$$M= F+S'i[1+s]+T+Ad+[Q+X]/l-U$$

$$= 29,750+48,851*0.0100[1+0.0500]$$
$$+3,385+7,899*0.3000$$
$$+[35,351+34,196]/0.9900$$
$$64,720$$

$$= \underline{41,548} \text{ millions USD}$$

Corporate IFRS-GAAP (B/S-I/S),ISBN-13: **978-1721066308**, ISBN-10: **1721066306**

Law-8312:

If both (**d**= D/A= Dividend Portion or Payout= 30.00%), (**X**= Xpress or Current Debt= 34,196 millions USD), (**T**= Tax Paid= 3,385 millions USD), (**A**= After Tax Income= 7,899 millions USD), (**i**= I/S= Interest Portion= 1.00%), (**U**= Utilized or Starting Capital= 64,720 millions USD), (**Q**= Quoted Longterm Liabilities= 35,351 millions USD), (**I**= L/E= Leverage or Gearing Ratio= 99.00%), (**S'**= Sales of Past Year= 48,851 millions USD), (**s**= [S/S']-1= Sales Growth= 5.00%), and (**M**= Margin of Contribution= 41,548 millions USD), then it's (**F**= Fixed Cost planned), is:

$$F= U+M-S'i[1+s]-T-Ad-[Q+X]/I$$

$$= 64,720+41,548-48,851*0.0100$$
$$[1+0.0500]-3,385-7,899$$
$$*0.3000-[35,351+34,196]$$
$$/0.9900$$

$$= \underline{29,750} \text{ millions USD}$$

Corporate IFRS-GAAP (B/S-I/S),ISBN-13: **978-1721066308**, ISBN-10: **1721066306**

Law-8313:

If both (**d**= D/A=cDividend Portion or Payout= 30.00%), (**X**= Xpress or Current Debt= 34,196 millions USD), (**T**= Tax Paid= 3,385 millions USD), (**A**= After Tax Income= 7,899 millions USD), (**i**= I/S= Interest Portion= 1.00%), (**U**= Utilized or Starting Capital= 64,720 millions USD), (**Q**= Quoted Longterm Liabilities= 35,351 millions USD), (**l**= L/E= Leverage or Gearing Ratio= 99.00%), (**F**= Fixed Cost= 29,740 millions USD), (**s**= [S/S']-1= Sales Growth= 5.00%), and (**M**= Margin of Contribution= 41,548 millions USD), then it's (**S'**= Sales Past), must be:

$$S' = \{U+M-F-T-Ad-[Q+X]/l\}/\{i[1+s]\}$$
$$= \{64,720+41,548-29,750-3,385$$
$$-7,899*0.3000-[35,351$$
$$+34,196]/0.9900\}/\{0.0100$$
$$[1+0.0500]\}$$
$$= \underline{48,851} \text{ millions USD}$$

Corporate IFRS-GAAP (B/S-I/S),ISBN-13: **978-1721066308**, ISBN-10: **1721066306**

Law-8314:

> If both (**d**= D/A= Dividend Portion or Payout= 30.00%), (**X**= Xpress or Current Debt= 34,196 millions USD), (**T**= Tax Paid= 3,385 millions USD), (**A**= After Tax Income= 7,899 millions USD), (**S'**= Sales of Past Year= 48,851 millions USD), (**U**= Utilized or Starting Capital= 64,720 millions USD), (**Q**= Quoted Longterm Liabilities= 35,351 millions USD), (**I**= L/E= Leverage or Gearing Ratio= 99.00%), (**F**= Fixed Cost= 29,740 millions USD), (**s**= [S/S']-1= Sales Growth= 5.00%), and (**M**= Margin of Contribution= 41,548 millions USD), then it's (**i**= Interest Portion planned), is:

$$i= \{U+M-F-T-Ad-[Q+X]/I\}/\{S'[1+s]\}$$

$$= \{64,720+41,548-29,750-3,385$$
$$-7,899*0.3000-[35,351$$
$$+34,196]/0.9900\}/\{48,851$$
$$[1+0.0500]\}$$

$$= 1.00\%$$

Corporate IFRS-GAAP (B/S-I/S),ISBN-13: **978-1721066308**, ISBN-10: **1721066306**

Law-8315:

If both (**d**= D/A= Dividend Portion or Payout= 30.00%), (**X**= Xpress or Current Debt= 34,196 millions USD), (**T**= Tax Paid= 3,385 millions USD), (**A**= After Tax Income= 7,899 millions USD), (**$'**= Sales of Past Year= 48,851 millions USD), (**U**= Utilized or Starting Capital= 64,720 millions USD), (**Q**= Quoted Longterm Liabilities= 35,351 millions USD), (**I**= L/E= Leverage or Gearing Ratio= 99.00%), (**F**= Fixed Cost= 29,740 millions USD), (**i**= I/S= Interest Portion= 1.00%), and (**M**= Margin of Contribution= 41,548 millions USD), then it's (**$**= Sales Growth planned), is:

$$\$= \{U+M-F-T-Ad-[Q+X]/I\}/[\$'i]-1$$

$$= \{64,720+41,548-29,750-3,385$$
$$-7,899*0.3000-[35,351$$
$$+34,196]/0.9900\}/[48,851$$
$$*0.0100]-1$$

$$= \underline{5.00\%}$$

Corporate IFRS-GAAP (B/S-I/S),ISBN-13: **978-1721066308**, ISBN-10: **1721066306**

<u>Law-8316</u>:

If both (**d**= D/A= Dividend Portion or Payout= <u>30.00%</u>), (**X**= Xpress or Current Debt= <u>34,196</u> millions USD), (**s**= [S/S']-1= Sales Growth= <u>5.00%</u>), (**A**= After Tax Income= <u>7,899</u> millions USD), (**S'**= Sales of Past Year= <u>48,851</u> millions USD), (**U**= Utilized or Starting Capital= <u>64,720</u> millions USD), (**Q**= Quoted Longterm Liabilities= <u>35,351</u> millions USD), (**I**= L/E= Leverage or Gearing Ratio= <u>99.00%</u>), (**F**= Fixed Cost= <u>29,740</u> millions USD), (**i**= I/S= Interest Portion= <u>1.00%</u>), and (**M**= Margin of Contribution= <u>41,548</u> millions USD), then it's (**T**= Tax Paid planned), is:

$$T= U+M-F-S'i[1+s]-Ad-[Q+X]/I$$
$$= 64,720+41,548-29,750-48,851$$
$$*0.0100[1+0.0500]-7,899$$
$$*0.3000-[35,351+34,196]$$
$$/0.9900$$
$$= \underline{3,385} \text{ millions USD}$$

Corporate IFRS-GAAP (B/S-I/S),ISBN-13: **978-1721066308**, ISBN-10: **1721066306**

Law-8317:

If both (**d**= D/A= Dividend Portion or Payout=
30.00%), (**X**= Xpress or Current Debt= 34,196
millions USD), (**s**= [S/S']-1= Sales Growth= 5.00%),
(**T**= Tax Paid= 3,385 millions USD), (**S'**= Sales of
Past Year= 48,851 millions USD), (**U**= Utilized or
Starting Capital= 64,720 millions USD), (**Q**= Quoted
Longterm Liabilities= 35,351 millions USD), (**I**=
L/E= Leverage or Gearing Ratio= 99.00%), (**F**=
Fixed Cost= 29,740 millions USD), (**i**= I/S= Interest
Portion= 1.00%), and (**M**= Margin of Contribution=
41,548 millions USD), then it's (**A**= After Tax
Income planned), is:

$$A= \{U+M-F-S'i[1+s]-T-[Q+X]/I\}/d$$

$$= \{64,720+41,548-29,750-48,851$$
$$*0.0100[1+0.0500]-3,385$$
$$-[35,351+34,196]/0.9900\}$$
$$/0.3000$$

$$= \underline{7,899} \text{ millions USD}$$

Corporate IFRS-GAAP (B/S-I/S),ISBN-13: **978-1721066308**, ISBN-10: **1721066306**

Law-8318:

If both (**A**= After Tax Income= 7,899 millions USD),
(**X**= Xpress or Current Debt= 34,196 millions USD),
(**s**= [S/S']-1= Sales Growth= 5.00%), (**T**= Tax Paid=
3,385 millions USD), (**S'**= Sales of Past Year= 48,851
millions USD), (**U**= Utilized or Starting Capital=
64,720 millions USD), (**Q**= Quoted Longterm
Liabilities= 35,351 millions USD), (**l**= L/E=
Leverage or Gearing Ratio= 99.00%), (**F**= Fixed
Cost= 29,740 millions USD), (**i**= I/S= Interest
Portion= 1.00%), and (**M**= Margin of Contribution=
41,548 millions USD), then it's (**d**= Dividend Portion
or Payout planned), is:

$$d= \{U+M-F-S'i[1+s]-T-[Q+X]/l\}/A$$
$$= \{64,720+41,548-29,750-48,851$$
$$*0.0100[1+0.0500]-3,385$$
$$-[35,351+34,196]/0.9900\}$$
$$/7,899$$

$$= \underline{30.00\%}$$

Corporate IFRS-GAAP (B/S-I/S),ISBN-13: **978-1721066308**, ISBN-10: **1721066306**

Law-8319:

If both (**A**= After Tax Income= 7,899 millions USD),
(**d**= D/A= Dividend Portion or Payout= 30.00%), (**s**=
[S/S']-1= Sales Growth= 5.00%), (**T**= Tax Paid=
3,385 millions USD), (**S'**= Sales of Past Year= 48,851
millions USD), (**U**= Utilized or Starting Capital=
64,720 millions USD), (**Q**= Quoted Longterm
Liabilities= 35,351 millions USD), (**I**= L/E=
Leverage or Gearing Ratio= 99.00%), (**F**= Fixed
Cost= 29,740 millions USD), (**i**= I/S= Interest
Portion= 1.00%), and (**M**= Margin of Contribution=
41,548 millions USD), then it's (**X**= Xpress or
Current Debt planned), is:

$$X = I\{U+M-F-S'i[1+s]-T-Ad\}-Q$$
$$= 0.9900\{64,720+41,548-29,750$$
$$-48,851*0.0100[1+0.0500]$$
$$-3,385-7,899*0.3000\}$$
$$-35,351$$
$$= 34,196 \text{ millions USD}$$

Corporate IFRS-GAAP (B/S-I/S),ISBN-13: **978-1721066308**, ISBN-10: **1721066306**

Law-8320:

If both (**A**= After Tax Income= 7,899 millions USD), (**c**= C/X= Current Ratio= 1.20 times), (**q**= [C-P]/X= Quick or Acid Test Ratio= 1.01 times), (**P**= Procured Inventories= 6,497 millions USD), (**d**= D/A= Dividend Portion or Payout= 30.00%), (**s**= [S/S']-1= Sales Growth= 5.00%), (**T**= Tax Paid= 3,385 millions USD), (**S'**= Sales of Past Year= 48,851 millions USD), (**U**= Utilized or Starting Capital= 64,720 millions USD), (**Q**= Quoted Longterm Liabilities= 35,351 millions USD), (**l**= L/E= Leverage or Gearing Ratio= 99.00%), (**F**= Fixed Cost= 29,740 millions USD), (**i**= I/S= Interest Portion= 1.00%), and (**M**= Margin of Contribution= 41,548 millions USD), then it's (**Q**= Quoted Longterm Liabilities planned), is:

$$Q= l\,\{U+M-F-S'i[1+s]-T-Ad\}-P/[c-q]$$
$$= 0.9900\{64,720+41,548-29,750$$
$$-48,851*0.0100[1+0.0500]$$
$$-3,385-7,899*0.3000\}-6,497$$
$$/[1.2000-1.0100]$$
$$= 35,351 \text{ millions USD}$$

Corporate IFRS-GAAP (B/S-I/S),ISBN-13: **978-1721066308**, ISBN-10: **1721066306**

Law-8321:

If both (**A**= After Tax Income= 7,899 millions USD), (**c**= C/X= Current Ratio= 1.20 times), (**q**= [C-P]/X= Quick or Acid Test Ratio= 1.01 times), (**P**= Procured Inventories= 6,497 millions USD), (**d**= D/A= Dividend Portion or Payout= 30.00%), (**s**= [S/S']-1= Sales Growth= 5.00%), (**T**= Tax Paid= 3,385 millions USD), (**S'**= Sales of Past Year= 48,851 millions USD), (**U**= Utilized or Starting Capital= 64,720 millions USD), (**Q**= Quoted Longterm Liabilities= 35,351 millions USD), (**F**= Fixed Cost= 29,740 millions USD), (**i**= I/S= Interest Portion= 1.00%), and (**M**= Margin of Contribution= 41,548 millions USD), then it's (**I** = Leverage or Gearing Ratio planned), is:

$$I = \{Q+P/[c\text{-}q]\}/\{U+M\text{-}F\text{-}S\text{'}i[1+s]\text{-}T\text{-}Ad\}$$
$$= \{35,351+6,497/[1.2000\text{-}1.0100]\}$$
$$/\{64,720+41,548\text{-}29,750$$
$$-48,851*0.0100[1+0.0500]$$
$$-3,385\text{-}7,899*0.3000\}$$
$$= \underline{99.00\%}$$

Corporate IFRS-GAAP (B/S-I/S),ISBN-13: **978-1721066308**, ISBN-10: **1721066306**

Law-8322:

If both (**A**= After Tax Income= 7,899 millions USD),
(**c**= C/X= Current Ratio= 1.20 times), (**q**= [C-P]/X=
Quick or Acid Test Ratio= 1.01 times), (**P**= Procured
Inventories= 6,497 millions USD), (**d**= D/A=
Dividend Portion or Payout= 30.00%), (**s**= [S/S']-1=
Sales Growth= 5.00%), (**T**= Tax Paid= 3,385 millions
USD), (**S'**= Sales of Past Year= 48,851 millions
USD), (**l**= L/E= Leverage or Gearing Ratio=
99.00%), (**Q**= Quoted Longterm Liabilities= 35,351
millions USD), (**F**= Fixed Cost= 29,740 millions
USD), (**i**= I/S= Interest Portion= 1.00%), and (**M**=
Margin of Contribution= 41,548 millions USD), then
it's (**U**= Utilized or Starting Capital planned), is:

$$U = F + S'i[1+s] + T + Ad + \{Q + P/[c-q]\}/l - M$$
$$= 29,750 + 48,851 * 0.0100[1+0.0500]$$
$$+ 3,385 + 7,899 * 0.3000$$
$$+ \{35,351 + 6,497/[1.2000$$
$$- 1.0100]\}/0.9900 - 41,548$$
$$= \underline{64,720} \text{ millions USD}$$

Corporate IFRS-GAAP (B/S-I/S),ISBN-13: **978-1721066308**, ISBN-10: **1721066306**

Law-8323:

If both (**A**= After Tax Income= 7,899 millions USD),
(**c**= C/X= Current Ratio= 1.20 times), (**q**= [C-P]/X=
Quick or Acid Test Ratio= 1.01 times), (**P**= Procured
Inventories= 6,497 millions USD), (**d**= D/A=
Dividend Portion or Payout= 30.00%), (**s**= [S/S']-1=
Sales Growth= 5.00%), (**T**= Tax Paid= 3,385 millions
USD), (**S'**= Sales of Past Year= 48,851 millions
USD), (**l**= L/E= Leverage or Gearing Ratio=
99.00%), (**Q**= Quoted Longterm Liabilities= 35,351
millions USD), (**F**= Fixed Cost= 29,740 millions
USD), (**i**= I/S= Interest Portion= 1.00%), and (**U**=
Utilized or Starting Capital= 64,720 millions USD),
then it's (**M**= Margin of Contribution planned), is:

$$M= F+S'i[1+s]+T+Ad+\{Q+P/[c-q]\}/l - U$$
$$= 29,750+48,851*0.0100[1+0.0500]$$
$$+3,385+7,899*0.3000$$
$$|\{35,351+6,497/[1.2000$$
$$-1.0100]\}/0.9900-64,720$$
$$= 41,548 \text{ millions USD}$$

Corporate IFRS-GAAP (B/S-I/S),ISBN-13: **978-1721066308**, ISBN-10: **1721066306**

<u>Law-8324</u>:

If both (**A**= After Tax Income= <u>7,899</u> millions USD), (**c**= C/X= Current Ratio= <u>1.20</u> times), (**q**= [C-P]/X= Quick or Acid Test Ratio= <u>1.01</u> times), (**P**= Procured Inventories= <u>6,497</u> millions USD), (**d**= D/A= Dividend Portion or Payout= <u>30.00%</u>), (**s**= [S/S']-1= Sales Growth= <u>5.00%</u>), (**T**= Tax Paid= <u>3,385</u> millions USD), (**S'**= Sales of Past Year= <u>48,851</u> millions USD), (**I**= L/E= Leverage or Gearing Ratio= <u>99.00%</u>), (**Q**= Quoted Longterm Liabilities= <u>35,351</u> millions USD), (**M**= Margin of Contribution= <u>41,548</u> millions USD), (**i**= I/S= Interest Portion= <u>1.00%</u>), and (**U**= Utilized or Starting Capital= <u>64,720</u> millions USD), then it's (**F**= Fixed Cost planned), is:

$$F= U+M-S'i[1+s]-T-Ad-\{Q+P/[c-q]\}/I$$
$$= 64,720+41,548-48,851*0.0100$$
$$[1+0.0500]-3,385-7,899$$
$$*0.3000-\{35,351+6,497$$
$$/[1.2000-1.0100]\}/0.9900$$
$$= \underline{29,750} \text{ millions USD}$$

Corporate IFRS-GAAP (B/S-I/S),ISBN-13: **978-1721066308**, ISBN-10: **1721066306**

Law-8325:

If both (**A**= After Tax Income= 7,899 millions USD),
(**c**= C/X= Current Ratio= 1.20 times), (**q**= [C-P]/X=
Quick or Acid Test Ratio= 1.01 times), (**P**= Procured
Inventories= 6,497 millions USD), (**d**= D/A=
Dividend Portion or Payout= 30.00%), (**s**= [S/S']-1=
Sales Growth= 5.00%), (**T**= Tax Paid= 3,385 millions
USD), (**F**= Fixed Cost= 29,740 millions USD), (**l**=
L/E= Leverage or Gearing Ratio= 99.00%), (**Q**=
Quoted Longterm Liabilities= 35,351 millions USD),
(**M**= Margin of Contribution= 41,548 millions USD),
(**i**= I/S= Interest Portion= 1.00%), and (**U**= Utilized
or Starting Capital= 64,720 millions USD), then it's
(**S'**= Sales past), must be:

$$S'= (U+M-F-T-Ad-\{Q+P/[c-q]\}/l\,)/\{i[1+s]\}$$

$$= (64,720+41,548-29,750-3,385$$
$$-7,899*0.3000-\{35,351$$
$$+6,497/[1.2000-1.0100]\}$$
$$/0.9900)/\{0.0100$$
$$[1+0.0500]\}$$

$$= 48,851 \text{ millions USD}$$

Corporate IFRS-GAAP (B/S-I/S),ISBN-13: **978-1721066308**, ISBN-10: **1721066306**

Law-8326:

If both (**A**= After Tax Income= 7,899 millions USD), (**c**= C/X= Current Ratio= 1.20 times), (**q**= [C-P]/X= Quick or Acid Test Ratio= 1.01 times), (**P**= Procured Inventories= 6,497 millions USD), (**d**= D/A= Dividend Portion or Payout= 30.00%), (**s**= [S/S']-1= Sales Growth= 5.00%), (**T**= Tax Paid= 3,385 millions USD), (**F**= Fixed Cost= 29,740 millions USD), (**l**= L/E= Leverage or Gearing Ratio= 99.00%), (**Q**= Quoted Longterm Liabilities= 35,351 millions USD), (**M**= Margin of Contribution= 41,548 millions USD), (**S'**= Sales of Past Year= 48,851 millions USD), and (**U**= Utilized or Starting Capital= 64,720 millions USD), then it's (**i**= I/S= Interest Portion planned), must be:

$$i= (U+M-F-T-Ad-\{Q+P/[c-q]\}/l)/\{S'[1+s]\}$$
$$= (64,720+41,548-29,750-3,385$$
$$-7,899*0.3000-\{35,351$$
$$+6,497/[1.2000-1.0100]\}$$
$$/0.9900)/\{48,851$$
$$[1+0.0500]\}$$
$$= 1.00\%$$

Law-8327:

If both (**A**= After Tax Income= 7,899 millions USD), (**c**= C/X= Current Ratio= 1.20 times), (**q**= [C-P]/X= Quick or Acid Test Ratio= 1.01 times), (**P**= Procured Inventories= 6,497 millions USD), (**d**= D/A= Dividend Portion or Payout= 30.00%), (**i**= I/S= Interest Portion= 1.00%), (**T**= Tax Paid= 3,385 millions USD), (**F**= Fixed Cost= 29,740 millions USD), (**l**= L/E= Leverage or Gearing Ratio= 99.00%), (**Q**= Quoted Longterm Liabilities= 35,351 millions USD), (**M**= Margin of Contribution= 41,548 millions USD), (**S'**= Sales of Past Year= 48,851 millions USD), and (**U**= Utilized or Starting Capital= 64,720 millions USD), then it's (**s**= Sales growth planned), is:

$$s= (U+M-F-T-Ad-\{Q+P/[c-q]\}/l)/[S'i]-1$$
$$= (64,720+41,548-29,750-3,385$$
$$-7,899*0.3000-\{35,351$$
$$+6,497/[1.2000-1.0100]\}$$
$$/0.9900)/[48,851*0.0100]-1$$
$$= 5.00\%$$

<u>Law-8328</u>:

If both (**A**= After Tax Income= <u>7,899</u> millions USD), (**c**= C/X= Current Ratio= <u>1.20</u> times), (**q**= [C-P]/X= Quick or Acid Test Ratio= <u>1.01</u> times), (**P**= Procured Inventories= <u>6,497</u> millions USD), (**d**= D/A= Dividend Portion or Payout= <u>30.00%</u>), (**i**= I/S= Interest Portion= <u>1.00%</u>), (**s**= [S/S']-1= Sales Growth= <u>5.00%</u>), (**F**= Fixed Cost= <u>29,740</u> millions USD), (**/**= L/E= Leverage or Gearing Ratio= <u>99.00%</u>), (**Q**= Quoted Longterm Liabilities= <u>35,351</u> millions USD), (**M**= Margin of Contribution= <u>41,548</u> millions USD), (**S'**= Sales of Past Year= <u>48,851</u> millions USD), and (**U**= Utilized or Starting Capital= <u>64,720</u> millions USD), then it's (**T**= Tax paid planned), is:

$$T= U+M-F-S'i[1+s]-Ad-\{Q+P/[c-q]\}/\textit{l}$$
$$= (64,720+41,548-29,750-48,851$$
$$*0.0100[1+0.0500]-7,899$$
$$*0.3000-\{35,351+6,497$$
$$/[1.2000-1.0100]\}/0.9900$$
$$= \underline{3,385} \text{ millions USD}$$

Corporate IFRS-GAAP (B/S-I/S),ISBN-13: **978-1721066308**, ISBN-10: **1721066306**

<u>Law-8329</u>:

If both (**T**= Tax Paid= <u>3,385</u> millions USD), (**c**= C/X= Current Ratio= <u>1.20</u> times), (**q**= [C-P]/X= Quick or Acid Test Ratio= <u>1.01</u> times), (**P**= Procured Inventories= <u>6,497</u> millions USD), (**d**= D/A= Dividend Portion or Payout= <u>30.00%</u>), (**i**= I/S= Interest Portion= <u>1.00%</u>), (**s**= [S/S']-1= Sales Growth= <u>5.00%</u>), (**F**= Fixed Cost= <u>29,740</u> millions USD), (**l**= L/E= Leverage or Gearing Ratio= <u>99.00%</u>), (**Q**= Quoted Longterm Liabilities= <u>35,351</u> millions USD), (**M**= Margin of Contribution= <u>41,548</u> millions USD), (**S'**= Sales of Past Year= <u>48,851</u> millions USD), and (**U**= Utilized or Starting Capital= <u>64,720</u> millions USD), then it's (**A**= After Tax Income planned), is:

$$A= (U+M-F-T-S'i[1+s]-\{Q+P/[c-q]\}/l)/d$$
$$= (64,720+41,548-29,750-3,385$$
$$48,851*0.0100[1+0.0500]$$
$$-\{35,351+6,497/[1.2000$$
$$-1.0100]\}/0.9900)/0.3000$$
$$= \underline{7,899} \text{ millions USD}$$

Corporate IFRS-GAAP (B/S-I/S),ISBN-13: **978-1721066308**, ISBN-10: **1721066306**

Law-8330:

If both (**T**= Tax Paid= 3,385 millions USD), (**c**= C/X= Current Ratio= 1.20 times), (**q**= [C-P]/X= Quick or Acid Test Ratio= 1.01 times), (**P**= Procured Inventories= 6,497 millions USD), (**A**= After Tax Income= 7,899 millions USD), (**i**= I/S= Interest Portion= 1.00%), (**s**= [S/S']-1= Sales Growth= 5.00%), (**F**= Fixed Cost= 29,740 millions USD), (**l**= L/E= Leverage or Gearing Ratio= 99.00%), (**Q**= Quoted Longterm Liabilities= 35,351 millions USD), (**M**= Margin of Contribution= 41,548 millions USD), (**S'**= Sales of Past Year= 48,851 millions USD), and (**U**= Utilized or Starting Capital= 64,720 millions USD), then it's (**d**= Dividend Portion or Payout planned), is:

$$\mathbf{d}= (\mathbf{U}+\mathbf{M}-\mathbf{F}-\mathbf{T}-\mathbf{S'i}[1+\mathbf{s}]-\{\mathbf{Q}+\mathbf{P}/[\mathbf{c}-\mathbf{q}]\}/\mathbf{l})/\mathbf{A}$$
$$= (64,720+41,548-29,750-3,385$$
$$-48,851*0.0100[1+0.0500]$$
$$-\{35,351+6,497/[1.2000$$
$$-1.0100]\}/0.9900)/7,899$$
$$= 30.00\%$$

Law-8331:

If both (**T**= Tax Paid= 3,385 millions USD), (**c**= C/X= Current Ratio= 1.20 times), (**q**= [C-P]/X= Quick or Acid Test Ratio= 1.01 times), (**d**= D/A= Dividend Portion or Payout= 30.00%), (**A**= After Tax Income= 7,899 millions USD), (**i**= I/S= Interest Portion= 1.00%), (**s**= [S/S']-1= Sales Growth= 5.00%), (**F**= Fixed Cost= 29,740 millions USD), (**l**= L/E= Leverage or Gearing Ratio= 99.00%), (**Q**= Quoted Longterm Liabilities= 35,351 millions USD), (**M**= Margin of Contribution= 41,548 millions USD), (**S'**= Sales of Past Year= 48,851 millions USD), and (**U**= Utilized or Starting Capital= 64,720 millions USD), then it's (**P**= Procured Inventories planned), is:

$$P= [c-q](l\{U+M-F-T-S'i[1+s]-Ad\}-Q)$$
$$= [1.2000-1.0100](0.9900\{64,720$$
$$+41,548-29,750-3,385$$
$$-48,851*0.0100[1+0.0500]$$
$$- 7,899*0.3000\}-35,351)$$
$$= \underline{6,497} \text{ millions USD}$$

Corporate IFRS-GAAP (B/S-I/S),ISBN-13: **978-1721066308**, ISBN-10: **1721066306**

Law-8332:

If both (**T**= Tax Paid= 3,385 millions USD), (**P**= Procured Inventories= 6,497 millions USD), (**q**= Quick or Acid Test Ratio= [C-P]/X= 1.01 times), (**d**= D/A= Dividend Portion or Payout= 30.00%), (**A**= After Tax Income= 7,899 millions USD), (**i**= I/S= Interest Portion= 1.00%), (**s**= [S/S']-1= Sales Growth= 5.00%), (**F**= Fixed Cost= 29,740 millions USD), (**I**= L/E= Leverage or Gearing Ratio= 99.00%), (**Q**= Quoted Longterm Liabilities= 35,351 millions USD), (**M**= Margin of Contribution= 41,548 millions USD), (**S'**= Sales of Past Year= 48,851 millions USD), and (**U**= Utilized or Starting Capital= 64,720 millions USD), then it's (**c**= Current Ratio planned), is:

$$c= q+P/(I\{U+M-F-T-S'i[1+s]-Ad\}-Q)$$
$$= 1.0100+6,497/(0.9900\{64,720$$
$$+41,548-29,750-3,385$$
$$-48,851*0.0100[1+0.0500]$$
$$-7,899*0.3000\}-35,351)$$
$$= \underline{1.20}\ times$$

Corporate IFRS-GAAP (B/S-I/S),ISBN-13: **978-1721066308**, ISBN-10: **1721066306**

<u>Law-8333</u>:

If both (**T**= Tax Paid= <u>3,385</u> millions USD), (**P**= Procured Inventories= <u>6,497</u> millions USD), (**c**= C/X= Current Ratio= <u>1.20</u> times), (**d**= D/A= Dividend Portion or Payout= <u>30.00%</u>), (**A**= After Tax Income= <u>7,899</u> millions USD), (**i**= I/S= Interest Portion= <u>1.00%</u>), (**s**= [S/S']-1= Sales Growth= <u>5.00%</u>), (**F**= Fixed Cost= <u>29,740</u> millions USD), (**l**= L/E= Leverage or Gearing Ratio= <u>99.00%</u>), (**Q**= Quoted Longterm Liabilities= <u>35,351</u> millions USD), (**M**= Margin of Contribution= <u>41,548</u> millions USD), (**S'**= Sales of Past Year= <u>48,851</u> millions USD), and (**U**= Utilized or Starting Capital= <u>64,720</u> millions USD), then it's (**q**= Quick or Acid Test Ratio planned), is:

$$q= c\text{-}P/(l\ \{U+M\text{-}F\text{-}T\text{-}S'i[1+s]\text{-}Ad\}\text{-}Q)$$

$$= 1.2000\text{-}6,497/(0.9900\{64,720$$
$$+41,548\text{-}29,750\text{-}3,385$$
$$-48,851*0.0100[1+0.0500]$$
$$-7,899*0.3000\}\text{-}35,351)$$

$$= \underline{1.01}\ \text{times}$$

Corporate IFRS-GAAP (B/S-I/S),ISBN-13: **978-1721066308**, ISBN-10: **1721066306**

Law-8334:

If both (**T**= Tax Paid= 3,385 millions USD), (**p**= 360
P/V= Procured Inventory Days= 240 Days), (**V**=
Variable Cost= 9,746 millions USD), (**c**= C/X=
Current Ratio= 1.20 times), (**d**= D/A= Dividend
Portion or Payout= 30.00%), (**A**= After Tax Income=
7,899 millions USD), (**i**= I/S= Interest Portion=
1.00%), (**s**= [S/S']-1= Sales Growth= 5.00%), (**F**=
Fixed Cost= 29,740 millions USD), (**Q**= Quoted
Longterm Liabilities= 35,351 millions USD), (**q**=
Quick or Acid Test Ratio= [C-P]/X= 1.01 times),
(**M**= Margin of Contribution= 41,548 millions USD),
(**S'**= Sales of Past Year= 48,851 millions USD), and
(**U**= Utilized or Starting Capital= 64,720 millions
USD), then it's (**I** = Leverage or Gearing Ratio
planned), is:

$$I = (\mathbf{Q}+\mathbf{Vp}/\{360[\mathbf{c}\text{-}\mathbf{q}]\})$$
$$/\{\mathbf{U}+\mathbf{M}\text{-}\mathbf{F}\text{-}\mathbf{T}\text{-}\mathbf{S'i}[1+\mathbf{s}]\text{-}\mathbf{Ad}\}$$
$$= (35,351+9,746*240/\{360[1.2000$$
$$-1.0100]\})/\{64,720+41,548$$
$$-29,750-3,385-48,851$$
$$*0.0100[1+0.0500]-7,899$$
$$*0.3000\}$$
$$= \underline{99.00\%}$$

Corporate IFRS-GAAP (B/S-I/S),ISBN-13: **978-1721066308**, ISBN-10: **1721066306**

Law-8335:

If both (**T**= Tax Paid= 3,385 millions USD), (**p**= 360 P/V= Procured Inventory Days= 240 Days), (**V**= Variable Cost= 9,746 millions USD), (**c**= C/X= Current Ratio= 1.20 times), (**d**= D/A= Dividend Portion or Payout= 30.00%), (**A**= After Tax Income= 7,899 millions USD), (**i**= I/S= Interest Portion= 1.00%), (**s**= [S/S']-1= Sales Growth= 5.00%), (**F**= Fixed Cost= 29,740 millions USD), (**Q**= Quoted Longterm Liabilities= 35,351 millions USD), (**q**= Quick or Acid Test Ratio= [C-P]/X= 1.01 times), (**U**= Utilized or Starting Capital= 64,720 millions USD), (**S'**= Sales of Past Year= 48,851 millions USD), and (**l**= L/E= Leverage or Gearing Ratio= 99.00%), then it's (**M**= Margin of Contribution planned), is:

$$M= F+T+S'i[1+s]+Ad$$
$$+(Q+Vp/\{360[c-q]\})/l-U$$

$$= 29,750+3,385+48,851*0.0100$$
$$[1+0.0500]+7,899*0.3000$$
$$+(35,351+9,746*240/\{360$$
$$[1.2000-1.0100]\})/0.9900$$
$$-64,720$$

$$= 41,548 \text{ millions USD}$$

Corporate IFRS-GAAP (B/S-I/S),ISBN-13: **978-1721066308**, ISBN-10: **1721066306**

Law-8336:

If both (**T**= Tax Paid= 3,385 millions USD), (**p**= 360 P/V= Procured Inventory Days= 240 Days), (**V**= Variable Cost= 9,746 millions USD), (**c**= C/X= Current Ratio= 1.20 times), (**d**= D/A= Dividend Portion or Payout= 30.00%), (**A**= After Tax Income= 7,899 millions USD), (**i**= I/S= Interest Portion= 1.00%), (**s**= [S/S']-1= Sales Growth= 5.00%), (**M**= Margin of Contribution= 41,548 millions USD), (**Q**= Quoted Longterm Liabilities= 35,351 millions USD), (**q**= [C-P]/X= Quick or Acid Test Ratio= 1.01 times), (**U**= Utilized or Starting Capital= 64,720 millions USD), (**S'**= Sales of Past Year= 48,851 millions USD), and (**F**= L/E= Leverage or Gearing Ratio= 99.00%), then it's (**F**= Fixed Cost planned), is:

$$F= U+M-T-S'i[1+s]-Ad$$
$$-(Q+Vp/\{360[c-q]\})/I$$
$$= 64,720+41,548-3,385-48,851$$
$$*0.0100[1+0.0500]-7,899$$
$$*0.3000-(35,351+9,746$$
$$*240/\{360[1.2000$$
$$-1.0100]\})/0.9900$$
$$= 29,750 \text{ millions USD}$$

Corporate IFRS-GAAP (B/S-I/S),ISBN-13: **978-1721066308**, ISBN-10: **1721066306**

Law-8337:

If both (**T**= Tax Paid= 3,385 millions USD), (**p**= 360 P/V= Procured Inventory Days= 240 Days), (**V**= Variable Cost= 9,746 millions USD), (**c**= C/X= Current Ratio= 1.20 times), (**d**= D/A= Dividend Portion or Payout= 30.00%), (**A**= After Tax Income= 7,899 millions USD), (**i**= I/S= Interest Portion= 1.00%), (**s**= [S/S']-1= Sales Growth= 5.00%), (**M**= Margin of Contribution= 41,548 millions USD), (**Q**= Quoted Longterm Liabilities= 35,351 millions USD), (**q**= [C-P]/X= Quick or Acid Test Ratio= 1.01 times), (**U**= Utilized or Starting Capital= 64,720 millions USD), (**F**= Fixed Cost= 29,740 millions USD), and (**I**= L/E= Leverage or Gearing Ratio= 99.00%), then it's (**S'**= Sales of Past), must be:

$$S' = [U+M-T-F-Ad-(Q+Vp/\{360[c-q]\})/I]$$
$$/\{i[1+s]\}$$
$$= [64,720+41,548-3,385-29,750$$
$$-7,899*0.3000-(35,351$$
$$+9,746*240/\{360[1.2000$$
$$-1.0100]\})/0.9900]$$
$$/\{0.0100[1+0.0500]\}$$
$$= 48,851 \text{ millions USD}$$

Corporate IFRS-GAAP (B/S-I/S),ISBN-13: **978-1721066308**, ISBN-10: **1721066306**

Law-8338:

If both (**T**= Tax Paid= 3,385 millions USD), (**p**= 360 P/V= Procured Inventory Days= 240 Days), (**V**= Variable Cost= 9,746 millions USD), (**c**= C/X= Current Ratio= 1.20 times), (**d**= D/A= Dividend Portion or Payout= 30.00%), (**A**= After Tax Income= 7,899 millions USD), (**$'**= Sales of Past Year= 48,851 millions USD), (**s**= [S/S']-1= Sales Growth= 5.00%), (**M**= Margin of Contribution= 41,548 millions USD), (**Q**= Quoted Longterm Liabilities= 35,351 millions USD), (**q**= [C-P]/X= Quick or Acid Test Ratio= 1.01 times), (**U**= Utilized or Starting Capital= 64,720 millions USD), (**F**= Fixed Cost= 29,740 millions USD), and (**I**= L/E= Leverage or Gearing Ratio= 99.00%), then it's (**i**= Interest Portion planned), is:

$$i= [U+M-T-F-Ad-(Q+Vp/\{360[c-q]\})/I]$$
$$/\{\$'[1+s]\}$$
$$= [64,720+41,548-3,385-29,750$$
$$-7,899*0.3000-(35,351$$
$$+9,746*240/\{360[1.2000$$
$$-1.0100]\})/0.9900]$$
$$/\{48,851[1+0.0500]\}$$
$$= 1.00\%$$

Corporate IFRS-GAAP (B/S-I/S),ISBN-13: **978-1721066308**, ISBN-10: **1721066306**

Law-8339:

If both (**T**= Tax Paid= 3,385 millions USD), (**p**= 360 P/V= Procured Inventory Days= 240 Days), (**V**= Variable Cost= 9,746 millions USD), (**c**= C/X= Current Ratio= 1.20 times), (**d**= D/A= Dividend Portion or Payout= 30.00%), (**A**= After Tax Income= 7,899 millions USD), (**S'**= Sales of Past Year= 48,851 millions USD), (**i**= I/S= Interest Portion= 1.00%), (**M**= Margin of Contribution= 41,548 millions USD), (**Q**= Quoted Longterm Liabilities= 35,351 millions USD), (**q**= [C-P]/X= Quick or Acid Test Ratio= 1.01 times), (**U**= Utilized or Starting Capital= 64,720 millions USD), (**F**= Fixed Cost= 29,740 millions USD), and (**l**= L/E= Leverage or Gearing Ratio= 99.00%), then it's (**s**= Sales Growth planned), is:

$$s= [U+M-T-F-Ad-(Q+Vp/\{360[c-q]\})/l]$$
$$/[S'i]-1$$
$$= [64,720+41,548-3,385-29,750$$
$$-7,899*0.3000-(35,351$$
$$+9,746*240/\{360[1.2000$$
$$-1.0100]\})/0.9900]$$
$$/[48,851*0.0100]-1$$
$$= 5.00\%$$

Corporate IFRS-GAAP (B/S-I/S),ISBN-13: **978-1721066308**, ISBN-10: **1721066306**

Law-8340:

If both (**s**= [S/S']-1= Sales Growth= 5.00%), (**p**= 360
P/V= Procured Inventory Days= 240 Days), (**V**=
Variable Cost= 9,746 millions USD), (**c**= C/X=
Current Ratio= 1.20 times), (**d**= D/A= Dividend
Portion or Payout= 30.00%), (**A**= After Tax Income=
7,899 millions USD), (**S'**= Sales of Past Year= 48,851
millions USD), (**i**= I/S= Interest Portion= 1.00%),
(**M**= Margin of Contribution= 41,548 millions USD),
(**Q**= Quoted Longterm Liabilities= 35,351 millions
USD), (**q**= [C-P]/X= Quick or Acid Test Ratio= 1.01
times), (**U**= Utilized or Starting Capital= 64,720
millions USD), (**F**= Fixed Cost= 29,740 millions
USD), and (**l**= L/E= Leverage or Gearing Ratio=
99.00%), then it's (**T**= Tax Paid planned), is:

$$T= U+M-S'i[1+s]-F-Ad$$
$$-(Q+Vp/\{360[c-q]\})/l$$
$$= 64,720+41,548-48,851*0.0100$$
$$[1+0.0500]-29,750-7,899$$
$$*0.3000-(35,351+9,746$$
$$*240/\{360[1.2000$$
$$-1.0100]\})/0.9900$$
$$= 3,385 \text{ millions USD}$$

Corporate IFRS-GAAP (B/S-I/S),ISBN-13: **978-1721066308**, ISBN-10: **1721066306**

Law-8341:

If both (**s**= [S/S']-1= Sales Growth= 5.00%), (**p**= 360 P/V= Procured Inventory Days= 240 Days), (**V**= Variable Cost= 9,746 millions USD), (**c**= C/X= Current Ratio= 1.20 times), (**d**= D/A= Dividend Portion or Payout= 30.00%), (**T**= Tax Paid= 3,385 millions USD), (**$'**= Sales of Past Year= 48,851 millions USD), (**i**= I/S= Interest Portion= 1.00%), (**M**= Margin of Contribution= 41,548 millions USD), (**Q**= Quoted Longterm Liabilities= 35,351 millions USD), (**q**= [C-P]/X= Quick or Acid Test Ratio= 1.01 times), (**U**= Utilized or Starting Capital= 64,720 millions USD), (**F**= Fixed Cost= 29,740 millions USD), and (**l**= L/E= Leverage or Gearing Ratio= 99.00%), then it's (**A**= After Tax Incomet planned), is:

$$A= [U+M-S'i[1+s]-F-T$$
$$-(Q+Vp/\{360[c-q]\})/l]/d$$
$$= [64,720+41,548-48,851*0.0100$$
$$[1+0.0500]-29,750-3,385$$
$$-(35,351+9,746*240/\{360$$
$$[1.2000-1.0100]\})$$
$$/0.9900]/0.3000$$
$$= 7,899 \text{ millions USD}$$

Corporate IFRS-GAAP (B/S-I/S),ISBN-13: **978-1721066308**, ISBN-10: **1721066306**

Law-8342:

If both (**s**= [S/S']-1= Sales Growth= <u>5.00%</u>), (**p**= 360
P/V= Procured Inventory Days= <u>240</u> Days), (**V**=
Variable Cost= <u>9,746</u> millions USD), (**c**= C/X=
Current Ratio= <u>1.20</u> times), (**A**= After Tax Income=
<u>7,899</u> millions USD), (**T**= Tax Paid= <u>3,385</u> millions
USD), (**S'**= Sales of Past Year= <u>48,851</u> millions
USD), (**i**= I/S= Interest Portion= <u>1.00%</u>), (**M**= Margin
of Contribution= <u>41,548</u> millions USD), (**Q**= Quoted
Longterm Liabilities= <u>35,351</u> millions USD), (**q**=
Quick or Acid Test Ratio= [C-P]/X= <u>1.01</u> times), (**U**=
Utilized or Starting Capital= <u>64,720</u> millions USD),
(**F**= Fixed Cost= <u>29,740</u> millions USD), and (**I**= L/E=
Leverage or Gearing Ratio= <u>99.00%</u>), then it's (**d**=
D/A= Dividend Portion or Payout planned), is:

$$d = [U+M-S'i[1+s]-F-T$$
$$-(Q+Vp/\{360[c-q]\})/I]/A$$
$$= [64,720+41,548-48,851*0.0100$$
$$[1+0.0500]-29,750-3,385$$
$$-(35,351+9,746*240/\{360$$
$$[1.2000-1.0100]\})$$
$$/0.9900]/7,899$$

$$= \underline{30.00\%}$$

Corporate IFRS-GAAP (B/S-I/S),ISBN-13: **978-1721066308**, ISBN-10: **1721066306**

Law-8343:

If both (s= [S/S']-1= Sales Growth= 5.00%), (p= 360 P/V= Procured Inventory Days= 240 Days), (d= D/A= Dividend Portion or Payout= 30.00%), (c= C/X= Current Ratio= 1.20 times), (A= After Tax Income= 7,899 millions USD), (T= Tax Paid= 3,385 millions USD), (S'= Sales of Past Year= 48,851 millions USD), (i= I/S= Interest Portion= 1.00%), (M= Margin of Contribution= 41,548 millions USD), (Q= Quoted Longterm Liabilities= 35,351 millions USD), (q= [C-P]/X= Quick or Acid Test Ratio= 1.01 times), (U= Utilized or Starting Capital= 64,720 millions USD), (F= Fixed Cost= 29,740 millions USD), and (I= L/E= Leverage or Gearing Ratio= 99.00%), then it's (V= Variable Cost planned), is:

$$V= 360[c\text{-}q](I\{U+M\text{-}S'i[1+s]\text{-}F\text{-}T\text{-}Ad\}\text{-}Q)/p$$
$$= 360[1.2000\text{-}1.0100](0.9900$$
$$\{64,720+41,548\text{-}48,851$$
$$*0.0100[1+0.0500]\text{-}29,750$$
$$\text{-}3,385\text{-}7,899*0.3000\}$$
$$\text{-}35,351)/240$$
$$= \underline{9,746} \text{ millions USD}$$

Corporate IFRS-GAAP (B/S-I/S),ISBN-13: **978-1721066308**, ISBN-10: **1721066306**

Law-8344:

If both (**s**= [S/S']-1= Sales Growth= 5.00%), (**V**= Variable Cost= 9,746 millions USD), (**d**= D/A= Dividend Portion or Payout= 30.00%), (**c**= C/X= Current Ratio= 1.20 times), (**A**= After Tax Income= 7,899 millions USD), (**T**= Tax Paid= 3,385 millions USD), (**S'**= Sales of Past Year= 48,851 millions USD), (**i**= I/S= Interest Portion= 1.00%), (**M**= Margin of Contribution= 41,548 millions USD), (**Q**= Quoted Longterm Liabilities= 35,351 millions USD), (**q**= [C-P]/X= Quick or Acid Test Ratio= 1.01 times), (**U**= Utilized or Starting Capital= 64,720 millions USD), (**F**= Fixed Cost= 29,740 millions USD), and (**I**= L/E= Leverage or Gearing Ratio= 99.00%), then it's (**p**= Procured Inventory Days planned), is:

$$\mathbf{p} = 360[\mathbf{c\text{-}q}](\mathbf{I}\ \{\mathbf{U\text{+}M\text{-}S'i}[1\text{+}\mathbf{s}]\text{-}\mathbf{F\text{-}T\text{-}Ad}\}\text{-}\mathbf{Q})/\mathbf{V}$$

$$= 360[1.2000\text{-}1.0100](0.9900$$
$$\{64,720\text{+}41,548\text{-}48,851$$
$$*0.0100[1\text{+}0.0500]\text{-}29,750$$
$$\text{-}3,385\text{-}7,899*0.3000\}$$
$$\text{-}35,351)/9,746$$

$$= \underline{240}\ \text{days}$$

Corporate IFRS-GAAP (B/S-I/S),ISBN-13: **978-1721066308**, ISBN-10: **1721066306**

<u>Law-8345</u>:

If both (s= [S/S']-1= Sales Growth= <u>5.00%</u>), (V= Variable Cost= <u>9,746</u> millions USD), (d= D/A= Dividend Portion or Payout= <u>30.00%</u>), (p= 360 P/V= Procured Inventory Days= <u>240</u> Days), (A= After Tax Income= <u>7,899</u> millions USD), (T= Tax Paid= <u>3,385</u> millions USD), (S'= Sales of Past Year= <u>48,851</u> millions USD), (i= I/S= Interest Portion= <u>1.00%</u>), (M= Margin of Contribution= <u>41,548</u> millions USD), (Q= Quoted Longterm Liabilities= <u>35,351</u> millions USD), (q= [C-P]/X= Quick or Acid Test Ratio= <u>1.01</u> times), (U= Utilized or Starting Capital= <u>64,720</u> millions USD), (F= Fixed Cost= <u>29,740</u> millions USD), and (l= L/E= Leverage or Gearing Ratio= <u>99.00%</u>), then it's (c= Curent Ratio planned), is:

$$c = q+Vp/[360(l\{U+M-S'i[1+s]-F-T-Ad\}-Q)]$$
$$= 1.0100+9,746*240/[360(0.9900$$
$$\{64,720+41,548-48,851$$
$$*0.0100[1+0.0500]-29,750$$
$$-3,385-7,899*0.3000\}$$
$$-35,351)]$$
$$= \underline{1.20} \text{ times}$$

Corporate IFRS-GAAP (B/S-I/S),ISBN-13: **978-1721066308**, ISBN-10: **1721066306**

Law-8346:

If both ($s=$ [S/S']-1= Sales Growth= 5.00%), ($V=$
Variable Cost= 9,746 millions USD), ($d=$ D/A=
Dividend Portion or Payout= 30.00%), ($p=$ 360 P/V=
Procured Inventory Days= 240 Days), ($A=$ After Tax
Income= 7,899 millions USD), ($T=$ Tax Paid= 3,385
millions USD), ($S'=$ Sales of Past Year= 48,851
millions USD), ($i=$ I/S= Interest Portion= 1.00%),
($M=$ Margin of Contribution= 41,548 millions USD),
($Q=$ Quoted Longterm Liabilities= 35,351 millions
USD), ($c=$ C/X= Current Ratio= 1.20 times), ($U=$
Utilized or Starting Capital= 64,720 millions USD),
($F=$ Fixed Cost= 29,740 millions USD), and ($I=$ L/E=
Leverage or Gearing Ratio= 99.00%), then it's ($q=$
Quick or Acid Test Ratio planned), is:

$$q= c\text{-}Vp/[360(I\{U+M\text{-}S'i[1+s]\text{-}F\text{-}T\text{-}Ad\}\text{-}Q)]$$
$$= 1.2000\text{-}9{,}746*240/[360(0.9900$$
$$\{64{,}720+41{,}548\text{-}48{,}851$$
$$*0.0100[1+0.0500]\text{-}29{,}750$$
$$\text{-}3{,}385\text{-}7{,}899*0.3000\}$$
$$\text{-}35{,}351)]$$
$$= 1.01 \text{ times}$$

Law-8347:

If both (**s**= [S/S']-1= Sales Growth= 5.00%), (**v**= V/S= Variable Portion= 19.00%), (**$**= Sales or Revenues= 51,294 millions USD), (**d**= D/A= Dividend Portion or Payout= 30.00%), (**p**= 360 P/V= Procured Inventory Days= 240 Days), (**A**= After Tax Income= 7,899 millions USD), (**T**= Tax Paid= 3,385 millions USD), (**$'**= Sales of Past Year= 48,851 millions USD), (**i**= I/S= Interest Portion= 1.00%), (**M**= Margin of Contribution= 41,548 millions USD), (**q**= [C-P]/X= Quick or Acid Test Ratio= 1.01 times), (**c**= C/X= Current Ratio= 1.20 times), (**U**= Utilized or Starting Capital= 64,720 millions USD), (**F**= Fixed Cost= 29,740 millions USD), and (**l**= L/E= Leverage or Gearing Ratio= 99.00%), then it's (**Q**= Quoted Longterm Liabilities planned), is:

$$Q = l\{U+M-S'i[1+s]-F-T-Ad\}-Svp/\{360[c-q]\}$$
$$= 0.9900\{64,720+41,548-48,851$$
$$*0.0100[1+0.0500]-29,750$$
$$-3,385-7,899*0.3000\}$$
$$-51,294*0.1900*240/\{360$$
$$[1.2000-1.0100]\}$$
$$= 35,351 \text{ millions USD}$$

Corporate IFRS-GAAP (B/S-I/S),ISBN-13: **978-1721066308**, ISBN-10: **1721066306**

Law-8348:

If both (s= [S/S']-1= Sales Growth= 5.00%), (v= V/S= Variable Portion= 19.00%), ($\$$= Sales or Revenues= 51,294 millions USD), (d= D/A= Dividend Portion or Payout= 30.00%), (p= 360 P/V= Procured Inventory Days= 240 Days), (A= After Tax Income= 7,899 millions USD), (T= Tax Paid= 3,385 millions USD), ($\$'$= Sales of Past Year= 48,851 millions USD), (i= I/S= Interest Portion= 1.00%), (M= Margin of Contribution= 41,548 millions USD), (q= [C-P]/X= Quick or Acid Test Ratio= 1.01 times), (c= C/X= Current Ratio= 1.20 times), (U= Utilized or Starting Capital= 64,720 millions USD), (F= Fixed Cost= 29,740 millions USD), and (Q= Quoted Longterm Liabilities= 35,351 millions USD), then it's (I= Leverage or Gearing Ratio planned), is:

$$I = (Q + \$vp/\{360[c-q]\})$$
$$/\{U+M-\$'i[1+s]-F-T-Ad\}$$
$$= (35,351+51,294*0.1900*240/\{360$$
$$[1.2000-1.0100]\})/\{64,720$$
$$+41,548-48,851*0.0100$$
$$[1+0.0500]-29,750-3,385$$
$$-7,899*0.3000\}$$

$$= \underline{99.00\%}$$

Corporate IFRS-GAAP (B/S-I/S),ISBN-13: **978-1721066308**, ISBN-10: **1721066306**

Law-8349:

If both (**s**= [S/S']-1= Sales Growth= 5.00%), (**v**= V/S= Variable Portion= 19.00%), (**$**= Sales or Revenues= 51,294 millions USD), (**d**= D/A= Dividend Portion or Payout= 30.00%), (**p**= 360 P/V= Procured Inventory Days= 240 Days), (**A**= After Tax Income= 7,899 millions USD), (**T**= Tax Paid= 3,385 millions USD), (**$'**= Sales of Past Year= 48,851 millions USD), (**i**= I/S= Interest Portion= 1.00%), (**M**= Margin of Contribution= 41,548 millions USD), (**q**= [C-P]/X= Quick or Acid Test Ratio= 1.01 times), (**c**= C/X= Current Ratio= 1.20 times), (**l**= L/E= Leverage or Gearing Ratio= 99.00%), (**F**= Fixed Cost= 29,740 millions USD), and (**Q**= Quoted Longterm Liabilities= 35,351 millions USD), then it's (**U**= Utilized or Starting Capital planned), is:

$$U = \$'i[1+s]+F+T+Ad$$
$$+(Q+\$vp/\{360[c-q]\})/l - M$$
$$= 48{,}851*0.0100[1+0.0500]+29{,}750$$
$$+3{,}385+7{,}899*0.3000$$
$$+(35{,}351+51{,}294*0.1900$$
$$*240/\{360[1.2000-1.0100]\})$$
$$/0.9900-41{,}548$$
$$= 64{,}720 \text{ millions USD}$$

Corporate IFRS-GAAP (B/S-I/S),ISBN-13: **978-1721066308**, ISBN-10: **1721066306**

<u>Law-8350</u>:

If both (**s**= [S/S']-1= Sales Growth= <u>5.00%</u>), (**v**= V/S= Variable Portion= <u>19.00%</u>), (**S**= Sales or Revenues= <u>51,294</u> millions USD), (**d**= D/A= Dividend Portion or Payout= <u>30.00%</u>), (**p**= 360 P/V= Procured Inventory Days= <u>240</u> Days), (**A**= After Tax Income= <u>7,899</u> millions USD), (**T**= Tax Paid= <u>3,385</u> millions USD), (**S'**= Sales of Past Year= <u>48,851</u> millions USD), (**i**= I/S= Interest Portion= <u>1.00%</u>), (**U**= Utilized or Starting Capital= <u>64,720</u> millions USD), (**q**= [C-P]/X= Quick or Acid Test Ratio= <u>1.01</u> times), (**c**= C/X= Current Ratio= <u>1.20</u> times), (**l**= L/E= Leverage or Gearing Ratio= <u>99.00%</u>), (**F**= Fixed Cost= <u>29,740</u> millions USD), and (**Q**= Quoted Longterm Liabilities= <u>35,351</u> millions USD), then it's (**M**= Margin of Contribution planned), is:

$$M= S'i[1+s]+F+T+Ad$$
$$+(Q+Svp/\{360[c-q]\})/l -U$$
$$= 48,851*0.0100[1+0.0500]+29,750$$
$$+3,385+7,899*0.3000$$
$$+(35,351+51,294*0.1900$$
$$*240/\{360[1.2000-1.0100]\})$$
$$/ 0.9900-64,720$$

$$= \underline{41,548} \text{ millions USD}$$

Corporate IFRS-GAAP (B/S-I/S),ISBN-13: **978-1721066308**, ISBN-10: **1721066306**

Law-8351:

If both (s= [S/S']-1= Sales Growth= 5.00%), (v= V/S= Variable Portion= 19.00%), (S= Sales or Revenues= 51,294 millions USD), (d= D/A= Dividend Portion or Payout= 30.00%), (p= 360 P/V= Procured Inventory Days= 240 Days), (A= After Tax Income= 7,899 millions USD), (T= Tax Paid= 3,385 millions USD), (S'= Sales of Past Year= 48,851 millions USD), (i= I/S= Interest Portion= 1.00%), (U= Utilized or Starting Capital= 64,720 millions USD), (q= [C-P]/X= Quick or Acid Test Ratio= 1.01 times), (c= C/X= Current Ratio= 1.20 times), (l= L/E= Leverage or Gearing Ratio= 99.00%), (M= Margin of Contribution= 41,548 millions USD), and (Q= Quoted Longterm Liabilities= 35,351 millions USD), then it's (F= Fixed Cost planned), is:

$$F= U+M-S'i[1+s]-T-Ad-(Q+Svp/\{360[c-q]\})/l$$
$$= 64,720+41,548-48,851*0.0100$$
$$[1+0.0500]- 3,385-7,899$$
$$*0.3000 -(35,351+51,294$$
$$*0.1900*240/\{360[1.2000$$
$$-1.0100]\})/0.9900$$
$$= 29,750 \text{ millions USD}$$

Corporate IFRS-GAAP (B/S-I/S),ISBN-13: **978-1721066308**, ISBN-10: **1721066306**

Law-8352:

If both (**s**= [S/S']-1= Sales Growth= 5.00%), (**v**= V/S= Variable Portion= 19.00%), (**S**= Sales or Revenues= 51,294 millions USD), (**d**= D/A= Dividend Portion or Payout= 30.00%), (**p**= 360 P/V= Procured Inventory Days= 240 Days), (**A**= After Tax Income= 7,899 millions USD), (**T**= Tax Paid= 3,385 millions USD), (**F**= Fixed Cost= 29,740 millions USD), (**i**= I/S= Interest Portion= 1.00%), (**U**= Utilized or Starting Capital= 64,720 millions USD), (**q**= [C-P]/X= Quick or Acid Test Ratio= 1.01 times), (**c**= C/X= Current Ratio= 1.20 times), (**l**= L/E= Leverage or Gearing Ratio= 99.00%), (**M**= Margin of Contribution= 41,548 millions USD), and (**Q**= Quoted Longterm Liabilities= 35,351 millions USD), then it's (**S'**= Sales Past), must be:

$$S'= [U+M-F-T-Ad-(Q+Svp/\{360[c-q]\})/l]$$
$$/\{i[1+s]\}$$
$$= [64,720+41,548-29,750-3,385$$
$$-7,899*0.3000-(35,351$$
$$+51,294*0.1900*240/\{360$$
$$[1.2000-1.0100]\})/0.9900]$$
$$/\{0.0100[1+0.0500]\}$$
$$= 48,851 \text{ millions USD}$$

Corporate IFRS-GAAP (B/S-I/S),ISBN-13: **978-1721066308**, ISBN-10: **1721066306**

Law-8353:

If both (**s**= [S/S']-1= Sales Growth= 5.00%), (**v**= V/S= Variable Portion= 19.00%), (**$**= Sales or Revenues= 51,294 millions USD), (**d**= D/A= Dividend Portion or Payout= 30.00%), (**p**= 360 P/V= Procured Inventory Days= 240 Days), (**A**= After Tax Income= 7,899 millions USD), (**T**= Tax Paid= 3,385 millions USD), (**F**= Fixed Cost= 29,740 millions USD), (**$'**= Sales of Past Year= 48,851 millions USD), (**U**= Utilized or Starting Capital= 64,720 millions USD), (**q**= [C-P]/X= Quick or Acid Test Ratio= 1.01 times), (**c**= C/X= Current Ratio= 1.20 times), (**l**= L/E= Leverage or Gearing Ratio= 99.00%), (**M**= Margin of Contribution= 41,548 millions USD), and (**Q**= Quoted Longterm Liabilities= 35,351 millions USD), then it's (**i**= Interest Portion planned), is:

$$\mathbf{i= [U+M-F-T-Ad-(Q+\$vp/\{360\lfloor c-q\rfloor\})/l\,]}$$
$$/\{\mathbf{\$'}[1+\mathbf{s}]\}$$
$$= [64,720+41,548-29,750-3,385$$
$$-7,899*0.3000-(35,351$$
$$+51,294*0.1900*240/\{360$$
$$[1.2000-1.0100]\})/0.9900]$$
$$/\{48,851[1+0.0500]\}$$
$$= 1.00\%$$

Corporate IFRS-GAAP (B/S-I/S),ISBN-13: **978-1721066308**, ISBN-10: **1721066306**

Law-8354:

If both (**i**= I/S= Interest Portion= 1.00%), (**v**= V/S= Variable Portion= 19.00%), (**$**= Sales or Revenues= 51,294 millions USD), (**d**= D/A= Dividend Portion or Payout= 30.00%), (**p**= 360 P/V= Procured Inventory Days= 240 Days), (**A**= After Tax Income= 7,899 millions USD), (**T**= Tax Paid= 3,385 millions USD), (**F**= Fixed Cost= 29,740 millions USD), (**$'**= Sales of Past Year= 48,851 millions USD), (**U**= Utilized or Starting Capital= 64,720 millions USD), (**q**= Quick or Acid Test Ratio= [C-P]/X= 1.01 times), (**c**= C/X= Current Ratio= 1.20 times), (**l**= L/E= Leverage or Gearing Ratio= 99.00%), (**M**= Margin of Contribution= 41,548 millions USD), and (**Q**= Quoted Longterm Liabilities= 35,351 millions USD), then it's (**s**= Sales Growth planned), is:

$$s= [U+M-F-T-Ad-(Q+\$vp/\{360[c-q]\})/l]$$
$$/[\$'i]-1$$
$$= [64,720+41,548-29,750-3,385$$
$$-7,899*0.3000-(35,351$$
$$+51,294*0.1900*240/\{360$$
$$[1.2000-1.0100]\})/0.9900]$$
$$/[48,851*0.0100]-1$$
$$= 5.00\%$$

Corporate IFRS-GAAP (B/S-I/S),ISBN-13: **978-1721066308**, ISBN-10: **1721066306**

Law-8355:

If both (i= I/S= Interest Portion= 1.00%), (v= V/S= Variable Portion= 19.00%), ($\$$= Sales or Revenues= 51,294 millions USD), (d= D/A= Dividend Portion or Payout= 30.00%), (p= 360 P/V= Procured Inventory Days= 240 Days), (A= After Tax Income= 7,899 millions USD), (s= [S/S']-1= Sales Growth= 5.00%), (F= Fixed Cost= 29,740 millions USD), ($\$'$= Sales of Past Year= 48,851 millions USD), (U= Utilized or Starting Capital= 64,720 millions USD), (q= Quick or Acid Test Ratio= [C-P]/X= 1.01 times), (c= C/X= Current Ratio= 1.20 times), (l= L/E= Leverage or Gearing Ratio= 99.00%), (M= Margin of Contribution= 41,548 millions USD), and (Q= Quoted Longterm Liabilities= 35,351 millions USD), then it's (T= Tax Paid planned), is:

$$T= U+M-F-\$'i[1+s]-Ad$$
$$-(Q+\$vp/\{360[c-q]\})/l$$
$$= 64,720+41,548-29,750-48,851$$
$$*0.0100[1+0.0500]-7,899$$
$$*0.3000-(35,351+51,294$$
$$*0.1900*240/\{360[1.2000$$
$$-1.0100]\})/0.9900$$
$$= 3,385 \text{ millions USD}$$

Corporate IFRS-GAAP (B/S-I/S),ISBN-13: **978-1721066308**, ISBN-10: **1721066306**

Law-8356:

If both (**i**= I/S= Interest Portion= 1.00%), (**v**= V/S= Variable Portion= 19.00%), (**S**= Sales or Revenues= 51,294 millions USD), (**d**= D/A= Dividend Portion or Payout= 30.00%), (**p**= 360 P/V= Procured Inventory Days= 240 Days), (**T**= Tax Paid= 3,385 millions USD), (**s**= [S/S']-1= Sales Growth= 5.00%), (**F**= Fixed Cost= 29,740 millions USD), (**S'**= Sales of Past Year= 48,851 millions USD), (**U**= Utilized or Starting Capital= 64,720 millions USD), (**q**= [C-P]/X= Quick or Acid Test Ratio= 1.01 times), (**c**= C/X= Current Ratio= 1.20 times), (**l**= L/E= Leverage or Gearing Ratio= 99.00%), (**M**= Margin of Contribution= 41,548 millions USD), and (**Q**= Quoted Longterm Liabilities= 35,351 millions USD), then it's (**A**= After Tax Income planned), is :

$$A = [U+M-F-S'i[1+s]-T$$
$$-(Q+Svp/\{360[c-q]\})/l\ ]/d$$
$$= [64,720+41,548-29,750-48,851$$
$$*0.0100[1+0.0500]-3,385$$
$$-(35,351+51,294*0.1900$$
$$*240/\{360[1.2000-1.0100]\})$$
$$/0.9900]/0.3000$$
$$= 7,899 \text{ millions USD}$$

Corporate IFRS-GAAP (B/S-I/S),ISBN-13: **978-1721066308**, ISBN-10: **1721066306**

Law-8357:

If both (**i**= I/S= Interest Portion= 1.00%), (**v**= V/S= Variable Portion= 19.00%), (**S**= Sales or Revenues= 51,294 millions USD), (**A**= After Tax Income= 7,899 millions USD), (**p**= 360 P/V= Procured Inventory Days= 240 Days), (**T**= Tax Paid= 3,385 millions USD), (**s**= [S/S']-1= Sales Growth= 5.00%), (**F**= Fixed Cost= 29,740 millions USD), (**S'**= Sales of Past Year= 48,851 millions USD), (**U**= Utilized or Starting Capital= 64,720 millions USD), (**q**= Quick or Acid Test Ratio= [C-P]/X= 1.01 times), (**c**= C/X= Current Ratio= 1.20 times), (**F**= L/E= Leverage or Gearing Ratio= 99.00%), (**M**= Margin of Contribution= 41,548 millions USD), and (**Q**= Quoted Longterm Liabilities= 35,351 millions USD), then it's (**d**= Dividend Portion or Payout planned), is:

$$d= [U+M-F-S'i[1+s]-T$$
$$-(Q+Svp/\{360[c-q]\})/F]/A$$
$$= [64,720+41,548-29,750-48,851$$
$$*0.0100[1+0.0500]-3,385$$
$$-(35,351+51,294*0.1900$$
$$*240/\{360[1.2000-1.0100]\})$$
$$/0.9900]/7,899$$
$$= 30.00\%$$

Corporate IFRS-GAAP (B/S-I/S),ISBN-13: **978-1721066308**, ISBN-10: **1721066306**

Law-8358:

If both (i= I/S= Interest Portion= 1.00%), (v= V/S= Variable Portion= 19.00%), (d= D/A= Dividend Portion or Payout= 30.00%), (A= After Tax Income= 7,899 millions USD), (p= 360 P/V= Procured Inventory Days= 240 Days), (T= Tax Paid= 3,385 millions USD), (s= [S/S']-1= Sales Growth= 5.00%), (F= Fixed Cost= 29,740 millions USD), (S'= Sales of Past Year= 48,851 millions USD), (U= Utilized or Starting Capital= 64,720 millions USD), (q= Quick or Acid Test Ratio= [C-P]/X= 1.01 times), (c= C/X= Current Ratio= 1.20 times), (l= L/E= Leverage or Gearing Ratio= 99.00%), (M= Margin of Contribution= 41,548 millions USD), and (Q= Quoted Longterm Liabilities= 35,351 millions USD), then it's (S= Sales or Revenues planned), is:

$$S= 360[c-q](l \{U+M-F-S'i[1+s]-T-Ad-Q) /[vp]$$

$$= 360[1.2000-1.0100](0.9900\{64,720 +41,548-29,750-48,851 *0.0100[1+0.0500]-3,385 -7,899*0.3000\}-35,351) /[0.1900*240]$$

$$= 51,294 \text{ millions USD}$$

Corporate IFRS-GAAP (B/S-I/S),ISBN-13: **978-1721066308**, ISBN-10: **1721066306**

Law-8359:

If both (**$**= Sales or Revenues= 51,294 millions USD), (**i**= I/S= Interest Portion= 1.00%), (**d**= D/A= Dividend Portion or Payout= 30.00%), (**A**= After Tax Income= 7,899 millions USD), (**p**= 360 P/V= Procured Inventory Days= 240 Days), (**T**= Tax Paid= 3,385 millions USD), (**s**= [S/S']-1= Sales Growth= 5.00%), (**F**= Fixed Cost= 29,740 millions USD), (**$'**= Sales of Past Year= 48,851 millions USD), (**U**= Utilized or Starting Capital= 64,720 millions USD), (**q**= [C-P]/X= Quick or Acid Test Ratio= 1.01 times), (**c**= C/X= Current Ratio= 1.20 times), (**l**= L/E= Leverage or Gearing Ratio= 99.00%), (**M**= Margin of Contribution= 41,548 millions USD), and (**Q**= Quoted Longterm Liabilities= 35,351 millions USD), then it's (**v**= Variable Portion planned), is:

$$v= 360[c\text{-}q](l \{U+M\text{-}F\text{-}S'i[1+s]\text{-}T\text{-}Ad\text{-}Q)$$
$$/\lfloor Sp \rfloor$$

$$= 360[1.2000\text{-}1.0100](0.9900$$
$$\{64,720+41,548\text{-}29,750$$
$$-48,851*0.0100[1+0.0500]$$
$$-3,385\text{-}7,899*0.3000\}$$
$$-35,351)/[51,294*240]$$

$$= 19.00\%$$

Corporate IFRS-GAAP (B/S-I/S),ISBN-13: **978-1721066308**, ISBN-10: **1721066306**

Law-8360:

If both (**$**= Sales or Revenues= 51,294 millions USD), (**i**= I/S= Interest Portion= 1.00%), (**d**= D/A= Dividend Portion or Payout= 30.00%), (**A**= After Tax Income= 7,899 millions USD), (**v**= V/S= Variable Portion= 19.00%), (**T**= Tax Paid= 3,385 millions USD), (**s**= [S/S']-1= Sales Growth= 5.00%), (**F**= Fixed Cost= 29,740 millions USD), (**$'**= Sales of Past Year= 48,851 millions USD), (**U**= Utilized or Starting Capital= 64,720 millions USD), (**q**= Quick or Acid Test Ratio= [C-P]/X= 1.01 times), (**c**= C/X= Current Ratio= 1.20 times), (**l**= L/E= Leverage or Gearing Ratio= 99.00%), (**M**= Margin of Contribution= 41,548 millions USD), and (**Q**= Quoted Longterm Liabilities= 35,351 millions USD), then it's (**p**= Procured Inventory Days planned), is:

$$p= 360[\textbf{c-q}](\textbf{l}\ \{\textbf{U+M-F-\$'i}[1+\textbf{s}]\textbf{-T-Ad-Q})/[\textbf{\$v}]$$

$$= 360[1.2000-1.0100](0.9900$$
$$\{64,720+41,548-29,750$$
$$-48,851*0.0100[1+0.0500]$$
$$-3,385-7,899*0.3000\}$$
$$-35,351)/[51,294*0.1900]$$

$$= \underline{240}\ \text{days}$$

Corporate IFRS-GAAP (B/S-I/S),ISBN-13: **978-1721066308**, ISBN-10: **1721066306**

Law-8361:

If both (**$**= Sales or Revenues= 51,294 millions USD), (**i**= I/S= Interest Portion= 1.00%), (**d**= D/A= Dividend Portion or Payout= 30.00%), (**A**= After Tax Income= 7,899 millions USD), (**v**= V/S= Variable Portion= 19.00%), (**T**= Tax Paid= 3,385 millions USD), (**s**= [S/S']-1= Sales Growth= 5.00%), (**F**= Fixed Cost= 29,740 millions USD), (**$'**= Sales of Past Year= 48,851 millions USD), (**U**= Utilized or Starting Capital= 64,720 millions USD), (**q**= Quick or Acid Test Ratio= [C-P]/X= 1.01 times), (**p**= 360 P/V= Procured Inventory Days= 240 Days), (**l**= L/E= Leverage or Gearing Ratio= 99.00%), (**M**= Margin of Contribution= 41,548 millions USD), and (**Q**= Quoted Longterm Liabilities= 35,351 millions USD), then it's (**c**= Current Ratio planned), is:

$$c= q+\$vp/[360(\textbf{\textit{l}}\,\{U+M-F-\$'i[1+s]-T-Ad-Q)]$$
$$= 1.0100+51,294*0.1900*240/[360$$
$$(0.9900\{64,720+41,548$$
$$-29,750-48,851*0.0100$$
$$[1+0.0500]-3,385-7,899$$
$$*0.3000\}-35,351)]$$
$$= 1.20 \text{ times}$$

Corporate IFRS-GAAP (B/S-I/S),ISBN-13: **978-1721066308**, ISBN-10: **1721066306**

Law-8362:

If both (**$**= Sales or Revenues= 51,294 millions USD), (**i**= I/S= Interest Portion= 1.00%), (**d**= D/A= Dividend Portion or Payout= 30.00%), (**A**= After Tax Income= 7,899 millions USD), (**v**= V/S= Variable Portion= 19.00%), (**T**= Tax Paid= 3,385 millions USD), (**s**= [S/S']-1= Sales Growth= 5.00%), (**F**= Fixed Cost= 29,740 millions USD), (**$'**= Sales of Past Year= 48,851 millions USD), (**U**= Utilized or Starting Capital= 64,720 millions USD), (**c**= C/X= Current Ratio= 1.20 times), (**p**= 360 P/V= Procured Inventory Days= 240 Days), (**l**= L/E= Leverage or Gearing Ratio= 99.00%), (**M**= Margin of Contribution= 41,548 millions USD), and (**Q**= Quoted Longterm Liabilities= 35,351 millions USD), then it's (**q**= Quick or Acid Test Ratio planned), is:

$$q= c\text{-}\$vp/[360(l\ \{U+M\text{-}F\text{-}\$'i[1+s]\text{-}T\text{-}Ad\text{-}Q)]$$
$$= 1.2000\text{-}51,294*0.1900*240/[360$$
$$(0.9900\{64,720+41,548$$
$$\text{-}29,750\text{-}48,851*0.0100$$
$$[1+0.0500]\text{-}3,385\text{-}7,899$$
$$*0.3000\}\text{-}35,351)]$$
$$= 1.01 \text{ times}$$

Corporate IFRS-GAAP (B/S-I/S),ISBN-13: **978-1721066308**, ISBN-10: **1721066306**

Law-8363:

If both (**i**= I/S= Interest Portion= 1.00%), (**d**= D/A= Dividend Portion or Payout= 30.00%), (**A**= After Tax Income= 7,899 millions USD), (**v**= V/S= Variable Portion= 19.00%), (**T**= Tax Paid= 3,385 millions USD), (**s**= [S/S']-1= Sales Growth= 5.00%), (**F**= Fixed Cost= 29,740 millions USD), (**S'**= Sales of Past Year= 48,851 millions USD), (**U**= Utilized or Starting Capital= 64,720 millions USD), (**c**= C/X= Current Ratio= 1.20 times), (**p**= 360 P/V= Procured Inventory Days= 240 Days), (**l**= L/E= Leverage or Gearing Ratio= 99.00%), (**M**= Margin of Contribution= 41,548 millions USD), and (**q**= Quick or Acid Test Ratio= [C-P]/X= 1.01 times), then it's (**Q**= Quoted Longterm Liabilities planned), is:

$$Q= l \{U+M-F-S'i[1+s]-T-Ad\}$$
$$-S'vp[1+s]/\{360[c-q]\}$$
$$= 0.9900\{64,720+41,548-29,750$$
$$-48,851*0.0100[1+0.0500]$$
$$-3,385-7,899*0.3000\}$$
$$-48,851*0.1900*240$$
$$[1+0.0500]/\{360$$
$$[1.2000-1.0100]\}$$
$$= 35,351 \text{ millions USD}$$

Corporate IFRS-GAAP (B/S-I/S),ISBN-13: **978-1721066308**, ISBN-10: **1721066306**

Law-8364:

If both (**i**= I/S= Interest Portion= 1.00%), (**d**= D/A= Dividend Portion or Payout= 30.00%), (**A**= After Tax Income= 7,899 millions USD), (**v**= V/S= Variable Portion= 19.00%), (**T**= Tax Paid= 3,385 millions USD), (**s**= [S/S']-1= Sales Growth= 5.00%), (**F**= Fixed Cost= 29,740 millions USD), (**S'**= Sales of Past Year= 48,851 millions USD), (**U**= Utilized or Starting Capital= 64,720 millions USD), (**c**= C/X= Current Ratio= 1.20 times), (**p**= 360 P/V= Procured Inventory Days= 240 Days), (**Q**= Quoted Longterm Liabilities= 35,351 millions USD), (**M**= Margin of Contribution= 41,548 millions USD), and (**q**= Quick or Acid Test Ratio= [C-P]/X= 1.01 times), then it's (**I** = Leverage or Gearing Ratio planned), is:

$$I = (Q+S'vp[1+s]/\{360[c-q]\})$$
$$/\{U+M-F-S'i[1+s]-T-Ad\}$$

$$= (35,351+48,851*0.1900*240$$
$$[1+0.0500]/\{360[1.2000$$
$$-1.0100]\})/\{64,720+41,548$$
$$-29,750-48,851*0.0100$$
$$[1+0.0500]-3,385$$
$$-7,899*0.3000\}$$

$$= 99.00\%$$

Corporate IFRS-GAAP (B/S-I/S),ISBN-13: **978-1721066308**, ISBN-10: **1721066306**

Law-8365:

If both (**i**= I/S= Interest Portion= 1.00%), (**d**= D/A= Dividend Portion or Payout= 30.00%), (**A**= After Tax Income= 7,899 millions USD), (**v**= V/S= Variable Portion= 19.00%), (**T**= Tax Paid= 3,385 millions USD), (**s**= [S/S']-1= Sales Growth= 5.00%), (**F**= Fixed Cost= 29,740 millions USD), (**S'**= Sales of Past Year= 48,851 millions USD), (**I**= L/E= Leverage or Gearing Ratio= 99.00%), (**c**= C/X= Current Ratio= 1.20 times), (**p**= 360 P/V= Procured Inventory Days= 240 Days), (**Q**= Quoted Longterm Liabilities= 35,351 millions USD), (**M**= Margin of Contribution= 41,548 millions USD), and (**q**= [C-P]/X= Quick or Acid Test Ratio= 1.01 times), then it's (**U**= Utilized or Starting Capital planned), is:

$$U= F+S'i[1+s]+T+Ad+(Q+S'vp[1+s]$$
$$/\{360[c-q]\})/I-M$$

$$= 29,750+48,851*0.0100[1+0.0500]$$
$$+3,385+7,899*0.3000$$
$$+(35,351+48,851*0.1900$$
$$*240[1+0.0500]/\{360[1.2000$$
$$-1.0100]\})/0.9900-41,548$$

$$= 64,720 \text{ millions USD}$$

Corporate IFRS-GAAP (B/S-I/S),ISBN-13: **978-1721066308**, ISBN-10: **1721066306**

<u>Law-8366</u>:

If both (**i**= I/S= Interest Portion= <u>1.00%</u>), (**d**= D/A= Dividend Portion or Payout= <u>30.00%</u>), (**A**= After Tax Income= <u>7,899</u> millions USD), (**v**= V/S= Variable Portion= <u>19.00%</u>), (**T**= Tax Paid= <u>3,385</u> millions USD), (**s**= [S/S']-1= Sales Growth= <u>5.00%</u>), (**S'**= Sales of Past Year= <u>48,851</u> millions USD), (**l**= L/E= Leverage or Gearing Ratio= <u>99.00%</u>), (**c**= C/X= Current Ratio= <u>1.20</u> times), (**p**= 360 P/V= Procured Inventory Days= <u>240</u> Days), (**F**= Fixed Cost= <u>29,740</u> millions USD), (**Q**= Quoted Longterm Liabilities= <u>35,351</u> millions USD), (**U**= Utilized or Starting Capital= <u>64,720</u> millions USD), and (**q**= [C-P]/X= Quick or Acid Test Ratio= <u>1.01</u> times), then it's (**M**= Margin of Contribution planned), is:

$$M= F+S'i[1+s]+T+Ad+(Q+S'vp[1+s]$$
$$/\{360[c-q]\})/l-U$$
$$= 29,750+48,851*0.0100[1+0.0500]$$
$$+3,385+7,899*0.3000$$
$$+(35,351+48,851*0.1900$$
$$*240[1+0.0500]/\{360$$
$$[1.2000-1.0100]\})$$
$$/0.9900-64,720$$
$$= \underline{41,548} \text{ millions USD}$$

Corporate IFRS-GAAP (B/S-I/S),ISBN-13: **978-1721066308**, ISBN-10: **1721066306**

Law-8367:

If both (**i**= I/S= Interest Portion= 1.00%), (**d**= D/A= Dividend Portion or Payout= 30.00%), (**A**= After Tax Income= 7,899 millions USD), (**v**= V/S= Variable Portion= 19.00%), (**T**= Tax Paid= 3,385 millions USD), (**s**= [S/S']-1= Sales Growth= 5.00%), (**S'**= Sales of Past Year= 48,851 millions USD), (**l**= L/E= Leverage or Gearing Ratio= 99.00%), (**c**= C/X= Current Ratio= 1.20 times), (**p**= 360 P/V= Procured Inventory Days= 240 Days), (**M**= Margin of Contribution= 41,548 millions USD), (**Q**= Quoted Longterm Liabilities= 35,351 millions USD), (**U**= Utilized or Starting Capital= 64,720 millions USD), and (**q**= [C-P]/X= Quick or Acid Test Ratio= 1.01 times), then it's (**F**= Fixed Cost planned), is:

$$F= U+M-S'i[1+s]-T-Ad-(Q+S'vp[1+s]$$
$$/\{360[c-q]\})/l$$

$$= 64,720+41,548-48,851*0.0100$$
$$[1+0.0500]-3,385-7,899$$
$$*0.3000 -(35,351+48,851$$
$$*0.1900*240[1+0.0500]$$
$$/\{360[1.2000-1.0100]\})$$
$$/0.9900$$

$$= \underline{29,750} \text{ millions USD}$$

Corporate IFRS-GAAP (B/S-I/S),ISBN-13: **978-1721066308**, ISBN-10: **1721066306**

Law-8368:

If both (**i**= I/S= Interest Portion= 1.00%), (**d**= D/A= Dividend Portion or Payout= 30.00%), (**A**= After Tax Income= 7,899 millions USD), (**v**= V/S= Variable Portion= 19.00%), (**T**= Tax Paid= 3,385 millions USD), (**s**= [S/S']-1= Sales Growth= 5.00%), (**F**= Fixed Cost= 29,740 millions USD), (**l**= L/E= Leverage or Gearing Ratio= 99.00%), (**c**= C/X= Current Ratio= 1.20 times), (**p**= 360 P/V= Procured Inventory Days= 240 Days), (**M**= Margin of Contribution= 41,548 millions USD), (**Q**= Quoted Longterm Liabilities= 35,351 millions USD), (**U**= Utilized or Starting Capital= 64,720 millions USD), and (**q**= [C-P]/X= Quick or Acid Test Ratio= 1.01 times), then it's (**S'**= Sales of Past), must be:

$$S' = [U+M-F-T-Ad-(Q+S'vp[1+s]$$
$$/\{360[c-q]\})/l]/\{i[1+s]\}$$
$$= [64,720+41,548-29,750-3,385$$
$$-7,899*0.3000-(35,351$$
$$+48,851*0.1900*240$$
$$[1+0.0500]/\{360[1.2000$$
$$-1.0100]\})/0.9900]$$
$$/\{0.0100[1+0.0500]\}$$
$$= 48,851 \text{ millions USD}$$

Corporate IFRS-GAAP (B/S-I/S),ISBN-13: **978-1721066308**, ISBN-10: **1721066306**

Law-8368:

If both (**i**= I/S= Interest Portion= 1.00%), (**d**= D/A= Dividend Portion or Payout= 30.00%), (**A**= After Tax Income= 7,899 millions USD), (**v**= V/S= Variable Portion= 19.00%), (**T**= Tax Paid= 3,385 millions USD), (**s**= [S/S']-1= Sales Growth= 5.00%), (**F**= Fixed Cost= 29,740 millions USD), (**I**= L/E= Leverage or Gearing Ratio= 99.00%), (**c**= C/X= Current Ratio= 1.20 times), (**p**= 360 P/V= Procured Inventory Days= 240 Days), (**M**= Margin of Contribution= 41,548 millions USD), (**Q**= Quoted Longterm Liabilities= 35,351 millions USD), (**U**= Utilized or Starting Capital= 64,720 millions USD), and (**q**= [C-P]/X= Quick or Acid Test Ratio= 1.01 times), then it's (**S'**= Sales of Past), must be:

$$S'= [U+M-F-T-Ad-(Q+S'vp[1+s]$$
$$/\{360[c-q]\})/I\ ]/\{i[1+s]\}$$
$$= [64,720+41,548-29,750-3,385$$
$$-7,899*0.3000-(35,351$$
$$+48,851*0.1900*240$$
$$[1+0.0500]/\{360[1.2000$$
$$-1.0100]\})/0.9900]$$
$$/\{0.0100[1+0.0500]\}$$
$$= 48,851 \text{ millions USD}$$

Corporate IFRS-GAAP (B/S-I/S),ISBN-13: **978-1721066308**, ISBN-10: **1721066306**

Law-8369:

If both (S'= Sales of Past Year= 48,851 millions USD), (d= D/A= Dividend Portion or Payout= 30.00%), (A= After Tax Income= 7,899 millions USD), (v= V/S= Variable Portion= 19.00%), (T= Tax Paid= 3,385 millions USD), (s= [S/S']-1= Sales Growth= 5.00%), (F= Fixed Cost= 29,740 millions USD), (I= L/E= Leverage or Gearing Ratio= 99.00%), (c= C/X= Current Ratio= 1.20 times), (p= 360 P/V= Procured Inventory Days= 240 Days), (M= Margin of Contribution= 41,548 millions USD), (Q= Quoted Longterm Liabilities= 35,351 millions USD), (U= Utilized or Starting Capital= 64,720 millions USD), and (q= [C-P]/X= Quick or Acid Test Ratio= 1.01 times), then it's (i= Interest Portion planned), is:

$$i= [U+M-F-T-Ad-(Q+S'vp[1+s]$$
$$/\{360[c-q]\})/I\ ]/\{S'[1+s]\}$$
$$= [64,720+41,548-29,750-3,385$$
$$-7,899*0.3000-(35,351$$
$$+48,851*0.1900*240$$
$$[1+0.0500]/\{360[1.2000$$
$$-1.0100]\})/0.9900]$$
$$/\{48,851[1+0.0500]\}$$
$$= 1.00\%$$

Corporate IFRS-GAAP (B/S-I/S),ISBN-13: **978-1721066308**, ISBN-10: **1721066306**

Law-8370:

If both ($\textbf{S'}$= Sales of Past Year= 48,851 millions USD), ($\textbf{d}$= D/A= Dividend Portion or Payout= 30.00%), ($\textbf{A}$= After Tax Income= 7,899 millions USD), ($\textbf{v}$= V/S= Variable Portion= 19.00%), ($\textbf{T}$= Tax Paid= 3,385 millions USD), ($\textbf{i}$= I/S= Interest Portion= 1.00%), ($\textbf{F}$= Fixed Cost= 29,740 millions USD), ($\textbf{l}$= L/E= Leverage or Gearing Ratio= 99.00%), ($\textbf{c}$= C/X= Current Ratio= 1.20 times), ($\textbf{p}$= 360 P/V= Procured Inventory Days= 240 Days), ($\textbf{M}$= Margin of Contribution= 41,548 millions USD), ($\textbf{Q}$= Quoted Longterm Liabilities= 35,351 millions USD), ($\textbf{U}$= Utilized or Starting Capital= 64,720 millions USD), and ($\textbf{q}$= [C-P]/X= Quick or Acid Test Ratio= 1.01 times), then it's ($\textbf{s}$= Sales Growth planned), is:

$$s= [U+M-F-T-Ad-(Q+S'vp[1+s]$$
$$/\{360[c-q]\})/l\,]/[S'i]-1$$
$$= [64,720+41,548-29,750-3,385$$
$$-7,899*0.3000-(35,351$$
$$+48,851*0.1900*240$$
$$[1+0.0500]/\{360[1.2000$$
$$-1.0100]\})/0.9900]$$
$$/[48,851*0.0100]-1$$
$$= 5.00\%$$

Corporate IFRS-GAAP (B/S-I/S),ISBN-13: **978-1721066308**, ISBN-10: **1721066306**

Law-8371:

If both (**S'**= Sales of Past Year= 48,851 millions USD), (**d**= D/A= Dividend Portion or Payout= 30.00%), (**A**= After Tax Income= 7,899 millions USD), (**v**= V/S= Variable Portion= 19.00%), (**s**= [S/S']-1= Sales Growth= 5.00%), (**i**= I/S= Interest Portion= 1.00%), (**F**= Fixed Cost= 29,740 millions USD), (**l**= L/E= Leverage or Gearing Ratio= 99.00%), (**c**= C/X= Current Ratio= 1.20 times), (**p**= 360 P/V= Procured Inventory Days= 240 Days), (**M**= Margin of Contribution= 41,548 millions USD), (**Q**= Quoted Longterm Liabilities= 35,351 millions USD), (**U**= Utilized or Starting Capital= 64,720 millions USD), and (**q**= [C-P]/X= Quick or Acid Test Ratio= 1.01 times), then it's (**T**= Tax Paid planned), is:

$$T= U+M-F-S'i[1+s]-Ad-(Q+S'up[1+s]$$
$$/\{360[c-q]\})/l$$
$$= 64,720+41,548-29,750-48,851$$
$$*0.0100[1+0.0500]$$
$$-7,899*0.3000-(35,351$$
$$+48,851*0.1900*240$$
$$[1+0.0500]/\{360[1.2000$$
$$-1.0100]\})/0.9900$$
$$= 3,385 \text{ millions USD}$$

Corporate IFRS-GAAP (B/S-I/S),ISBN-13: **978-1721066308**, ISBN-10: **1721066306**

<u>Law-8372</u>:

If both (**$'**= Sales of Past Year= <u>48,851</u> millions USD), (**d**= D/A= Dividend Portion or Payout= <u>30.00%</u>), (**T**= Tax Paid= <u>3,385</u> millions USD), (**v**= V/S= Variable Portion= <u>19.00%</u>), (**s**= [S/S']-1= Sales Growth= <u>5.00%</u>), (**i**= I/S= Interest Portion= <u>1.00%</u>), (**F**= Fixed Cost= <u>29,740</u> millions USD), (**l**= L/E= Leverage or Gearing Ratio= <u>99.00%</u>), (**c**= C/X= Current Ratio= <u>1.20</u> times), (**p**= 360 P/V= Procured Inventory Days= <u>240</u> Days), (**M**= Margin of Contribution= <u>41,548</u> millions USD), (**Q**= Quoted Longterm Liabilities= <u>35,351</u> millions USD), (**U**= Utilized or Starting Capital= <u>64,720</u> millions USD), and (**q**= [C-P]/X= Quick or Acid Test Ratio= <u>1.01</u> times), then it's (**A**= After Tax Income planned), is:

$$A= [U+M-F-S'i[1+s]-T-(Q+S'vp[1+s]$$
$$/\{360[c-q]\})/l]/d$$
$$= [64,720+41,548-29,750-48,851$$
$$*0.0100[1+0.0500]-3,385$$
$$-(35,351+48,851*0.1900$$
$$*240[1+0.0500]/\{360[1.2000$$
$$-1.0100]\})/0.9900]/0.3000$$
$$= \underline{7,899} \text{ millions USD}$$

Corporate IFRS-GAAP (B/S-I/S),ISBN-13: **978-1721066308**, ISBN-10: **1721066306**

Law-8373:

If both (**$'$**= Sales of Past Year= 48,851 millions
USD), (**A**= After Tax Income= 7,899 millions USD),
(**T**= Tax Paid= 3,385 millions USD), (**v**= V/S=
Variable Portion= 19.00%), (**s**= [S/S']-1= Sales
Growth= 5.00%), (**i**= I/S= Interest Portion= 1.00%),
(**F**= Fixed Cost= 29,740 millions USD), (**l**= L/E=
Leverage or Gearing Ratio= 99.00%), (**c**= C/X=
Current Ratio= 1.20 times), (**p**= 360 P/V= Procured
Inventory Days= 240 Days), (**M**= Margin of
Contribution= 41,548 millions USD), (**Q**= Quoted
Longterm Liabilities= 35,351 millions USD), (**U**=
Utilized or Starting Capital= 64,720 millions USD),
and (**q**= [C-P]/X= Quick or Acid Test Ratio= 1.01
times), then it's (**d**= Dividend Portion or Payout
planned), is:

$$d= [U+M-F-S'i[1+s]-T-(Q+S'vp[1+s]$$
$$/\{360[c-q]\})/l\,]/A$$
$$= [64,720+41,548-29,750-48,851$$
$$*0.0100[1+0.0500]-3,385$$
$$-(35,351+48,851*0.1900$$
$$*240[1+0.0500]/\{360$$
$$[1.2000-1.0100]\})$$
$$/0.9900]/7,899$$
$$= 30.00\%$$

Corporate IFRS-GAAP (B/S-I/S),ISBN-13: **978-1721066308**, ISBN-10: **1721066306**

Law-8374:

If both ($\textbf{S'}$= Sales of Past Year= 48,851 millions USD), (**A**= After Tax Income= 7,899 millions USD), (**T**= Tax Paid= 3,385 millions USD), (**d**= D/A= Dividend Portion or Payout= 30.00%), (**s**= [S/S']-1= Sales Growth= 5.00%), (**i**= I/S= Interest Portion= 1.00%), (**F**= Fixed Cost= 29,740 millions USD), (**l**= L/E= Leverage or Gearing Ratio= 99.00%), (**c**= C/X= Current Ratio= 1.20 times), (**p**= 360 P/V= Procured Inventory Days= 240 Days), (**M**= Margin of Contribution= 41,548 millions USD), (**Q**= Quoted Longterm Liabilities= 35,351 millions USD), (**U**= Utilized or Starting Capital= 64,720 millions USD), and (**q**= [C-P]/X= Quick or Acid Test Ratio= 1.01 times), then it's (**v**= Variable Portion planned), is:

$$\textbf{v}= 360[\textbf{c}-\textbf{q}](\textbf{l} \{\textbf{U}+\textbf{M}-\textbf{F}-\textbf{S'i}[1+\textbf{s}]-\textbf{T}-\textbf{Ad}\}-\textbf{Q})$$
$$/\{\textbf{S'p}[1+\textbf{s}]\}$$
$$= 360[1.2000-1.0100](0.9900$$
$$\{64,720+41,548-29,750$$
$$-48,851*0.0100[1+0.0500]$$
$$-3,385-7,899*0.3000\}$$
$$-35,351)/\{48,851$$
$$*240[1+0.0500]\}$$
$$= 19.00\%$$

Corporate IFRS-GAAP (B/S-I/S),ISBN-13: **978-1721066308**, ISBN-10: **1721066306**

<u>Law-8375</u>:

If both (**S'**= Sales of Past Year= <u>48,851</u> millions USD), (**A**= After Tax Income= <u>7,899</u> millions USD), (**T**= Tax Paid= <u>3,385</u> millions USD), (**d**= D/A= Dividend Portion or Payout= <u>30.00%</u>), (**s**= [S/S']-1= Sales Growth= <u>5.00%</u>), (**i**= I/S= Interest Portion= <u>1.00%</u>), (**F**= Fixed Cost= <u>29,740</u> millions USD), (**l**= L/E= Leverage or Gearing Ratio= <u>99.00%</u>), (**c**= C/X= Current Ratio= <u>1.20</u> times), (**v**= V/S= Variable Portion= <u>19.00%</u>), (**M**= Margin of Contribution= <u>41,548</u> millions USD), (**Q**= Quoted Longterm Liabilities= <u>35,351</u> millions USD), (**U**= Utilized or Starting Capital= <u>64,720</u> millions USD), and (**q**= Quick or Acid Test Ratio= [C-P]/X= <u>1.01</u> times), then it's (**p**= Procured Inventory Days planned), is:

$$p= 360[c\text{-}q](l\ \{U\text{+}M\text{-}F\text{-}S'i[1\text{+}s]\text{-}T\text{-}Ad\}\text{-}Q)$$
$$/\{S'v[1\text{+}s]\}$$

$$= 360[1.2000\text{-}1.0100](0.9900$$
$$\{64,720\text{+}41,548\text{-}29,750$$
$$\text{-}48,851*0.0100[1\text{+}0.0500]$$
$$\text{-}3,385\text{-}7,899*0.3000\}$$
$$\text{-}35,351)/\{48,851$$
$$*0.1900[1\text{+}0.0500]\}$$

$$= \underline{240}\ \text{days}$$

Corporate IFRS-GAAP (B/S-I/S),ISBN-13: **978-1721066308**, ISBN-10: **1721066306**

<u>Law-8376</u>:

If both (**$'**= Sales of Past Year= <u>48,851</u> millions USD), (**A**= After Tax Income= <u>7,899</u> millions USD), (**T**= Tax Paid= <u>3,385</u> millions USD), (**d**= D/A= Dividend Portion or Payout= <u>30.00%</u>), (**s**= [S/S']-1= Sales Growth= <u>5.00%</u>), (**i**= I/S= Interest Portion= <u>1.00%</u>), (**F**= Fixed Cost= <u>29,740</u> millions USD), (***l***= L/E= Leverage or Gearing Ratio= <u>99.00%</u>), (**p**= 360 P/V= Procured Inventory Days= <u>240</u> Days), (**v**= V/S= Variable Portion= <u>19.00%</u>), (**M**= Margin of Contribution= <u>41,548</u> millions USD), (**Q**= Quoted Longterm Liabilities= <u>35,351</u> millions USD), (**U**= Utilized or Starting Capital= <u>64,720</u> millions USD), and (**q**= [C-P]/X= Quick or Acid Test Ratio= <u>1.01</u> times), then it's (**c**= Current Ratio planned), is:

$$c= q+\text{$'}vp[1+s]/[360(\textit{l} \{U+M-F-\text{$'}i[1+s]$$
$$-T-Ad\}-Q)]$$
$$= 1.0100+48,851*0.1900*240$$
$$[1+0.0500]/[360(0.9900$$
$$\{64,720+41,548-29,750$$
$$-48,851*0.0100[1+0.0500]$$
$$-3,385-7,899*0.3000\}$$
$$-35,351)]$$
$$= \underline{1.20} \text{ times}$$

Law-8377:

If both (**$'**= Sales of Past Year= 48,851 millions USD), (**A**= After Tax Income= 7,899 millions USD), (**T**= Tax Paid= 3,385 millions USD), (**d**= D/A= Dividend Portion or Payout= 30.00%), (**s**= [S/S']-1= Sales Growth= 5.00%), (**i**= I/S= Interest Portion= 1.00%), (**F**= Fixed Cost= 29,740 millions USD), (**I**= L/E= Leverage or Gearing Ratio= 99.00%), (**p**= 360 P/V= Procured Inventory Days= 240 Days), (**v**= V/S= Variable Portion= 19.00%), (**M**= Margin of Contribution= 41,548 millions USD), (**Q**= Quoted Longterm Liabilities= 35,351 millions USD), (**U**= Utilized or Starting Capital= 64,720 millions USD), and (**c**= C/X= Current Ratio= 1.20 times), then it's (**q**= Quick or Acid Test Ratio planned), is:

$$q= c\text{-}\$'vp[1+s]/[360(I \{U+M-F-\$'i[1+s]$$
$$-T\text{-}Ad\}-Q)]$$
$$= 1.2000\text{-}48,851*0.1900*240$$
$$[1+0.0500]/[360(0.9900$$
$$\{64,720+41,548\text{-}29,750$$
$$-48,851*0.0100[1+0.0500]$$
$$-3,385\text{-}7,899*0.3000\}$$
$$-35,351)]$$
$$= 1.01 \text{ times}$$

Corporate IFRS-GAAP (B/S-I/S),ISBN-13: **978-1721066308**, ISBN-10: **1721066306**

Law-8378:

If both (**$'**= Sales of Past Year= 48,851 millions USD), (**t**= T/B= Tax Rate= 30.00%), (**D**= Dividend Paid= 2,370 millions USD), (**i**= I/S= Interest Portion= 1.00%), (**F**= Fixed Cost= 29,740 millions USD), (**I**= L/E= Leverage or Gearing Ratio= 99.00%), (**M**= Margin of Contribution= 41,548 millions USD), (**X**= Xpress or Current Debt= 34,196 millions USD), (**U**= Utilized or Starting Capital= 64,720 millions USD), and (**s**= [S/S']-1= Sales Growth= 5.00%), then it's (**Q**= Quoted Longterm Liabilities planned), is:

$$Q = I\,(U + \{M - F - S'i[1+s][1-t] - D) - X$$
$$= 0.9900(64{,}720 + \{41{,}548 - 29{,}750$$
$$-48{,}851 * 0.0100[1 + 0.0500]$$
$$[1 - 0.3000] - 2{,}370) - 34{,}196$$
$$= \underline{35{,}351} \text{ millions USD}$$

Corporate IFRS-GAAP (B/S-I/S),ISBN-13: **978-1721066308**, ISBN-10: **1721066306**

Law-8379:

If both ($\textbf{S'}$= Sales of Past Year= 48,851 millions USD), ($\textbf{t}$= T/B= Tax Rate= 30.00%), ($\textbf{D}$= Dividend Paid= 2,370 millions USD), ($\textbf{i}$= I/S= Interest Portion= 1.00%), ($\textbf{F}$= Fixed Cost= 29,740 millions USD), ($\textbf{Q}$= Quoted Longterm Liabilities= 35,351 millions USD), ($\textbf{M}$= Margin of Contribution= 41,548 millions USD), ($\textbf{X}$= Xpress or Current Debt= 34,196 millions USD), ($\textbf{U}$= Utilized or Starting Capital= 64,720 millions USD), and ($\textbf{s}$= [S/S']-1= Sales Growth= 5.00%), then it's ($\textbf{I}$= Leverage or Gearing Ratio planned), is:

$$\textbf{I} = [\textbf{Q}+\textbf{X}]/(\textbf{U}+\{\textbf{M}-\textbf{F}-\textbf{S'i}[1+\textbf{s}][1-\textbf{t}]-\textbf{D})$$
$$= [35,351+34,196]/(64,720+\{41,548$$
$$-29,750-48,851*0.0100$$
$$[1+0.0500][1-0.3000]-2,370)$$
$$= 99.00\%$$

Corporate IFRS-GAAP (B/S-I/S),ISBN-13: **978-1721066308**, ISBN-10: **1721066306**

Law-8380:

If both ($**S'**$= Sales of Past Year= 48,851 millions USD), ($**t**$= T/B= Tax Rate= 30.00%), ($**D**$= Dividend Paid= 2,370 millions USD), ($**i**$= I/S= Interest Portion= 1.00%), ($**F**$= Fixed Cost= 29,740 millions USD), ($**Q**$= Quoted Longterm Liabilities= 35,351 millions USD), ($**M**$= Margin of Contribution= 41,548 millions USD), ($**X**$= Xpress or Current Debt= 34,196 millions USD), ($**I**$= L/E= Leverage or Gearing Ratio= 99.00%), and ($**s**$= [S/S']-1= Sales Growth= 5.00%), then it's ($**U**$= Utilized or Starting Capital planned), is:

$$U= D+[Q+X]/I -\{M-F-S'i[1+s][1-t]$$

$$= 2,370+[35,351+34,196]/0.9900$$
$$-\{41,548-29,750-48,851$$
$$*0.0100[1+0.0500]$$
$$[1-0.3000]$$

$$= 64,720 \text{ millions USD}$$

Corporate IFRS-GAAP (B/S-I/S),ISBN-13: **978-1721066308**, ISBN-10: **1721066306**

Law-8381:

If both (**$'**= Sales of Past Year= 48,851 millions USD), (**t**= T/B= Tax Rate= 30.00%), (**D**= Dividend Paid= 2,370 millions USD), (**i**= I/S= Interest Portion= 1.00%), (**F**= Fixed Cost= 29,740 millions USD), (**Q**= Quoted Longterm Liabilities= 35,351 millions USD), (**U**= Utilized or Starting Capital= 64,720 millions USD), (**X**= Xpress or Current Debt= 34,196 millions USD), (**I**= L/E= Leverage or Gearing Ratio= 99.00%), and (**s**= [S/S']-1= Sales Growth= 5.00%), then it's (**M**= Margin of Contribution planned), is:

$$M= F+S'i[1+s]+\{D+[Q+X]/I-U\}/[1-t]$$
$$= 29,750+48,851*0.0100[1+0.0500]$$
$$+\{2,370+[35,351+34,196]$$
$$/0.9900-64,720\}/[1-0.3000]$$
$$= 41,548 \text{ millions USD}$$

Corporate IFRS-GAAP (B/S-I/S),ISBN-13: **978-1721066308**, ISBN-10: **1721066306**

Law-8382:

If both (**S'**= Sales of Past Year= 48,851 millions
USD), (**t**= T/B= Tax Rate= 30.00%), (**D**= Dividend
Paid= 2,370 millions USD), (**i**= I/S= Interest Portion=
1.00%), (**M**= Margin of Contribution= 41,548
millions USD), (**Q**= Quoted Longterm Liabilities=
35,351 millions USD), (**U**= Utilized or Starting
Capital= 64,720 millions USD), (**X**= Xpress or
Current Debt= 34,196 millions USD), (**I**= L/E=
Leverage or Gearing Ratio= 99.00%), and (**s**= [S/S']-
1= Sales Growth= 5.00%), then it's (**F**= Fixed Cost
planned), is:

$$\textbf{F= M-S'i}[1+\textbf{s}]-(\textbf{D}+\{[\textbf{Q}+\textbf{X}]/\textbf{I}-\textbf{U}\})/[1-\textbf{t}]$$
$$= 41,548-48,851*0.0100[1-0.0500]$$
$$-(2,370+\{[35,351+34,196]$$
$$/0.9900-64,720\})/[1-0.3000]$$
$$= 29,750 \text{ millions USD}$$

Corporate IFRS-GAAP (B/S-I/S),ISBN-13: **978-1721066308**, ISBN-10: **1721066306**

<u>Law-8383</u>:

If both (**F**= Fixed Cost= <u>29,740</u> millions USD), (**t**= T/B= Tax Rate= <u>30.00%</u>), (**D**= Dividend Paid= <u>2,370</u> millions USD), (**i**= I/S= Interest Portion= <u>1.00%</u>), (**M**= Margin of Contribution= <u>41,548</u> millions USD), (**Q**= Quoted Longterm Liabilities= <u>35,351</u> millions USD), (**U**= Utilized or Starting Capital= <u>64,720</u> millions USD), (**X**= Xpress or Current Debt= <u>34,196</u> millions USD), (**l**= L/E= Leverage or Gearing Ratio= <u>99.00%</u>), and (**s**= [S/S']-1= Sales Growth= <u>5.00%</u>), then it's (**S'**= Sales Past), must be:

$$S'= [M-F-(D+\{[Q+X]/l-U\})/[1-t]]/\{i[1+s]\}$$
$$= [41,548-29,750-(2,370+\{[35,351+34,196]/0.9900-64,720\})$$
$$/[1-0.3000]]$$
$$/\{0.0100[1-0.0500]\}$$
$$= \underline{48,851} \text{ millions USD}$$

Corporate IFRS-GAAP (B/S-I/S),ISBN-13: **978-1721066308**, ISBN-10: **1721066306**

<u>Law-8384</u>:

If both (**F**= Fixed Cost= <u>29,740</u> millions USD), (**t**= T/B= Tax Rate= <u>30.00%</u>), (**D**= Dividend Paid= <u>2,370</u> millions USD), (**S'**= Sales of Past Year= <u>48,851</u> millions USD), (**M**= Margin of Contribution= <u>41,548</u> millions USD), (**Q**= Quoted Longterm Liabilities= <u>35,351</u> millions USD), (**U**= Utilized or Starting Capital= <u>64,720</u> millions USD), (**X**= Xpress or Current Debt= <u>34,196</u> millions USD), (**I**= L/E= Leverage or Gearing Ratio= <u>99.00%</u>), and (**s**= [S/S']-1= Sales Growth= <u>5.00%</u>), then it's (**i**= Interest Portion planned), is:

$$\textbf{i} = \textbf{[M-F-(D}+\{[\textbf{Q}+\textbf{X}]/\textbf{I}-\textbf{U}\})/[1-\textbf{t}]\textbf{]}/\{\textbf{S'}[1+\textbf{s}]\}$$

$$= \textbf{[}41,548-29,750-(2,370+\{[35,351$$
$$+34,196]/0.9900-64,720\})$$
$$/[1-0.3000]\textbf{]}/\{48,851$$
$$[1-0.0500]\}$$

$$= \underline{1.00\%}$$

Corporate IFRS-GAAP (B/S-I/S),ISBN-13: **978-1721066308**, ISBN-10: **1721066306**

Law-8385:

If both (**F**= Fixed Cost= 29,740 millions USD), (**t**= T/B= Tax Rate= 30.00%), (**D**= Dividend Paid= 2,370 millions USD), (**S'**= Sales of Past Year= 48,851 millions USD), (**M**= Margin of Contribution= 41,548 millions USD), (**Q**= Quoted Longterm Liabilities= 35,351 millions USD), (**U**= Utilized or Starting Capital= 64,720 millions USD), (**X**= Xpress or Current Debt= 34,196 millions USD), (**I**= L/E= Leverage or Gearing Ratio= 99.00%), and (**i**= I/S= Interest Portion= 1.00%), then it's (**s**= Sales Growth planned), is:

$$s= [M-F-(D+\{[Q+X]/I-U\})/[1-t]]/[S'i]-1$$
$$= [41,548-29,750-(2,370+\{[35,351$$
$$+34,196]/0.9900-64,720\})$$
$$/[1-0.3000]]/[48,851$$
$$*0.0100]-1$$
$$= 5.00\%$$

Corporate IFRS-GAAP (B/S-I/S),ISBN-13: **978-1721066308**, ISBN-10: **1721066306**

<u>Law-8386</u>:

If both (**F**= Fixed Cost= <u>29,740</u> millions USD), (**s**= [S/S']-1= Sales Growth= <u>5.00%</u>), (**D**= Dividend Paid= <u>2,370</u> millions USD), (**S'**= Sales of Past Year= <u>48,851</u> millions USD), (**M**= Margin of Contribution= <u>41,548</u> millions USD), (**Q**= Quoted Longterm Liabilities= <u>35,351</u> millions USD), (**U**= Utilized or Starting Capital= <u>64,720</u> millions USD), (**X**= Xpress or Current Debt= <u>34,196</u> millions USD), (**I**= L/E= Leverage or Gearing Ratio= <u>99.00%</u>), and (**i**= I/S= Interest Portion= <u>1.00%</u>), then it's (**t**= Tax Rate planned), is:

$$t= 1-(D+\{[Q+X]/I-U\})/\{M-F-S'i[1+s]\}$$
$$= 1-(2,370+\{[35,351+34,196]/0.9900$$
$$-64,720\})/\{41,548-29,750$$
$$-48,851*0.0100[1+0.0500]\}$$
$$= 30.00\%$$

Corporate IFRS-GAAP (B/S-I/S),ISBN-13: **978-1721066308**, ISBN-10: **1721066306**

<u>Law-8387</u>:

If both (**F**= Fixed Cost= <u>29,740</u> millions USD), (**s**= [S/S']-1= Sales Growth= <u>5.00%</u>), (**t**= T/B= Tax Rate= <u>30.00%</u>), (**S'**= Sales of Past Year= <u>48,851</u> millions USD), (**M**= Margin of Contribution= <u>41,548</u> millions USD), (**Q**= Quoted Longterm Liabilities= <u>35,351</u> millions USD), (**U**= Utilized or Starting Capital= <u>64,720</u> millions USD), (**X**= Xpress or Current Debt= <u>34,196</u> millions USD), (**I**= L/E= Leverage or Gearing Ratio= <u>99.00%</u>), and (**i**= I/S= Interest Portion= <u>1.00%</u>), then it's (**D**= Dividend Paid planned), is:

$$D= U+\{M-F-S'i[1+s]\}[1-t]-[Q+X]/I$$
$$= 64,720+\{41,548-29,750-48,851$$
$$*0.0100[1+0.0500]\}$$
$$[1-0.3000]-[35,351$$
$$+34,196]/0.9900$$
$$= 2,370 \text{ millions USD}$$

Corporate IFRS-GAAP (B/S-I/S),ISBN-13: **978-1721066308**, ISBN-10: **1721066306**

Law-8388:

If both (**F**= Fixed Cost= 29,740 millions USD), (**s**= [S/S']-1= Sales Growth= 5.00%), (**t**= T/B= Tax Rate= 30.00%), (**S'**= Sales of Past Year= 48,851 millions USD), (**M**= Margin of Contribution= 41,548 millions USD), (**Q**= Quoted Longterm Liabilities= 35,351 millions USD), (**U**= Utilized or Starting Capital= 64,720 millions USD), (**D**= Dividend Paid= 2,370 millions USD), (**l**= L/E= Leverage or Gearing Ratio= 99.00%), and (**i**= I/S= Interest Portion= 1.00%), then it's (**X**= Xpress or Current Debt planned), is:

$$\mathbf{X} = \mathbf{\mathit{l}}\,(\mathbf{U} + \{\mathbf{M} \text{-} \mathbf{F} \text{-} \mathbf{S'i}[1 + \mathbf{s}]\}[1 \text{-} \mathbf{t}] \text{-} \mathbf{D}\} \text{-} \mathbf{Q}$$

$$= 0.9900(64{,}720 + \{41{,}548 \text{-} 29{,}750$$
$$\text{-}48{,}851 * 0.0100[1 + 0.0500]\}$$
$$[1\text{-}0.3000]\text{-}2{,}370\} \text{-} 35{,}351$$

$$= 34{,}196 \text{ millions USD}$$

Corporate IFRS-GAAP (B/S-I/S),ISBN-13: **978-1721066308**, ISBN-10: **1721066306**

Law-8389:

If both (**F**= Fixed Cost= 29,740 millions USD), (**s**= [S/S']-1= Sales Growth= 5.00%), (**t**= T/B= Tax Rate= 30.00%), (**S'**= Sales of Past Year= 48,851 millions USD), (**M**= Margin of Contribution= 41,548 millions USD), (**c**= C/X= Current Ratio= 1.20 times), (**q**= [C-P]/X= Quick or Acid Test Ratio= 1.01 times), (**P**= Procured Inventories= 6,497 millions USD), (**U**= Utilized or Starting Capital= 64,720 millions USD), (**D**= Dividend Paid= 2,370 millions USD), (**l**= L/E= Leverage or Gearing Ratio= 99.00%), and (**i**= I/S= Interest Portion= 1.00%), then it's (**Q**= Quoted Longterm Liabilities planned), is:

$$Q = l(U+\{M-F-S'i[1+s]\}[1-t]-D\}-P/[c-q]$$
$$= 0.9900(64,720+\{41,548-29,750$$
$$-48,851*0.0100[1+0.0500]\}$$
$$[1-0.3000]-2,370\}-6,497$$
$$/[1.2000-1.0100]$$
$$= 35,351 \text{ millions USD}$$

Law-8390:

If both (**F**= Fixed Cost= 29,740 millions USD), (**s**= [S/S']-1= Sales Growth= 5.00%), (**t**= T/B= Tax Rate= 30.00%), (**S'**= Sales of Past Year= 48,851 millions USD), (**M**= Margin of Contribution= 41,548 millions USD), (**c**= C/X= Current Ratio= 1.20 times), (**q**= [C-P]/X= Quick or Acid Test Ratio= 1.01 times), (**P**= Procured Inventories= 6,497 millions USD), (**U**= Utilized or Starting Capital= 64,720 millions USD), (**D**= Dividend Paid= 2,370 millions USD), (**Q**= Quoted Longterm Liabilities= 35,351 millions USD), and (**i**= I/S= Interest Portion= 1.00%), then it's (**I** = Leverage or Gearing Ratio planned), is:

$$I = \{Q+P/[c\text{-}q]\}/(U+\{M\text{-}F\text{-}S'i[1+s]\}[1\text{-}t]\text{-}D\}$$
$$= \{35,351+6,497/[1.2000\text{-}1.0100]\}$$
$$/(64,720+\{41,548\text{-}29,750$$
$$-48,851*0.0100[1+0.0500]\}$$
$$[1\text{-}0.3000]\text{-}2,370\}$$
$$= 99.00\%$$

Corporate IFRS-GAAP (B/S-I/S),ISBN-13: **978-1721066308**, ISBN-10: **1721066306**

<u>Law-8391</u>:

If both (**F**= Fixed Cost= <u>29,740</u> millions USD), (**s**= [S/S']-1= Sales Growth= <u>5.00%</u>), (**t**= T/B= Tax Rate= <u>30.00%</u>), (**S'**= Sales of Past Year= <u>48,851</u> millions USD), (**M**= Margin of Contribution= <u>41,548</u> millions USD), (**c**= C/X= Current Ratio= <u>1.20</u> times), (**q**= [C-P]/X= Quick or Acid Test Ratio= <u>1.01</u> times), (**P**= Procured Inventories= <u>6,497</u> millions USD), (**F**= L/E= Leverage or Gearing Ratio= <u>99.00%</u>), (**D**= Dividend Paid= <u>2,370</u> millions USD), (**Q**= Quoted Longterm Liabilities= <u>35,351</u> millions USD), and (**i**= I/S= Interest Portion= <u>1.00%</u>), then it's (**U**= Utilized or Starting Capital planned), is:

$$U= D+\{Q+P/[c\text{-}q]\}/\textit{I} -\{M\text{-}F\text{-}S'i[1+s]\}[1\text{-}t]$$
$$= 2,370+\{35,351+6,497/[1.2000$$
$$-1.0100]\}/0.9900-\{41,548$$
$$-29,750-48,851*0.0100$$
$$[1+0.0500]\}[1-0.3000]$$
$$= \underline{64,720} \text{ millions USD}$$

Corporate IFRS-GAAP (B/S-I/S),ISBN-13: **978-1721066308**, ISBN-10: **1721066306**

Law-8392:

If both (**F**= Fixed Cost= <u>29,740</u> millions USD), (**s**= [S/S']-1= Sales Growth= <u>5.00%</u>), (**t**= T/B= Tax Rate= <u>30.00%</u>), (**S'**= Sales of Past Year= <u>48,851</u> millions USD), (**U**= Utilized or Starting Capital= <u>64,720</u> millions USD), (**c**= C/X= Current Ratio= <u>1.20</u> times), (**q**= [C-P]/X= Quick or Acid Test Ratio= <u>1.01</u> times), (**P**= Procured Inventories= <u>6,497</u> millions USD), (**l**= L/E= Leverage or Gearing Ratio= <u>99.00%</u>), (**D**= Dividend Paid= <u>2,370</u> millions USD), (**Q**= Quoted Longterm Liabilities= <u>35,351</u> millions USD), and (**i**= I/S= Interest Portion= <u>1.00%</u>), then it's (**M**= Margin of Contribution planned), is:

$$M= F+S'i[1+s]+(D+\{Q+P/[c-q]\}/l\ -U)/[1-t]$$
$$= 29,750+48,851*0.0100[1+0.0500]$$
$$+(2,370+\{35,351+6,497$$
$$/[1.2000-1.0100]\}/0.9900$$
$$-64,720)/[1-0.3000]$$
$$= \underline{41,548} \text{ millions USD}$$

Corporate IFRS-GAAP (B/S-I/S),ISBN-13: **978-1721066308**, ISBN-10: **1721066306**

Law-8393:

If both (**M**= Margin of Contribution= 41,548 millions USD), (**s**= [S/S']-1= Sales Growth= 5.00%), (**t**= T/B= Tax Rate= 30.00%), (**S'**= Sales of Past Year= 48,851 millions USD), (**U**= Utilized or Starting Capital= 64,720 millions USD), (**c**= C/X= Current Ratio= 1.20 times), (**q**= [C-P]/X= Quick or Acid Test Ratio= 1.01 times), (**P**= Procured Inventories= 6,497 millions USD), (**I**= L/E= Leverage or Gearing Ratio= 99.00%), (**D**= Dividend Paid= 2,370 millions USD), (**Q**= Quoted Longterm Liabilities= 35,351 millions USD), and (**i**= I/S= Interest Portion= 1.00%), then it's (**F**= Fixed Cost planned), is:

$$F= M-S'i[1+s]-(D+\{Q+P/[c-q]\}/I-U)/[1-t]$$
$$= 41,548-48,851*0.0100[1+0.0500]$$
$$-(2,370+\{35,351+6,497$$
$$/[1.2000-1.0100]\}/0.9900$$
$$-64,720)/[1-0.3000]$$
$$= 29,750 \text{ millions USD}$$

Corporate IFRS-GAAP (B/S-I/S),ISBN-13: **978-1721066308**, ISBN-10: **1721066306**

<u>Law-8394</u>:

If both (**M**= Margin of Contribution= <u>41,548</u> millions USD), (**s**= [S/S']-1= Sales Growth= <u>5.00%</u>), (**t**= T/B= Tax Rate= <u>30.00%</u>), (**F**= Fixed Cost= <u>29,740</u> millions USD), (**U**= Utilized or Starting Capital= <u>64,720</u> millions USD), (**c**= C/X= Current Ratio= <u>1.20</u> times), (**q**= [C-P]/X= Quick or Acid Test Ratio= <u>1.01</u> times), (**P**= Procured Inventories= <u>6,497</u> millions USD), (**l**= L/E= Leverage or Gearing Ratio= <u>99.00%</u>), (**D**= Dividend Paid= <u>2,370</u> millions USD), (**Q**= Quoted Longterm Liabilities= <u>35,351</u> millions USD), and (**i**= I/S= Interest Portion= <u>1.00%</u>), then it's (**S'**= Sales Past), must be:

$$S' = [M-F-(D+\{Q+P/[c-q]\}/l\,-U)/[1-t]]$$
$$/\{i[1+s]\}$$
$$= [41,548-29,750-(2,370+\{35,351$$
$$+6,497/[1.2000-1.0100]\}$$
$$/0.9900-64,720)/[1-0.3000]]$$
$$/\{0.0100[1+0.0500]\}$$
$$= \underline{48,851} \text{ millions USD}$$

Corporate IFRS-GAAP (B/S-I/S),ISBN-13: **978-1721066308**, ISBN-10: **1721066306**

Law-8395:

If both (**M**= Margin of Contribution= 41,548 millions USD), (**s**= [S/S']-1= Sales Growth= 5.00%), (**t**= T/B= Tax Rate= 30.00%), (**F**= Fixed Cost= 29,740 millions USD), (**U**= Utilized or Starting Capital= 64,720 millions USD), (**c**= C/X= Current Ratio= 1.20 times), (**q**= [C-P]/X= Quick or Acid Test Ratio= 1.01 times), (**P**= Procured Inventories= 6,497 millions USD), (**I**= L/E= Leverage or Gearing Ratio= 99.00%), (**D**= Dividend Paid= 2,370 millions USD), (**Q**= Quoted Longterm Liabilities= 35,351 millions USD), and (**S'**= Sales of Past Year= 48,851 millions USD), then it's (**i**= Interest Portion planned), is:

$$i= [M-F-(D+\{Q+P/[c-q]\}/I-U)/[1-t]]$$
$$/\{S'[1+s]\}$$
$$= [41,548-29,750-(2,370+\{35,351$$
$$+6,497/[1.2000-1.0100]\}$$
$$/0.9900-64,720)/[1-0.3000]]$$
$$/\{48,851[1+0.0500]\}$$
$$= \underline{1.00\%}$$

Corporate IFRS-GAAP (B/S-I/S),ISBN-13: **978-1721066308**, ISBN-10: **1721066306**

Law-8396:

If both (**M**= Margin of Contribution= 41,548 millions USD), (**i**= I/S= Interest Portion= 1.00%), (**t**= T/B= Tax Rate= 30.00%), (**F**= Fixed Cost= 29,740 millions USD), (**U**= Utilized or Starting Capital= 64,720 millions USD), (**c**= C/X= Current Ratio= 1.20 times), (**q**= [C-P]/X= Quick or Acid Test Ratio= 1.01 times), (**P**= Procured Inventories= 6,497 millions USD), (**l**= L/E= Leverage or Gearing Ratio= 99.00%), (**D**= Dividend Paid= 2,370 millions USD), (**Q**= Quoted Longterm Liabilities= 35,351 millions USD), and (**$'**= Sales of Past Year= 48,851 millions USD), then it's (**$**= Sales Growth planned), is:

$$s= [M\text{-}F\text{-}(D+\{Q+P/[c\text{-}q]\}/l\text{-}U)/[1\text{-}t]]$$
$$/[\$'i]\text{-}1$$
$$= [41,548\text{-}29,750\text{-}(2,370+\{35,351$$
$$+6,497/[1.2000\text{-}1.0100]\}$$
$$/0.9900\text{-}64,720)/[1\text{-}0.3000]]$$
$$/[48,851*0.0100]\text{-}1$$
$$= \underline{5.00\%}$$

Corporate IFRS-GAAP (B/S-I/S),ISBN-13: **978-1721066308**, ISBN-10: **1721066306**

Law-8397:

If both (**M**= Margin of Contribution= 41,548 millions USD), (**i**= I/S= Interest Portion= 1.00%), (**s**= [S/S']-1= Sales Growth= 5.00%), (**F**= Fixed Cost= 29,740 millions USD), (**U**= Utilized or Starting Capital= 64,720 millions USD), (**c**= C/X= Current Ratio= 1.20 times), (**q**= [C-P]/X= Quick or Acid Test Ratio= 1.01 times), (**P**= Procured Inventories= 6,497 millions USD), (**I**= L/E= Leverage or Gearing Ratio= 99.00%), (**D**= Dividend Paid= 2,370 millions USD), (**Q**= Quoted Longterm Liabilities= 35,351 millions USD), and (**S'**= Sales of Past Year= 48,851 millions USD), then it's (**t**= Tax Rate planned), is:

$$t= 1-[(D+\{Q+P/[c-q]\}/I-U)]/\{M-F-S'i[1+s]\}$$

$$= 1-[(2,370+\{35,351+6,497/[1.2000$$
$$-1.0100]\}/0.9900-64,720)]$$
$$/\{41,548-29,750-48,851$$
$$*0.0100[1+0.0500]\}$$

$$= 30.00\%$$

Corporate IFRS-GAAP (B/S-I/S),ISBN-13: **978-1721066308**, ISBN-10: **1721066306**

Law-8398:

If both (**M**= Margin of Contribution= 41,548 millions USD), (**i**= I/S= Interest Portion= 1.00%), (**s**= [S/S']-1= Sales Growth= 5.00%), (**F**= Fixed Cost= 29,740 millions USD), (**U**= Utilized or Starting Capital= 64,720 millions USD), (**c**= C/X= Current Ratio= 1.20 times), (**q**= [C-P]/X= Quick or Acid Test Ratio= 1.01 times), (**P**= Procured Inventories= 6,497 millions USD), (**l**= L/E= Leverage or Gearing Ratio= 99.00%), (**t**= T/B= Tax Rate= 30.00%), (**Q**= Quoted Longterm Liabilities= 35,351 millions USD), and (**S'**= Sales of Past Year= 48,851 millions USD), then it's (**D**= Dividend Paid planned), is:

$$\mathbf{D}= \mathbf{U}+\{\mathbf{M}\text{-}\mathbf{F}\text{-}\mathbf{S'i}[1+\mathbf{s}]\}[1\text{-}\mathbf{t}]\text{-}\{\mathbf{Q}+\mathbf{P}/[\mathbf{c}\text{-}\mathbf{q}]\}/\mathbf{l}$$

$$= 64,720+\{41,548\text{-}29,750\text{-}48,851$$
$$*0.0100[1+0.0500]\}$$
$$[1\text{-}0.3000]\}\text{-}\{35,351+6,497$$
$$/[1.2000\text{-}1.0100]\}/0.9900$$

$$= \underline{2,370} \text{ millions USD}$$

Corporate IFRS-GAAP (B/S-I/S),ISBN-13: **978-1721066308**, ISBN-10: **1721066306**

<u>Law-8399</u>:

If both (**M**= Margin of Contribution= <u>41,548</u> millions USD), (**i**= I/S= Interest Portion= <u>1.00%</u>), (**s**= [S/S']-1= Sales Growth= <u>5.00%</u>), (**F**= Fixed Cost= <u>29,740</u> millions USD), (**U**= Utilized or Starting Capital= <u>64,720</u> millions USD), (**c**= C/X= Current Ratio= <u>1.20</u> times), (**q**= [C-P]/X= Quick or Acid Test Ratio= <u>1.01</u> times), (**D**= Dividend Paid= <u>2,370</u> millions USD), (**l**= L/E= Leverage or Gearing Ratio= <u>99.00%</u>), (**t**= T/B= Tax Rate= <u>30.00%</u>), (**Q**= Quoted Longterm Liabilities= <u>35,351</u> millions USD), and (**S'**= Sales of Past Year= <u>48,851</u> millions USD), then it's (**P**= Procured Inventories planned), is:

$$P= [c\text{-}q][\textit{l}\,(U+\{M\text{-}F\text{-}S'i[1+s]\}[1\text{-}t]\text{-}D)\text{-}Q]$$

$$= [1.2000\text{-}1.0100][0.9900(64,720$$
$$+\{41,548\text{-}29,750\text{-}48,851$$
$$*0.0100[1+0.0500]$$
$$[1\text{-}0.3000]\}\text{-}2,370)\text{-}35,351]$$

$$= \underline{6,497} \text{ millions USD}$$

Corporate IFRS-GAAP (B/S-I/S),ISBN-13: **978-1721066308**, ISBN-10: **1721066306**

Law-8400:

If both (**M**= Margin of Contribution= 41,548 millions USD), (**i**= I/S= Interest Portion= 1.00%), (**s**= [S/S']-1= Sales Growth= 5.00%), (**F**= Fixed Cost= 29,740 millions USD), (**U**= Utilized or Starting Capital= 64,720 millions USD), (**P**= Procured Inventories= 6,497 millions USD), (**q**= [C-P]/X= Quick or Acid Test Ratio= 1.01 times), (**D**= Dividend Paid= 2,370 millions USD), (**l**= L/E= Leverage or Gearing Ratio= 99.00%), (**t**= T/B= Tax Rate= 30.00%), (**Q**= Quoted Longterm Liabilities= 35,351 millions USD), and (**S'**= Sales of Past Year= 48,851 millions USD), then it's (**c**= Current Ratio planned), is:

$$c= q+P/[\textbf{\textit{l}}\,(U+\{M\text{-}F\text{-}S'i[1+s]\}[1\text{-}t]\text{-}D)\text{-}Q]$$

$$
\begin{aligned}
&= 1.0100+6,497/[0.9900(64,720 \\
&\quad +\{41,548\text{-}29,750\text{-}48,851 \\
&\quad *0.0100[1+0.0500] \\
&\quad [1\text{-}0.3000]\}\text{-}2,370)\text{-}35,351]
\end{aligned}
$$

$$= 1.20 \text{ times}$$

Corporate IFRS-GAAP (B/S-I/S),ISBN-13: **978-1721066308**, ISBN-10: **1721066306**

Law-8401:

If both (**M**= Margin of Contribution= 41,548 millions USD), (**i**= I/S= Interest Portion= 1.00%), (**s**= [S/S']-1= Sales Growth= 5.00%), (**F**= Fixed Cost= 29,740 millions USD), (**U**= Utilized or Starting Capital= 64,720 millions USD), (**P**= Procured Inventories= 6,497 millions USD), (**c**= C/X= Current Ratio= 1.20 times), (**D**= Dividend Paid= 2,370 millions USD), (**l**= L/E= Leverage or Gearing Ratio= 99.00%), (**t**= T/B= Tax Rate= 30.00%), (**Q**= Quoted Longterm Liabilities= 35,351 millions USD), and (**S'**= Sales of Past Year= 48,851 millions USD), then it's (**q**= Quick or Acid Test Ratio planned), is:

$$q = c\text{-}P/[\textbf{\textit{l}}\,(U+\{M\text{-}F\text{-}S'i[1+s]\}[1\text{-}t]\text{-}D)\text{-}Q]$$

$$= 1.2000\text{-}6,497/[0.9900(64,720$$
$$+\{41,548\text{-}29,750\text{-}48,851$$
$$*0.0100[1+0.0500]$$
$$[1\text{-}0.3000]\}\text{-}2,370)\text{-}35,351]$$

$$= 1.01 \text{ times}$$

Corporate IFRS-GAAP (B/S-I/S),ISBN-13: **978-1721066308**, ISBN-10: **1721066306**

Law-8402:

If both (**M**= Margin of Contribution= 41,548 millions USD), (**i**= I/S= Interest Portion= 1.00%), (**s**= [S/S']-1= Sales Growth= 5.00%), (**F**= Fixed Cost= 29,740 millions USD), (**U**= Utilized or Starting Capital= 64,720 millions USD), (**p**= 360 P/V= Procured Inventory Days= 240 Days), (**V**= Variable Cost= 9,746 millions USD), (**c**= C/X= Current Ratio= 1.20 times), (**D**= Dividend Paid= 2,370 millions USD), (**l**= L/E= Leverage or Gearing Ratio= 99.00%), (**t**= T/B= Tax Rate= 30.00%), (**q**= [C-P]/X= Quick or Acid Test Ratio= 1.01 times), and (**S'**= Sales of Past Year= 48,851 millions USD), then it's (**Q**= Quoted Longterm Liabilities planned), is:

$$Q = l(U+\{M-F-S'i[1+s]\}[1-t]-D)$$
$$-Vp/\{360[c-q]\}$$
$$= 0.9900(64,720+\{41,548-29,750$$
$$-48,851*0.0100[1+0.0500]$$
$$[1-0.3000]\}-2,370)-9,746$$
$$*240/\{360[1.2000-1.0100]\}$$
$$= 35,351 \text{ millions USD}$$

Corporate IFRS-GAAP (B/S-I/S),ISBN-13: **978-1721066308**, ISBN-10: **1721066306**

Law-8403:

If both (**M**= Margin of Contribution= 41,548 millions USD), (**i**= I/S= Interest Portion= 1.00%), (**s**= [S/S']-1= Sales Growth= 5.00%), (**F**= Fixed Cost= 29,740 millions USD), (**U**= Utilized or Starting Capital= 64,720 millions USD), (**p**= 360 P/V= Procured Inventory Days= 240 Days), (**V**= Variable Cost= 9,746 millions USD), (**c**= C/X= Current Ratio= 1.20 times), (**D**= Dividend Paid= 2,370 millions USD), (**Q**= Quoted Longterm Liabilities= 35,351 millions USD), (**t**= T/B= Tax Rate= 30.00%), (**q**= [C-P]/X= Quick or Acid Test Ratio= 1.01 times), and (**S'**= Sales of Past Year= 48,851 millions USD), then it's (**I** = Leverage or Gearing Ratio planned), is:

$$I = (\mathbf{Q}+\mathbf{Vp}/\{360[\mathbf{c}\text{-}\mathbf{q}]\})/(\mathbf{U}+\{\mathbf{M}\text{-}\mathbf{F}\text{-}\mathbf{S'i}[1+\mathbf{s}]\}$$
$$[1\text{-}\mathbf{t}]\text{-}\mathbf{D})$$
$$= (35,351+9,746*240/\{360[1.2000$$
$$-1.0100]\})/(64,720+\{41,548$$
$$-29,750-48,851*0.0100$$
$$[1+0.0500][1-0.3000]\}$$
$$-2,370)$$

$$= \underline{99.00\%}$$

Corporate IFRS-GAAP (B/S-I/S),ISBN-13: **978-1721066308**, ISBN-10: **1721066306**

Law-8404:

If both (**M**= Margin of Contribution= <u>41,548</u> millions USD), (**i**= I/S= Interest Portion= <u>1.00%</u>), (**s**= [S/S']-1= Sales Growth= <u>5.00%</u>), (**F**= Fixed Cost= <u>29,740</u> millions USD), (**Q**= Quoted Longterm Liabilities= <u>35,351</u> millions USD), (**p**= 360 P/V= Procured Inventory Days= <u>240</u> Days), (**V**= Variable Cost= <u>9,746</u> millions USD), (**c**= C/X= Current Ratio= <u>1.20</u> times), (**D**= Dividend Paid= <u>2,370</u> millions USD), (**I**= L/E= Leverage or Gearing Ratio= <u>99.00%</u>), (**t**= T/B= Tax Rate= <u>30.00%</u>), (**q**= [C-P]/X= Quick or Acid Test Ratio= <u>1.01</u> times), and (**S'**= Sales of Past Year= <u>48,851</u> millions USD), then it's (**U**= Utilized or Starting Capital planned), is:

$$U= D+(Q+Vp/\{360[c-q]\})/I$$
$$-\{M-F-S'i[1+s]\}[1-t]$$
$$= 2,370+(35,351+9,746*240/\{360$$
$$[1.2000-1.0100]\})/0.9900$$
$$-\{41,548-29,750-48,851$$
$$*0.0100[1+0.0500]\}$$
$$[1-0.3000]$$
$$= \underline{64,720} \text{ millions USD}$$

Corporate IFRS-GAAP (B/S-I/S),ISBN-13: **978-1721066308**, ISBN-10: **1721066306**

<u>Law-8405</u>:

If both (**U**= Utilized or Starting Capital= <u>64,720</u> millions USD), (**i**= I/S= Interest Portion= <u>1.00%</u>), (**s**= [S/S']-1= Sales Growth= <u>5.00%</u>), (**F**= Fixed Cost= <u>29,740</u> millions USD), (**Q**= Quoted Longterm Liabilities= <u>35,351</u> millions USD), (**p**= 360 P/V= Procured Inventory Days= <u>240</u> Days), (**V**= Variable Cost= <u>9,746</u> millions USD), (**c**= C/X= Current Ratio= <u>1.20</u> times), (**D**= Dividend Paid= <u>2,370</u> millions USD), (**l**= L/E= Leverage or Gearing Ratio= <u>99.00%</u>), (**t**= T/B= Tax Rate= <u>30.00%</u>), (**q**= [C-P]/X= Quick or Acid Test Ratio= <u>1.01</u> times), and (**S'**= Sales of Past Year= <u>48,851</u> millions USD), then it's (**M**= Margin of Contribution planned), is:

$$M = F + S'i[1+s] + [D + (Q + Vp/\{360[c-q]\})/l - U] \; / [1-t]$$

$$= 29,750 + 48,851*0.0100[1+0.0500] + [2,370 + (35,351 + 9,746 *240/\{360[1.2000 - 1.0100]\}) /0.9900 - 64,720]/[1 - 0.3000]$$

$$= \underline{41,548} \text{ millions USD}$$

Corporate IFRS-GAAP (B/S-I/S),ISBN-13: **978-1721066308**, ISBN-10: **1721066306**

Law-8406:

If both (**U**= Utilized or Starting Capital= 64,720 millions USD), (**i**= I/S= Interest Portion= 1.00%), (**s**= [S/S']-1= Sales Growth= 5.00%), (**M**= Margin of Contribution= 41,548 millions USD), (**Q**= Quoted Longterm Liabilities= 35,351 millions USD), (**p**= 360 P/V= Procured Inventory Days= 240 Days), (**V**= Variable Cost= 9,746 millions USD), (**c**= C/X= Current Ratio= 1.20 times), (**D**= Dividend Paid= 2,370 millions USD), (**l**= L/E= Leverage or Gearing Ratio= 99.00%), (**t**= T/B= Tax Rate= 30.00%), (**q**= [C-P]/X= Quick or Acid Test Ratio= 1.01 times), and (**S'**= Sales of Past Year= 48,851 millions USD), then it's (**F**= Fixed Cost planned), is:

$$F= M\text{-}S'i[1+s]\text{-}[D+(Q+Vp/\{360[c\text{-}q]\})/l\text{-}U]$$
$$/[1\text{-}t]$$
$$= 41{,}548\text{-}48{,}851*0.0100[1+0.0500]$$
$$-[2{,}370+(35{,}351|9{,}746$$
$$*240/\{360[1.2000\text{-}1.0100]\})$$
$$/0.9900\text{-}64{,}720]/[1\text{-}0.3000]$$
$$= 29{,}750 \text{ millions USD}$$

Corporate IFRS-GAAP (B/S-I/S),ISBN-13: **978-1721066308**, ISBN-10: **1721066306**

Law-8407:

If both (**U**= Utilized or Starting Capital= 64,720 millions USD), (**i**= I/S= Interest Portion= 1.00%), (**s**= [S/S']-1= Sales Growth= 5.00%), (**M**= Margin of Contribution= 41,548 millions USD), (**Q**= Quoted Longterm Liabilities= 35,351 millions USD), (**p**= 360 P/V= Procured Inventory Days= 240 Days), (**V**= Variable Cost= 9,746 millions USD), (**c**= C/X= Current Ratio= 1.20 times), (**D**= Dividend Paid= 2,370 millions USD), (**I**= L/E= Leverage or Gearing Ratio= 99.00%), (**t**= T/B= Tax Rate= 30.00%), (**q**= [C-P]/X= Quick or Acid Test Ratio= 1.01 times), and (**F**= Fixed Cost= 29,740 millions USD), then it's (**S'**= Sales Past), is:

$$S'= \{M-F-[D+(Q+Vp/\{360[c-q]\})/I-U]$$
$$/[1-t]\}/\{i[1+s]\}$$
$$= \{41,548-29,750-[2,370+(35,351$$
$$+9,746*240/\{360[1.2000$$
$$-1.0100]\})/0.9900-64,720]$$
$$/[1-0.3000]\}$$
$$/\{0.0100[1+0.0500]\}$$
$$= 48,851 \text{ millions USD}$$

Corporate IFRS-GAAP (B/S-I/S),ISBN-13: **978-1721066308**, ISBN-10: **1721066306**

<u>Law-8408</u>:

If both (**U**= Utilized or Starting Capital= <u>64,720</u> millions USD), (**S'**= Sales of Past Year= <u>48,851</u> millions USD), (**s**= [S/S']-1= Sales Growth= <u>5.00%</u>), (**M**= Margin of Contribution= <u>41,548</u> millions USD), (**Q**= Quoted Longterm Liabilities= <u>35,351</u> millions USD), (**p**= 360 P/V= Procured Inventory Days= <u>240</u> Days), (**V**= Variable Cost= <u>9,746</u> millions USD), (**c**= C/X= Current Ratio= <u>1.20</u> times), (**D**= Dividend Paid= <u>2,370</u> millions USD), (**I**= L/E= Leverage or Gearing Ratio= <u>99.00%</u>), (**t**= T/B= Tax Rate= <u>30.00%</u>), (**q**= [C-P]/X= Quick or Acid Test Ratio= <u>1.01</u> times), and (**F**= Fixed Cost= <u>29,740</u> millions USD), then it's (**i**= Interest Portion planned), is:

$$i= \{M\text{-}F\text{-}[D+(Q+Vp/\{360[c\text{-}q]\})/I\text{-}U]$$
$$/[1\text{-}t]\}/\{S'[1+s]\}$$
$$= \{41,548\text{-}29,750\text{-}[2,370+(35,351$$
$$+9,746^{*}240/\{360[1.2000$$
$$-1.0100]\})/0.9900\text{-}64,720]$$
$$/[1\text{-}0.3000]\}$$
$$/\{48,851[1+0.0500]\}$$
$$= \underline{1.00\%}$$

Corporate IFRS-GAAP (B/S-I/S),ISBN-13: **978-1721066308**, ISBN-10: **1721066306**

<u>Law-8409</u>:

If both (**U**= Utilized or Starting Capital= <u>64,720</u> millions USD), (**S'**= Sales of Past Year= <u>48,851</u> millions USD), (**i**= I/S= Interest Portion= <u>1.00%</u>), (**M**= Margin of Contribution= <u>41,548</u> millions USD), (**Q**= Quoted Longterm Liabilities= <u>35,351</u> millions USD), (**p**= 360 P/V= Procured Inventory Days= <u>240</u> Days), (**V**= Variable Cost= <u>9,746</u> millions USD), (**c**= C/X= Current Ratio= <u>1.20</u> times), (**D**= Dividend Paid= <u>2,370</u> millions USD), (**I**= L/E= Leverage or Gearing Ratio= <u>99.00%</u>), (**t**= T/B= Tax Rate= <u>30.00%</u>), (**q**= [C-P]/X= Quick or Acid Test Ratio= <u>1.01</u> times), and (**F**= Fixed Cost= <u>29,740</u> millions USD), then it's (**s**= Sales Growth planned), is:

$$s= \{M\text{-}F\text{-}[D+(Q+Vp/\{360[c\text{-}q]\})/I\text{-}U]$$
$$/[1\text{-}t]\}/[S'i]\text{-}1$$
$$= \{41,548\text{-}29,750\text{-}[2,370+(35,351$$
$$+9,746*240/\{360[1.2000$$
$$-1.0100]\})/0.9900\text{-}64,720]$$
$$/[1\text{-}0.3000]\}/[48,851$$
$$*0.0100]\text{-}1$$
$$= \underline{5.00}\%$$

Corporate IFRS-GAAP (B/S-I/S),ISBN-13: **978-1721066308**, ISBN-10: **1721066306**

Law-8410:

If both (**U**= Utilized or Starting Capital= 64,720 millions USD), (**S'**= Sales of Past Year= 48,851 millions USD), (**i**= I/S= Interest Portion= 1.00%), (**M**= Margin of Contribution= 41,548 millions USD), (**Q**= Quoted Longterm Liabilities= 35,351 millions USD), (**p**= 360 P/V= Procured Inventory Days= 240 Days), (**V**= Variable Cost= 9,746 millions USD), (**c**= C/X= Current Ratio= 1.20 times), (**D**= Dividend Paid= 2,370 millions USD), (**F**= L/E= Leverage or Gearing Ratio= 99.00%), (**s**= [S/S']-1= Sales Growth= 5.00%), (**q**= [C-P]/X= Quick or Acid Test Ratio= 1.01 times), and (**F**= Fixed Cost= 29,740 millions USD), then it's (**t**= Tax Rate planned), is:

$$t= 1-[\mathbf{D}+(\mathbf{Q}+\mathbf{Vp}/\{360[\mathbf{c}\text{-}\mathbf{q}]\})/\mathbf{I}\text{ -}\mathbf{U}]$$
$$/\{\mathbf{M}\text{-}\mathbf{F}\text{-}\mathbf{S'i}[1+\mathbf{s}]\}$$
$$= 1-[2,370+(35,351+9,746*240/\{360$$
$$[1.2000\text{-}1.0100]\})/0.9900$$
$$-64,720]/[1-0.3000]\}/\{41,548$$
$$-29,750\text{-}48,851*0.0100$$
$$[1+0.0500]\}$$
$$= 30.00\%$$

Corporate IFRS-GAAP (B/S-I/S),ISBN-13: **978-1721066308**, ISBN-10: **1721066306**

Law-8411:

If both (**U**= Utilized or Starting Capital= 64,720 millions USD), (**S'**= Sales of Past Year= 48,851 millions USD), (**i**= I/S= Interest Portion= 1.00%), (**M**= Margin of Contribution= 41,548 millions USD), (**Q**= Quoted Longterm Liabilities= 35,351 millions USD), (**p**= 360 P/V= Procured Inventory Days= 240 Days), (**V**= Variable Cost= 9,746 millions USD), (**c**= C/X= Current Ratio= 1.20 times), (**t**= T/B= Tax Rate= 30.00%), (**l**= L/E= Leverage or Gearing Ratio= 99.00%), (**s**= [S/S']-1= Sales Growth= 5.00%), (**q**= [C-P]/X= Quick or Acid Test Ratio= 1.01 times), and (**F**= Fixed Cost= 29,740 millions USD), then it's (**D**= Dividend Paid planned), is:

$$D= U+\{M\text{-}F\text{-}S'i[1+s]\}[1\text{-}t]$$
$$-(Q+Vp/\{360[c\text{-}q]\})/l$$
$$= 64,720+\{41,548\text{-}29,750\text{-}48,851$$
$$*0.0100[1+0.0500]\}$$
$$[1\text{-}0.3000]\text{-}(35,351+9,746$$
$$*240/\{360[1.2000\text{-}1.0100]\})$$
$$/0.9900$$
$$= \underline{2,370} \text{ millions USD}$$

Corporate IFRS-GAAP (B/S-I/S),ISBN-13: **978-1721066308**, ISBN-10: **1721066306**

Law-8412:

If both (**U**= Utilized or Starting Capital= 64,720 millions USD), (**S'**= Sales of Past Year= 48,851 millions USD), (**i**= I/S= Interest Portion= 1.00%), (**M**= Margin of Contribution= 41,548 millions USD), (**Q**= Quoted Longterm Liabilities= 35,351 millions USD), (**p**= 360 P/V= Procured Inventory Days= 240 Days), (**D**= Dividend Paid= 2,370 millions USD), (**c**= C/X= Current Ratio= 1.20 times), (**t**= T/B= Tax Rate= 30.00%), (**l**= L/E= Leverage or Gearing Ratio= 99.00%), (**s**= [S/S']-1= Sales Growth= 5.00%), (**q**= [C-P]/X= Quick or Acid Test Ratio= 1.01 times), and (**F**= Fixed Cost= 29,740 millions USD), then it's (**V**= Variable Cost planned), is:

$$V= 360[c-q][l\,(U+\{M-F-S'i[1+s]\}$$
$$[1-t]-D)-Q]/p$$
$$= 360[1.2000-1.0100][0.9900(64,720$$
$$+\{41,548-29,750-48,851$$
$$*0.0100[1+0.0500]\}$$
$$[1-0.3000]-2,370)$$
$$-35,351]/240$$
$$= 9,746 \text{ millions USD}$$

Corporate IFRS-GAAP (B/S-I/S),ISBN-13: **978-1721066308**, ISBN-10: **1721066306**

Law-8413:

If both (**U**= Utilized or Starting Capital= 64,720 millions USD), (**S'**= Sales of Past Year= 48,851 millions USD), (**i**= I/S= Interest Portion= 1.00%), (**M**= Margin of Contribution= 41,548 millions USD), (**Q**= Quoted Longterm Liabilities= 35,351 millions USD), (**V**= Variable Cost= 9,746 millions USD), (**D**= Dividend Paid= 2,370 millions USD), (**c**= C/X= Current Ratio= 1.20 times), (**t**= T/B= Tax Rate= 30.00%), (**I**= L/E= Leverage or Gearing Ratio= 99.00%), (**s**= [S/S']-1= Sales Growth= 5.00%), (**q**= [C-P]/X= Quick or Acid Test Ratio= 1.01 times), and (**F**= Fixed Cost= 29,740 millions USD), then it's (**p**= Procured Inventory Days planned), is:

$$p= 360[c-q][I\,(U+\{M-F-S'i[1+s]\}$$
$$[1-t]-D)-Q]/V$$
$$= 360[1.2000-1.0100][0.9900$$
$$(64,720+\{41,548-29,750$$
$$-48,851*0.0100[1+0.0500]\}$$
$$[1-0.3000]-2,370)-35,351]$$
$$/9,746$$

$$= \underline{240}\ \text{days}$$

Corporate IFRS-GAAP (B/S-I/S),ISBN-13: **978-1721066308**, ISBN-10: **1721066306**

<u>Law-8414</u>:

If both (**U**= Utilized or Starting Capital= <u>64,720</u> millions USD), (**S'**= Sales of Past Year= <u>48,851</u> millions USD), (**i**= I/S= Interest Portion= <u>1.00%</u>), (**M**= Margin of Contribution= <u>41,548</u> millions USD), (**Q**= Quoted Longterm Liabilities= <u>35,351</u> millions USD), (**V**= Variable Cost= <u>9,746</u> millions USD), (**D**= Dividend Paid= <u>2,370</u> millions USD), (**p**= 360 P/V= Procured Inventory Days= <u>240</u> Days), (**t**= T/B= Tax Rate= <u>30.00%</u>), (**I**= L/E= Leverage or Gearing Ratio= <u>99.00%</u>), (**s**= [S/S']-1= Sales Growth= <u>5.00%</u>), (**q**= [C-P]/X= Quick or Acid Test Ratio= <u>1.01</u> times), and (**F**= Fixed Cost= <u>29,740</u> millions USD), then it's (**c**= Current Ratio planned), is:

$$c= q+Vp/\{360[I/(U+\{M-F-S'i[1+s]\}$$
$$[1-t]-D)-Q]\}$$
$$= 1.0100+9,746*240/\{360[0.9900$$
$$(64,720+\{41,548-29,750$$
$$-48,851*0.0100[1+0.0500]\}$$
$$[1-0.3000]-2,370)-35,351]\}$$
$$= \underline{1.20}\ times$$

Corporate IFRS-GAAP (B/S-I/S),ISBN-13: **978-1721066308**, ISBN-10: **1721066306**

<u>Law-8415</u>:

If both (**U**= Utilized or Starting Capital= <u>64,720</u> millions USD), (**$'**= Sales of Past Year= <u>48,851</u> millions USD), (**i**= I/S= Interest Portion= <u>1.00%</u>), (**M**= Margin of Contribution= <u>41,548</u> millions USD), (**Q**= Quoted Longterm Liabilities= <u>35,351</u> millions USD), (**V**= Variable Cost= <u>9,746</u> millions USD), (**D**= Dividend Paid= <u>2,370</u> millions USD), (**p**= 360 P/V= Procured Inventory Days= <u>240</u> Days), (**t**= T/B= Tax Rate= <u>30.00%</u>), (**F**= L/E= Leverage or Gearing Ratio= <u>99.00%</u>), (**s**= [S/S']-1= Sales Growth= <u>5.00%</u>), (**c**= C/X= Current Ratio= <u>1.20</u> times), and (**F**= Fixed Cost= <u>29,740</u> millions USD), then it's (**q**= Quick or Acid Test Ratio planned), is:

$$q = c\text{-}Vp/\{360[\textbf{\textit{I}}\,(U+\{M\text{-}F\text{-}S\text{'}i[1+s]\}$$
$$[1\text{-}t]\text{-}D)\text{-}Q]\}$$

$$= 1.2000\text{-}9{,}746*240/\{360[0.9900$$
$$(64{,}720+\{41{,}548\text{-}29{,}750$$
$$-48{,}851*0.0100[1+0.0500]\}$$
$$[1\text{-}0.3000]\text{-}2{,}370)\text{-}35{,}351]\}$$

$$= \underline{1.01}\ \text{times}$$

Corporate IFRS-GAAP (B/S-I/S),ISBN-13: **978-1721066308**, ISBN-10: **1721066306**

Law-8416:

If both (**U**= Utilized or Starting Capital= 64,720 millions USD), (**v**= V/S= Variable Portion= 19.00%), (**S'**= Sales of Past Year= 48,851 millions USD), (**i**= I/S= Interest Portion= 1.00%), (**M**= Margin of Contribution= 41,548 millions USD), (**q**= [C-P]/X= Quick or Acid Test Ratio= 1.01 times), (**S**= Sales or Revenues= 51,294 millions USD), (**D**= Dividend Paid= 2,370 millions USD), (**p**= 360 P/V= Procured Inventory Days= 240 Days), (**t**= T/B= Tax Rate= 30.00%), (**l**= L/E= Leverage or Gearing Ratio= 99.00%), (**s**= [S/S']-1= Sales Growth= 5.00%), (**c**= C/X= Current Ratio= 1.20 times), and (**F**= Fixed Cost= 29,740 millions USD), then it's (**Q**= Quoted Longterm Liabilities planned), is:

$$Q = l\,(U+\{M\text{-}F\text{-}S\text{'}i[1+s]\}[1\text{-}t]\text{-}D)$$
$$\quad -Svp/\{360[c\text{-}q]\}$$
$$= 0.9900(64,720+\{41,548-29,750$$
$$\quad -48,851*0.0100[1+0.0500]\}$$
$$\quad [1-0.3000]-2,370)-51,294$$
$$\quad *0.1900*240/\{360[1.2000$$
$$\quad -1.0100]\}$$
$$= \underline{35,351} \text{ millions USD}$$

Corporate IFRS-GAAP (B/S-I/S),ISBN-13: **978-1721066308**, ISBN-10: **1721066306**

Law-8417:

If both (**U**= Utilized or Starting Capital= 64,720 millions USD), (**v**= V/S= Variable Portion= 19.00%), (**S'**= Sales of Past Year= 48,851 millions USD), (**i**= I/S= Interest Portion= 1.00%), (**M**= Margin of Contribution= 41,548 millions USD), (**q**= [C-P]/X= Quick or Acid Test Ratio= 1.01 times), (**S**= Sales or Revenues= 51,294 millions USD), (**D**= Dividend Paid= 2,370 millions USD), (**p**= 360 P/V= Procured Inventory Days= 240 Days), (**t**= T/B= Tax Rate= 30.00%), (**Q**= Quoted Longterm Liabilities= 35,351 millions USD), (**s**= [S/S']-1= Sales Growth= 5.00%), (**c**= C/X= Current Ratio= 1.20 times), and (**F**= Fixed Cost= 29,740 millions USD), then it's (**I** = Leverage or Gearing Ratio planned), is:

$$I = (Q+Svp/\{360[c-q]\})/(U+\{M-F-S'i[1+s]\}[1-t]-D)$$

$$= (35,351+51,294*0.1900*240/\{360[1.2000-1.0100]\})/(64,720+\{41,548-29,750-48,851*0.0100[1+0.0500]\}[1-0.3000]-2,370)$$

$$= \underline{99.00\%}$$

Corporate IFRS-GAAP (B/S-I/S),ISBN-13: **978-1721066308**, ISBN-10: **1721066306**

<u>Law-8418</u>:

If both (**Q**= Quoted Longterm Liabilities= <u>35,351</u> millions USD), (**v**= V/S= Variable Portion= <u>19.00%</u>), (**S'**= Sales of Past Year= <u>48,851</u> millions USD), (**i**= I/S= Interest Portion= <u>1.00%</u>), (**M**= Margin of Contribution= <u>41,548</u> millions USD), (**q**= [C-P]/X= Quick or Acid Test Ratio= <u>1.01</u> times), (**S**= Sales or Revenues= <u>51,294</u> millions USD), (**D**= Dividend Paid= <u>2,370</u> millions USD), (**p**= 360 P/V= Procured Inventory Days= <u>240</u> Days), (**t**= T/B= Tax Rate= <u>30.00%</u>), (**l**= L/E= Leverage or Gearing Ratio= <u>99.00%</u>), (**s**= [S/S']-1= Sales Growth= <u>5.00%</u>), (**c**= C/X= Current Ratio= <u>1.20</u> times), and (**F**= Fixed Cost= <u>29,740</u> millions USD), then it's (**U**= Utilized or Starting Capital planned), is:

$$U= D+(Q+Svp/\{360[c-q]\})/l$$
$$-\{M-F-S'i[1+s]\}[1-t]$$
$$= 2,370 +(35,351+51,294*0.1900$$
$$*240/\{360[1.2000-1.0100]\})$$
$$/0.9900-\{41,548-29,750$$
$$-48,851*0.0100[1+0.0500]\}$$
$$[1-0.3000]$$
$$= \underline{64,720} \text{ millions USD}$$

Corporate IFRS-GAAP (B/S-I/S),ISBN-13: **978-1721066308**, ISBN-10: **1721066306**

Law-8419:

If both (Q= Quoted Longterm Liabilities= 35,351 millions USD), (v= V/S= Variable Portion= 19.00%), (S'= Sales of Past Year= 48,851 millions USD), (i= I/S= Interest Portion= 1.00%), (U= Utilized or Starting Capital= 64,720 millions USD), (q= Quick or Acid Test Ratio= [C-P]/X= 1.01 times), (S= Sales or Revenues= 51,294 millions USD), (D= Dividend Paid= 2,370 millions USD), (p= 360P/V= Procured Inventory Days= 240 Days), (t= T/B= Tax Rate= 30.00%), (l= L/E= Leverage or Gearing Ratio= 99.00%), (s= [S/S']-1= Sales Growth= 5.00%), (c= C/X= Current Ratio= 1.20 times), and (F= Fixed Cost= 29,740 millions USD), then it's (M= Margin of Contribution planned), is:

$$M= F+S'i[1+s]+[D+(Q+Svp$$
$$/\{360[c-q]\})/l\ -U]/[1-t]$$
$$= 29{,}750+48{,}851*0.0100[1+0.0500]$$
$$+[2{,}370+(35{,}351+51{,}294$$
$$*0.1900*240/\{360[1.2000$$
$$-1.0100]\})/0.9900-64{,}720]$$
$$/[1-0.3000]$$
$$= 41{,}548 \text{ millions USD}$$

Corporate IFRS-GAAP (B/S-I/S),ISBN-13: **978-1721066308**, ISBN-10: **1721066306**

Law-8420:

If both (**Q**= Quoted Longterm Liabilities= 35,351 millions USD), (**v**= V/S= Variable Portion= 19.00%), (**S'**= Sales of Past Year= 48,851 millions USD), (**i**= I/S= Interest Portion= 1.00%), (**U**= Utilized or Starting Capital= 64,720 millions USD), (**q**= Quick or Acid Test Ratio= [C-P]/X= 1.01 times), (**S**= Sales or Revenues= 51,294 millions USD), (**D**= Dividend Paid= 2,370 millions USD), (**p**= 360 P/V= Procured Inventory Days= 240 Days), (**t**= T/B= Tax Rate= 30.00%), (**l**= L/E= Leverage or Gearing Ratio= 99.00%), (**s**= [S/S']-1= Sales Growth= 5.00%), (**c**= C/X= Current Ratio= 1.20 times), and (**M**= Margin of Contribution= 41,548 millions USD), then it's (**F**= Fixed Cost planned), is:

$$F= M\text{-}S'i[1+s]\text{-}[D+(Q+Svp$$
$$/\{360[c\text{-}q]\})/l\text{-}U]/[1\text{-}t]$$

$$= 41,548\text{-}48,851*0.0100[1+0.0500]$$
$$-[2,370+(35,351+51,294$$
$$*0.1900*240/\{360[1.2000$$
$$-1.0100]\})/0.9900\text{-}64,720]$$
$$/[1\text{-}0.3000]$$

$$= 29,750 \text{ millions USD}$$

Corporate IFRS-GAAP (B/S-I/S),ISBN-13: **978-1721066308**, ISBN-10: **1721066306**

Law-8421:

If both (**Q**= Quoted Longterm Liabilities= 35,351 millions USD), (**v**= V/S= Variable Portion= 19.00%), (**F**= Fixed Cost= 29,740 millions USD), (**i**= I/S= Interest Portion= 1.00%), (**U**= Utilized or Starting Capital= 64,720 millions USD), (**q**= [C-P]/X= Quick or Acid Test Ratio= 1.01 times), (**S**= Sales or Revenues= 51,294 millions USD), (**D**= Dividend Paid= 2,370 millions USD), (**p**= 360 P/V= Procured Inventory Days= 240 Days), (**t**= T/B= Tax Rate= 30.00%), (**l**= L/E= Leverage or Gearing Ratio= 99.00%), (**s**= [S/S']-1= Sales Growth= 5.00%), (**c**= C/X= Current Ratio= 1.20 times), and (**M**= Margin of Contribution= 41,548 millions USD), then it's (**S'**= Sales Past), must be:

$$S' = \{M-F-[D+(Q+Svp/\{360[c-q]\})/l - U]/[1-t]\}/\{i[1+s]\}$$

$$= \{41,548-29,750-[2,370+(35,351$$
$$+51,294*0.1900*240/\{360$$
$$[1.2000-1.0100]\})/0.9900$$
$$-64,720]/[1-0.3000]\}$$
$$/\{0.0100[1+0.0500]\}$$

$$= 48,851 \text{ millions USD}$$

Corporate IFRS-GAAP (B/S-I/S),ISBN-13: **978-1721066308**, ISBN-10: **1721066306**

Law-8422:

If both (**Q**= Quoted Longterm Liabilities= 35,351 millions USD), (**v**= V/S= Variable Portion= 19.00%), (**F**= Fixed Cost= 29,740 millions USD), (**S'**= Sales of Past Year= 48,851 millions USD), (**U**= Utilized or Starting Capital= 64,720 millions USD), (**q**= Quick or Acid Test Ratio= [C-P]/X= 1.01 times), (**S**= Sales or Revenues= 51,294 millions USD), (**D**= Dividend Paid= 2,370 millions USD), (**p**= 360 P/V= Procured Inventory Days= 240 Days), (**t**= T/B= Tax Rate= 30.00%), (**l**= L/E= Leverage or Gearing Ratio= 99.00%), (**s**= [S/S']-1= Sales Growth= 5.00%), (**c**= C/X= Current Ratio= 1.20 times), and (**M**= Margin of Contribution= 41,548 millions USD), then it's (**i**= Interest Portion planned), is:

$$i= \{M\text{-}F\text{-}[D+(Q+Svp/\{360[c\text{-}q]\})/l\text{-}U]$$
$$/[1\text{-}t]\}/\{S'[1+s]\}$$
$$= \{41,548\text{-}29,750\text{-}[2,370+(35,351$$
$$+51,294*0.1900*240/\{360$$
$$[1.2000\text{-}1.0100]\})/0.9900$$
$$-64,720]/[1\text{-}0.3000]\}/\{48,851$$
$$[1+0.0500]\}$$
$$= 1.00\%$$

Corporate IFRS-GAAP (B/S-I/S),ISBN-13: **978-1721066308**, ISBN-10: **1721066306**

Law-8423:

If both (**Q**= Quoted Longterm Liabilities= 35,351 millions USD), (**v**= V/S= Variable Portion= 19.00%), (**F**= Fixed Cost= 29,740 millions USD), (**S'**= Sales of Past Year= 48,851 millions USD), (**U**= Utilized or Starting Capital= 64,720 millions USD), (**q**= Quick or Acid Test Ratio= [C-P]/X= 1.01 times), (**S**= Sales or Revenues= 51,294 millions USD), (**D**= Dividend Paid= 2,370 millions USD), (**p**= 360 P/V= Procured Inventory Days= 240 Days), (**t**= T/B= Tax Rate= 30.00%), (**l**= L/E= Leverage or Gearing Ratio= 99.00%), (**i**= I/S= Interest Portion= 1.00%), (**c**= C/X= Current Ratio= 1.20 times), and (**M**= Margin of Contribution= 41,548 millions USD), then it's (**s**= Sales Growth planned), is:

$$s= \{M\text{-}F\text{-}[D+(Q+Svp/\{360[c\text{-}q]\})/l\text{-}U] /[1\text{-}t]\}/[S'i]\text{-}1$$

$$= \{41,548\text{-}29,750\text{-}[2,370+(35,351 +51,294*0.1900*240/\{360 [1.2000\text{-}1.0100]\})/0.9900 -64,720]/[1\text{-}0.3000]\} /[48,851*0.0100]\text{-}1$$

$$= 5.00\%$$

Corporate IFRS-GAAP (B/S-I/S),ISBN-13: **978-1721066308**, ISBN-10: **1721066306**

Law-8424:

If both (**Q**= Quoted Longterm Liabilities= <u>35,351</u> millions USD), (**v**= V/S= Variable Portion= <u>19.00%</u>), (**F**= Fixed Cost= <u>29,740</u> millions USD), (**S'**= Sales of Past Year= <u>48,851</u> millions USD), (**U**= Utilized or Starting Capital= <u>64,720</u> millions USD), (**q**= Quick or Acid Test Ratio= [C-P]/X= <u>1.01</u> times), (**S**= Sales or Revenues= <u>51,294</u> millions USD), (**D**= Dividend Paid= <u>2,370</u> millions USD), (**p**= 360 P/V= Procured Inventory Days= <u>240</u> Days), (**s**= [S/S']-1= Sales Growth= <u>5.00%</u>), (**l**= L/E= Leverage or Gearing Ratio= <u>99.00%</u>), (**i**= I/S= Interest Portion= <u>1.00%</u>), (**c**= C/X= Current Ratio= <u>1.20</u> times), and (**M**= Margin of Contribution= <u>41,548</u> millions USD), then it's (**t**= Tax Rate planned), is:

$$t = 1-[\mathbf{D}+(\mathbf{Q}+\mathbf{Svp}/\{360[\mathbf{c}-\mathbf{q}]\})/\mathbf{l}-\mathbf{U}]$$
$$/\{\mathbf{M}-\mathbf{F}-\mathbf{S'i}[1+\mathbf{s}]\}$$

$$= 1-[2,370+(35,351+51,294*0.1900$$
$$*240/\{360[1.2000-1.0100]\})$$
$$/0.9900-64,720]/\{41,548$$
$$-29,750-48,851*0.0100$$
$$[1+0.0500]\}$$

$$= \underline{30.00\%}$$

Corporate IFRS-GAAP (B/S-I/S),ISBN-13: **978-1721066308**, ISBN-10: **1721066306**

Law-8425:

If both (Q= Quoted Longterm Liabilities= 35,351 millions USD), (v= V/S= Variable Portion= 19.00%), (F= Fixed Cost= 29,740 millions USD), (S'= Sales of Past Year= 48,851 millions USD), (U= Utilized or Starting Capital= 64,720 millions USD), (q= Quick or Acid Test Ratio= [C-P]/X= 1.01 times), (S= Sales or Revenues= 51,294 millions USD), (t= T/B= Tax Rate= 30.00%), (p= 360 P/V= Procured Inventory Days= 240 Days), (s= [S/S']-1= Sales Growth= 5.00%), (l= L/E= Leverage or Gearing Ratio= 99.00%), (i= I/S= Interest Portion= 1.00%), (c= C/X= Current Ratio= 1.20 times), and (M= Margin of Contribution= 41,548 millions USD), then it's (D= Dividend Paid planned), is:

$$D= U+\{M-F-S'i[1+s]\}[1-d]$$
$$-(Q+Svp/\{360[c-q]\})/l$$
$$= 64,720+\{41,548-29,750-48,851$$
$$*0.0100[1+0.0500]\}$$
$$[1-0.3000]-(35,351+51,294$$
$$*0.1900*240/\{360[1.2000$$
$$-1.0100]\})/0.9900$$
$$= 2,370 \text{ millions USD}$$

Corporate IFRS-GAAP (B/S-I/S),ISBN-13: **978-1721066308**, ISBN-10: **1721066306**

Law-8426:

If both (**Q**= Quoted Longterm Liabilities= 35,351 millions USD), (**v**= V/S= Variable Portion= 19.00%), (**F**= Fixed Cost= 29,740 millions USD), (**S'**= Sales of Past Year= 48,851 millions USD), (**U**= Utilized or Starting Capital= 64,720 millions USD), (**q**= Quick or Acid Test Ratio= [C-P]/X= 1.01 times), (**D**= Dividend Paid= 2,370 millions USD), (**t**= T/B= Tax Rate= 30.00%), (**p**= 360 P/V= Procured Inventory Days= 240 Days), (**s**= [S/S']-1= Sales Growth= 5.00%), (**l**= L/E= Leverage or Gearing Ratio= 99.00%), (**i**= I/S= Interest Portion= 1.00%), (**c**= C/X= Current Ratio= 1.20 times), and (**M**= Margin of Contribution= 41,548 millions USD), then it's (**S**= Sales or Revenues planned), is:

$$S= 360[\textbf{c-q}][\textbf{l}\,(\textbf{U}+\{\textbf{M-F-S'i}[1+\textbf{s}]\}$$
$$[1\textbf{-t}]\textbf{-D})\textbf{-Q}]/[\textbf{vp}]$$
$$= 360[1.2000\text{-}1.0100][0.9900$$
$$(64{,}720+\{41{,}548\text{-}29{,}750$$
$$-48{,}851*0.0100[1+0.0500]\}$$
$$[1\text{-}0.3000]\text{-}2{,}370)\text{-}35{,}351]$$
$$/[0.1900*240]$$
$$= 51{,}294 \text{ millions USD}$$

Corporate IFRS-GAAP (B/S-I/S),ISBN-13: **978-1721066308**, ISBN-10: **1721066306**

<u>Law-8427</u>:

If both (**Q**= Quoted Longterm Liabilities= <u>35,351</u> millions USD), (**S**= Sales or Revenues= <u>51,294</u> millions USD), (**F**= Fixed Cost= <u>29,740</u> millions USD), (**S'**= Sales of Past Year= <u>48,851</u> millions USD), (**U**= Utilized or Starting Capital= <u>64,720</u> millions USD), (**q**= [C-P]/X= Quick or Acid Test Ratio= <u>1.01</u> times), (**D**= Dividend Paid= <u>2,370</u> millions USD), (**t**= T/B= Tax Rate= <u>30.00%</u>), (**p**= 360 P/V= Procured Inventory Days= <u>240</u> Days), (**s**= [S/S']-1= Sales Growth= <u>5.00%</u>), (**l**= L/E= Leverage or Gearing Ratio= <u>99.00%</u>), (**i**= I/S= Interest Portion= <u>1.00%</u>), (**c**= C/X= Current Ratio= <u>1.20</u> times), and (**M**= Margin of Contribution= <u>41,548</u> millions USD), then it's (**v**= Variable Portion planned), is:

$$v= 360[c\text{-}q][l\,(U+\{M\text{-}F\text{-}S'i[1+s]\}$$
$$[1\text{-}t]\text{-}D)\text{-}Q]/[Sp]$$
$$= 360[1.2000\text{-}1.0100][0.9900$$
$$(64,720+\{41,548\text{-}29,750$$
$$-48,851*0.0100[1+0.0500]\}$$
$$[1\text{-}0.3000]\text{-}2,370)\text{-}35,351]$$
$$/[51,294*240]$$
$$= \underline{19.00\%}$$

Corporate IFRS-GAAP (B/S-I/S),ISBN-13: **978-1721066308**, ISBN-10: **1721066306**

<u>Law-8428</u>:

If both (**Q**= Quoted Longterm Liabilities= <u>35,351</u> millions USD), (**$**= Sales or Revenues= <u>51,294</u> millions USD), (**F**= Fixed Cost= <u>29,740</u> millions USD), (**$'**= Sales of Past Year= <u>48,851</u> millions USD), (**U**= Utilized or Starting Capital= <u>64,720</u> millions USD), (**q**= [C-P]/X= Quick or Acid Test Ratio= <u>1.01</u> times), (**D**= Dividend Paid= <u>2,370</u> millions USD), (**t**= T/B= Tax Rate= <u>30.00%</u>), (**v**= V/S= Variable Portion= <u>19.00%</u>), (**s**= [S/S']-1= Sales Growth= <u>5.00%</u>), (**I**= L/E= Leverage or Gearing Ratio= <u>99.00%</u>), (**i**= I/S= Interest Portion= <u>1.00%</u>), (**c**= C/X= Current Ratio= <u>1.20</u> times), and (**M**= Margin of Contribution= <u>41,548</u> millions USD), then it's (**p**= Procured Inventory Days planned), is:

$$p= 360[c-q][I\,(U+\{M-F-S'i[1+s]\}$$
$$[1-t]-D)-Q]/[Sv]$$
$$= 360[1.2000-1.0100][0.9900$$
$$(64,720+\{41,548-29,750$$
$$-48,851*0.0100[1+0.0500]\}$$
$$[1-0.3000]-2,370)-35,351]$$
$$/[51,294*0.1900]$$
$$= \underline{240}\ days$$

Corporate IFRS-GAAP (B/S-I/S),ISBN-13: **978-1721066308**, ISBN-10: **1721066306**

<u>Law-8429</u>:

If both (**Q**= Quoted Longterm Liabilities= <u>35,351</u> millions USD), (**S**= Sales or Revenues= <u>51,294</u> millions USD), (**F**= Fixed Cost= <u>29,740</u> millions USD), (**S'**= Sales of Past Year= <u>48,851</u> millions USD), (**U**= Utilized or Starting Capital= <u>64,720</u> millions USD), (**q**= [C-P]/X= Quick or Acid Test Ratio= <u>1.01</u> times), (**D**= Dividend Paid= <u>2,370</u> millions USD), (**t**= T/B= Tax Rate= <u>30.00%</u>), (**v**= V/S= Variable Portion= <u>19.00%</u>), (**s**= [S/S']-1= Sales Growth= <u>5.00%</u>), (**l**= L/E= Leverage or Gearing Ratio= <u>99.00%</u>), (**i**= I/S= Interest Portion= <u>1.00%</u>), (**p**= 360 P/V= Procured Inventory Days= <u>240</u> Days), and (**M**= Margin of Contribution= <u>41,548</u> millions USD), then it's (**c**= Current Ratio planned), is:

$$c= q+Svp/\{360[l/(U+\{M-F-S'i[1+s]\} [1-t]-D)-Q]\}$$

$$= 1.0100+51,294*0.1900*240 /\{360[0.9900(64,720 +\{41,548-29,750-48,851 *0.0100[1+0.0500]\} [1-0.3000]-2,370)-35,351]\}$$

$$= \underline{1.20} \text{ times}$$

Corporate IFRS-GAAP (B/S-I/S),ISBN-13: **978-1721066308**, ISBN-10: **1721066306**

Law-8430:

If both (**Q**= Quoted Longterm Liabilities= 35,351 millions USD), (**$**= Sales or Revenues= 51,294 millions USD), (**F**= Fixed Cost= 29,740 millions USD), (**$'**= Sales of Past Year= 48,851 millions USD), (**U**= Utilized or Starting Capital= 64,720 millions USD), (**c**= C/X= Current Ratio= 1.20 times), (**D**= Dividend Paid= 2,370 millions USD), (**t**= T/B= Tax Rate= 30.00%), (**v**= V/S= Variable Portion= 19.00%), (**s**= [S/S']-1= Sales Growth= 5.00%), (**F**= L/E= Leverage or Gearing Ratio= 99.00%), (**i**= I/S= Interest Portion= 1.00%), (**p**= 360 P/V= Procured Inventory Days= 240 Days), and (**M**= Margin of Contribution= 41,548 millions USD), then it's (**q**= Quick or Acid Test Ratio planned), is:

$$q = c\text{-}\$vp/\{360[\textit{I}(U+\{M\text{-}F\text{-}\$'i[1+s]\}$$
$$[1\text{-}t]\text{-}D)\text{-}Q]\}$$
$$= 1.2000\text{-}51,294*0.1900*240/\{360$$
$$[0.9900(64,720+\{41,548$$
$$-29,750\text{-}48,851*0.0100$$
$$[1+0.0500]\}][1\text{-}0.3000]$$
$$-2,370)\text{-}35,351]\}$$
$$= \underline{1.01}\text{ times}$$

Corporate IFRS-GAAP (B/S-I/S),ISBN-13: **978-1721066308**, ISBN-10: **1721066306**

Law-8431:

If both (**q**= [C-P]/X= Quick or Acid Test Ratio= 1.01 times), (**F**= Fixed Cost= 29,740 millions USD), (**S'**= Sales of Past Year= 48,851 millions USD), (**U**= Utilized or Starting Capital= 64,720 millions USD), (**c**= C/X= Current Ratio= 1.20 times), (**D**= Dividend Paid= 2,370 millions USD), (**t**= T/B= Tax Rate= 30.00%), (**v**= V/S= Variable Portion= 19.00%), (**s**= [S/S']-1= Sales Growth= 5.00%), (**f**= L/E= Leverage or Gearing Ratio= 99.00%), (**i**= I/S= Interest Portion= 1.00%), (**p**= 360 P/V= Procured Inventory Days= 240 Days), and (**M**= Margin of Contribution= 41,548 millions USD), then it's (**Q**= Quoted Longterm Liabilities planned), is:

$$Q= f(U+\{M-F-S'i[1+s]\}[1-t]-D)$$
$$-S'vp[1+s]/\{360[c-q]\}$$
$$= 0.9900(64,720+\{41,548-29,750$$
$$-48,851*0.0100[1+0.0500]\}$$
$$[1-0.3000]-2,370)-48,851$$
$$*0.1900*240[1+0.0500]$$
$$/\{360[1.2000-1.0100]\}$$
$$= 35,351 \text{ millions USD}$$

Corporate IFRS-GAAP (B/S-I/S),ISBN-13: **978-1721066308**, ISBN-10: **1721066306**

<u>Law-8432</u>:

If both (**q**= [C-P]/X= Quick or Acid Test Ratio= <u>1.01</u>
times), (**F**= Fixed Cost= <u>29,740</u> millions USD), (**S'**=
Sales of Past Year= <u>48,851</u> millions USD), (**U**=
Utilized or Starting Capital= <u>64,720</u> millions USD),
(**c**= C/X= Current Ratio= <u>1.20</u> times), (**D**= Dividend
Paid= <u>2,370</u> millions USD), (**t**= T/B= Tax Rate=
<u>30.00%</u>), (**v**= V/S= Variable Portion= <u>19.00%</u>), (**s**=
[S/S']-1= Sales Growth= <u>5.00%</u>), (**Q**= Quoted
Longterm Liabilities= <u>35,351</u> millions USD), (**i**= I/S=
Interest Portion= <u>1.00%</u>), (**p**= 360 P/V= Procured
Inventory Days= <u>240</u> Days), and (**M**= Margin of
Contribution= <u>41,548</u> millions USD), then it's (**I** =
Leverage or Gearing Ratio planned), is:

$$I = (Q+S'vp[1+s]/\{360[c-q]\})/(U+\{M-F$$
$$-S'i[1+s]\}[1-t]-D)$$
$$= (35,351+48,851*0.1900*240$$
$$[1+0.0500]/\{360[1.2000$$
$$-1.0100]\})/(64,720+\{41,548$$
$$-29,750-48,851*0.0100$$
$$[1+0.0500]\}[1-0.3000]$$
$$-2,370)$$
$$= \underline{99.00\%}$$

Corporate IFRS-GAAP (B/S-I/S),ISBN-13: **978-1721066308**, ISBN-10: **1721066306**

<u>Law-8433</u>:

If both ($\mathbf{q}$= [C-P]/X= Quick or Acid Test Ratio= <u>1.01</u> times), ($\mathbf{F}$= Fixed Cost= <u>29,740</u> millions USD), ($\mathbf{S'}$= Sales of Past Year= <u>48,851</u> millions USD), ($\mathbf{l}$= L/E= Leverage or Gearing Ratio= <u>99.00%</u>), ($\mathbf{c}$= C/X= Current Ratio= <u>1.20</u> times), ($\mathbf{D}$= Dividend Paid= <u>2,370</u> millions USD), ($\mathbf{t}$= T/B= Tax Rate= <u>30.00%</u>), ($\mathbf{v}$= V/S= Variable Portion= <u>19.00%</u>), ($\mathbf{s}$= [S/S']-1= Sales Growth= <u>5.00%</u>), ($\mathbf{Q}$= Quoted Longterm Liabilities= <u>35,351</u> millions USD), ($\mathbf{i}$= I/S= Interest Portion= <u>1.00%</u>), ($\mathbf{p}$= 360 P/V= Procured Inventory Days= <u>240</u> Days), and ($\mathbf{M}$= Margin of Contribution= <u>41,548</u> millions USD), then it's ($\mathbf{U}$= Utilized or Starting Capital planned), is:

$$\mathbf{U}= \mathbf{D}+(\mathbf{Q}+\mathbf{S'vp}[1+\mathbf{s}]/\{360[\mathbf{c}\text{-}\mathbf{q}]\})/\mathbf{l}$$
$$-\{\mathbf{M}\text{-}\mathbf{F}\text{-}\mathbf{S'i}[1+\mathbf{s}]\}[1\text{-}\mathbf{t}]$$
$$= 2{,}370+(35{,}351+48{,}851*0.1900$$
$$*240[1+0.0500]/\{360[1.2000$$
$$-1.0100]\})/0.9900-\{41{,}548$$
$$-29{,}750-48{,}851*0.0100$$
$$[1+0.0500]\}[1-0.3000]$$
$$= \underline{64{,}720} \text{ millions USD}$$

Corporate IFRS-GAAP (B/S-I/S),ISBN-13: **978-1721066308**, ISBN-10: **1721066306**

Law-8434:

If both (q= [C-P]/X= Quick or Acid Test Ratio= <u>1.01</u> times), (**F**= Fixed Cost= <u>29,740</u> millions USD), (**$'**= Sales of Past Year= <u>48,851</u> millions USD), (**l**= L/E= Leverage or Gearing Ratio= <u>99.00%</u>), (**c**= C/X= Current Ratio= <u>1.20</u> times), (**D**= Dividend Paid= <u>2,370</u> millions USD), (**t**= T/B= Tax Rate= <u>30.00%</u>), (**v**= V/S= Variable Portion= <u>19.00%</u>), (**s**= [S/S']-1= Sales Growth= <u>5.00%</u>), (**Q**= Quoted Longterm Liabilities= <u>35,351</u> millions USD), (**i**= I/S= Interest Portion= <u>1.00%</u>), (**p**= 360 P/V= Procured Inventory Days= <u>240</u> Days), and (**U**= Utilized or Starting Capital= <u>64,720</u> millions USD), then it's (**M**= Margin of Contribution planned), is:

$$M= F+\$'i[1+s]+[D+(Q+\$'vp[1+s]$$
$$/\{360[c-q]\})/l-U]/[1-t]$$
$$= 29,750+48,851*0.0100[1+0.0500]$$
$$+[2,370+(35,351+48,851$$
$$*0.1900*240[1+0.0500]$$
$$/\{360[1.2000-1.0100]\})$$
$$/0.9900-64,720]/[1-0.3000]$$
$$= \underline{41,548} \text{ millions USD}$$

Corporate IFRS-GAAP (B/S-I/S),ISBN-13: **978-1721066308**, ISBN-10: **1721066306**

<u>Law-8435</u>:

If both (**q**= [C-P]/X= Quick or Acid Test Ratio= <u>1.01</u> times), (**M**= Margin of Contribution= <u>41,548</u> millions USD), (**S'**= Sales of Past Year= <u>48,851</u> millions USD), (**I**= L/E= Leverage or Gearing Ratio= <u>99.00%</u>), (**c**= C/X= Current Ratio= <u>1.20</u> times), (**D**= Dividend Paid= <u>2,370</u> millions USD), (**t**= T/B= Tax Rate= <u>30.00%</u>), (**v**= V/S= Variable Portion= <u>19.00%</u>), (**s**= [S/S']-1= Sales Growth= <u>5.00%</u>), (**Q**= Quoted Longterm Liabilities= <u>35,351</u> millions USD), (**i**= I/S= Interest Portion= <u>1.00%</u>), (**p**= 360 P/V= Procured Inventory Days= <u>240</u> Days), and (**U**= Utilized or Starting Capital= <u>64,720</u> millions USD), then it's (**F**= Fixed Cost planned), is:

$$F= M\text{-}S'i[1+s]\text{-}[D+(Q+S'vp[1+s]$$
$$/\{360[c\text{-}q]\})/I\text{-}U]/[1\text{-}t]$$
$$= 41,548\text{-}48,851*0.0100[1+0.0500]$$
$$\text{-}[2,370+(35,351+48,851$$
$$*0.1900*240[1+0.0500]$$
$$/\{360[1.2000\text{-}1.0100]\})$$
$$/0.9900\text{-}64,720]/[1\text{-}0.3000]$$
$$= \underline{29,750} \text{ millions USD}$$

Corporate IFRS-GAAP (B/S-I/S),ISBN-13: **978-1721066308**, ISBN-10: **1721066306**

Law-8436:

If both (**q**= [C-P]/X= Quick or Acid Test Ratio= 1.01 times), (**M**= Margin of Contribution= 41,548 millions USD), (**F**= Fixed Cost= 29,740 millions USD), (**l**= L/E= Leverage or Gearing Ratio= 99.00%), (**c**= C/X= Current Ratio= 1.20 times), (**D**= Dividend Paid= 2,370 millions USD), (**t**= T/B= Tax Rate= 30.00%), (**v**= V/S= Variable Portion= 19.00%), (**s**= [S/S']-1= Sales Growth= 5.00%), (**Q**= Quoted Longterm Liabilities= 35,351 millions USD), (**i**= I/S= Interest Portion= 1.00%), (**p**= 360 P/V= Procured Inventory Days= 240 Days), and (**U**= Utilized or Starting Capital= 64,720 millions USD), then it's (**S'**= Sales Past), must be:

$$S'= \{M\text{-}F\text{-}[D+(Q+S'vp[1+s]/\{360[c\text{-}q]\})/l\text{-}U]$$
$$/[1\text{-}t]\}/\{i[1+s]\}$$
$$= \{41,548\text{-}29,750\text{-}[2,370+(35,351$$
$$+48,851*0.1900*240$$
$$[1+0.0500]/\{360[1.2000$$
$$-1.0100]\})/0.9900\text{-}64,720]$$
$$/[1\text{-}0.3000]\}$$
$$/\{0.0100[1+0.0500]\}$$
$$= \underline{48,851} \text{ millions USD}$$

Corporate IFRS-GAAP (B/S-I/S),ISBN-13: **978-1721066308**, ISBN-10: **1721066306**

<u>Law-8437</u>:

If both (q= [C-P]/X= Quick or Acid Test Ratio= <u>1.01</u> times), (M= Margin of Contribution= <u>41,548</u> millions USD), (F= Fixed Cost= <u>29,740</u> millions USD), (I= L/E= Leverage or Gearing Ratio= <u>99.00%</u>), (c= C/X= Current Ratio= <u>1.20</u> times), (D= Dividend Paid= <u>2,370</u> millions USD), (t= T/B= Tax Rate= <u>30.00%</u>), (v= V/S= Variable Portion= <u>19.00%</u>), (s= [S/S']-1= Sales Growth= <u>5.00%</u>), (Q= Quoted Longterm Liabilities= <u>35,351</u> millions USD), (S'= Sales of Past Year= <u>48,851</u> millions USD), (p= 360 P/V= Procured Inventory Days= <u>240</u> Days), and (U= Utilized or Starting Capital= <u>64,720</u> millions USD), then it's (i= Interest Portion planned), must be:

$$i= \{M\text{-}F\text{-}[D+(Q+S'vp[1+s]/\{360[c\text{-}q]\})/I\text{-}U]$$
$$/[1\text{-}t]\}/\{S'[1+s]\}$$
$$= \{41,548\text{-}29,750\text{-}[2,370+(35,351$$
$$+48,851*0.1900*240$$
$$[1+0.0500]/\{360[1.2000$$
$$-1.0100]\})/0.9900\text{-}64,720]$$
$$/[1\text{-}0.3000]\}$$
$$/\{48,851[1+0.0500]\}$$
$$= \underline{1.00\%}$$

Corporate IFRS-GAAP (B/S-I/S),ISBN-13: **978-1721066308**, ISBN-10: **1721066306**

Law-8438:

If both (**q**= [C-P]/X= Quick or Acid Test Ratio= <u>1.01</u> times), (**M**= Margin of Contribution= <u>41,548</u> millions USD), (**F**= Fixed Cost= <u>29,740</u> millions USD), (**I**= L/E= Leverage or Gearing Ratio= <u>99.00%</u>), (**c**= C/X= Current Ratio= <u>1.20</u> times), (**D**= Dividend Paid= <u>2,370</u> millions USD), (**t**= T/B= Tax Rate= <u>30.00%</u>), (**v**= V/S= Variable Portion= <u>19.00%</u>), (**i**= I/S= Interest Portion= <u>1.00%</u>), (**Q**= Quoted Longterm Liabilities= <u>35,351</u> millions USD), (**S'**= Sales of Past Year= <u>48,851</u> millions USD), (**p**= 360 P/V= Procured Inventory Days= <u>240</u> Days), and (**U**= Utilized or Starting Capital= <u>64,720</u> millions USD), then it's (**s**= Sales Growth planned), must be:

$$s= \{M\text{-}F\text{-}[D+(Q+S\text{'}vp[1+s]/\{360[c\text{-}q]\})/I\text{-}U]$$
$$/[1\text{-}t]\}/[S\text{'}i]\text{-}1$$
$$= \{41,548\text{-}29,750\text{-}[2,370+(35,351$$
$$+48,851*0.1900*240$$
$$[1+0.0500]/\{360[1.2000$$
$$-1.0100]\})/0.9900\text{-}64,720]$$
$$/[1\text{-}0.3000]\}/[48,851$$
$$*0.0100]\text{-}1$$
$$= \underline{5.00\%}$$

Law-8439:

If both (**q**= [C-P]/X= Quick or Acid Test Ratio= 1.01 times), (**M**= Margin of Contribution= 41,548 millions USD), (**F**= Fixed Cost= 29,740 millions USD), (**I**= L/E= Leverage or Gearing Ratio= 99.00%), (**c**= C/X= Current Ratio= 1.20 times), (**D**= Dividend Paid= 2,370 millions USD), (**s**= [S/S']-1= Sales Growth= 5.00%), (**v**= V/S= Variable Portion= 19.00%), (**i**= I/S= Interest Portion= 1.00%), (**Q**= Quoted Longterm Liabilities= 35,351 millions USD), (**S'**= Sales of Past Year= 48,851 millions USD), (**p**= 360 P/V= Procured Inventory Days= 240 Days), and (**U**= Utilized or Starting Capital= 64,720 millions USD), then it's (**t**= Tax Rate planned), is:

$$t= 1-[D+(Q+S'vp[1+s]/\{360[c-q]\})/I-U]$$
$$/\{M-F-S'i[1+s]\}$$

$$= 1-[2,370+(35,351+48,851*0.1900$$
$$*240[1+0.0500]/\{360[1.2000$$
$$-1.0100]\})/0.9900-64,720]$$
$$/\{41,548-29,750-48,851$$
$$*0.0100[1+0.0500]\}$$

$$= \underline{30.00\%}$$

Corporate IFRS-GAAP (B/S-I/S),ISBN-13: **978-1721066308**, ISBN-10: **1721066306**

<u>Law-8440</u>:

If both (**q**= [C-P]/X= Quick or Acid Test Ratio= <u>1.01</u> times), (**M**= Margin of Contribution= <u>41,548</u> millions USD), (**F**= Fixed Cost= <u>29,740</u> millions USD), (**I**= L/E= Leverage or Gearing Ratio= <u>99.00%</u>), (**c**= C/X= Current Ratio= <u>1.20</u> times), (**t**= T/B= Tax Rate= <u>30.00%</u>), (**s**= [S/S']-1= Sales Growth= <u>5.00%</u>), (**v**= V/S= Variable Portion= <u>19.00%</u>), (**i**= I/S= Interest Portion= <u>1.00%</u>), (**Q**= Quoted Longterm Liabilities= <u>35,351</u> millions USD), (**S'**= Sales of Past Year= <u>48,851</u> millions USD), (**p**= 360 P/V= Procured Inventory Days= <u>240</u> Days), and (**U**= Utilized or Starting Capital= <u>64,720</u> millions USD), then it's (**D**= Dividend Paid planned), is:

$$D= U+\{M\text{-}F\text{-}S'i[1+s]\}[1\text{-}t]\text{-}(Q+S'vp[1+s]$$
$$/\{360[c\text{-}q]\})/I$$

$$= 64,720+\{41,548\text{-}29,750\text{-}48,851$$
$$*0.0100[1+0.0500]\}$$
$$[1\text{-}0.3000]\text{-}(35,351+48,851$$
$$*0.1900*240[1+0.0500]$$
$$/\{360[1.2000\text{-}1.0100]\})$$
$$/0.9900$$

$$= \underline{2,370} \text{ millions USD}$$

Corporate IFRS-GAAP (B/S-I/S),ISBN-13: **978-1721066308**, ISBN-10: **1721066306**

Law-8441:

If both (**q**= [C-P]/X= Quick or Acid Test Ratio= 1.01 times), (**M**= Margin of Contribution= 41,548 millions USD), (**F**= Fixed Cost= 29,740 millions USD), (**I**= L/E= Leverage or Gearing Ratio= 99.00%), (**c**= C/X= Current Ratio= 1.20 times), (**t**= T/B= Tax Rate= 30.00%), (**s**= [S/S']-1= Sales Growth= 5.00%), (**D**= Dividend Paid= 2,370 millions USD), (**i**= I/S= Interest Portion= 1.00%), (**Q**= Quoted Longterm Liabilities= 35,351 millions USD), (**S'**= Sales of Past Year= 48,851 millions USD), (**p**= 360 P/V= Procured Inventory Days= 240 Days), and (**U**= Utilized or Starting Capital= 64,720 millions USD), then it's (**v**= Variable Portion planned), is:

$$v= 360[c\text{-}q][I\,(U+\{M\text{-}F\text{-}S'i[1+s]\}[1\text{-}t]\text{-}D)\text{-}Q]$$
$$/\{S'p[1+s]\}$$

$$= 360[1.2000\text{-}1.0100][0.9900$$
$$(64,720+\{41,548\text{-}29,750$$
$$-48,851*0.0100[1+0.0500]\}$$
$$[1\text{-}0.3000]\text{-}2,370)\text{-}35,351]$$
$$/\{48,851*240[1+0.0500]\}$$

$$= \underline{19.00\%}$$

Corporate IFRS-GAAP (B/S-I/S),ISBN-13: **978-1721066308**, ISBN-10: **1721066306**

Law-8442:

If both (q= [C-P]/X= Quick or Acid Test Ratio= 1.01 times), (M= Margin of Contribution= 41,548 millions USD), (F= Fixed Cost= 29,740 millions USD), (I= L/E= Leverage or Gearing Ratio= 99.00%), (c= C/X= Current Ratio= 1.20 times), (t= T/B= Tax Rate= 30.00%), (s= [S/S']-1= Sales Growth= 5.00%), (D= Dividend Paid= 2,370 millions USD), (i= I/S= Interest Portion= 1.00%), (Q= Quoted Longterm Liabilities= 35,351 millions USD), (S'= Sales of Past Year= 48,851 millions USD), (v= V/S= Variable Portion= 19.00%), and (U= Utilized or Starting Capital= 64,720 millions USD), then it's (p= Procured Inventory planned planned), is:

$$p = 360[c\text{-}q][I\,(U+\{M\text{-}F\text{-}S'i[1+s]\}[1\text{-}t]\text{-}D)\text{-}Q]$$
$$/\{S'v[1+s]\}$$

$$= 360[1.2000\text{-}1.0100][0.9900(64,720$$
$$+\{41,548\text{-}29,750\text{-}48,851$$
$$*0.0100[1+0.0500]\}$$
$$[1\text{-}0.3000]\text{-}2,370)\text{-}35,351]$$
$$/\{48,851*0.1900[1+0.0500]\}$$

$$= \underline{240} \text{ days}$$

Corporate IFRS-GAAP (B/S-I/S),ISBN-13: **978-1721066308**, ISBN-10: **1721066306**

Law-8443:

If both (q= [C-P]/X= Quick or Acid Test Ratio= 1.01 times), (M= Margin of Contribution= 41,548 millions USD), (F= Fixed Cost= 29,740 millions USD), (I= L/E= Leverage or Gearing Ratio= 99.00%), (p= 360 P/V= Procured Inventory Days= 240 Days), (t= T/B= Tax Rate= 30.00%), (s= [S/S']-1= Sales Growth= 5.00%), (D= Dividend Paid= 2,370 millions USD), (i= I/S= Interest Portion= 1.00%), (Q= Quoted Longterm Liabilities= 35,351 millions USD), (S'= Sales of Past Year= 48,851 millions USD), (v= V/S= Variable Portion= 19.00%), and (U= Utilized or Starting Capital= 64,720 millions USD), then it's (c= Current Ratio planned), is:

$$c= q+S'vp[1+s]/\{360[c-q][I\,(U+\{M-F$$
$$-S'i[1+s]\}[1-t]-D)-Q]\}$$
$$= 1.0100+48,851*0.1900*240$$
$$[1+0.0500]/\{360[0.9900$$
$$(64,720+\{41,548-29,750$$
$$-48,851*0.0100[1+0.0500]\}$$
$$[1-0.3000]-2,370)-35,351]\}$$
$$= 1.20 \text{ times}$$

Corporate IFRS-GAAP (B/S-I/S),ISBN-13: **978-1721066308**, ISBN-10: **1721066306**

Law-8444:

If both (**c**= C/X= Current Ratio= 1.20 times), (**M**= Margin of Contribution= 41,548 millions USD), (**F**= Fixed Cost= 29,740 millions USD), (**I**= L/E= Leverage or Gearing Ratio= 99.00%), (**p**= 360 P/V= Procured Inventory Days= 240 Days), (**t**= T/B= Tax Rate= 30.00%), (**s**= [S/S']-1= Sales Growth= 5.00%), (**D**= Dividend Paid= 2,370 millions USD), (**i**= I/S= Interest Portion= 1.00%), (**Q**= Quoted Longterm Liabilities= 35,351 millions USD), (**S'**= Sales of Past Year= 48,851 millions USD), (**v**= V/S= Variable Portion= 19.00%), and (**U**= Utilized or Starting Capital= 64,720 millions USD), then it's (**q**= Quick or Acid Test Ratio planned), is:

$$q= c\text{-}S'vp[1+s]/\{360[c\text{-}q][I\,(U+\{M\text{-}F$$
$$-S'i[1+s]\}[1\text{-}t]\text{-}D)\text{-}Q]\}$$
$$= 1.2000\text{-}48,851*0.1900*240$$
$$[1+0.0500]/\{360[0.9900$$
$$(64,720+\{41,548\text{-}29,750$$
$$-48,851*0.0100[1+0.0500]\}$$
$$[1\text{-}0.3000]\text{-}2,370)\text{-}35,351]\}$$
$$= 1.01 \text{ times}$$

Corporate IFRS-GAAP (B/S-I/S),ISBN-13: **978-1721066308**, ISBN-10: **1721066306**

<u>Law-8445</u>:

If both (**M**= Margin of Contribution= <u>41,548</u> millions USD), (**S'**= Sales of Past Year= <u>48,851</u> millions USD), (**s**= [S/S']-1= Sales Growth= <u>5.00%</u>), (**F**= Fixed Cost= <u>29,740</u> millions USD), (**I**= L/E= Leverage or Gearing Ratio= <u>99.00%</u>), (**t**= T/B= Tax Rate= <u>30.00%</u>), (**d**= D/A= Dividend Portion or Payout= <u>30.00%</u>), (**A**= After Tax Income= <u>7,899</u> millions USD), (**i**= I/S= Interest Portion= <u>1.00%</u>), (**X**= Xpress or Current Debt= <u>34,196</u> millions USD), and (**U**= Utilized or Starting Capital= <u>64,720</u> millions USD), then it's (**Q**= Quoted Longterm Liabilities planned), is:

$$Q= I(U+\{M-F-S'i[1+s]\}[1-t]-Ad)-X$$
$$= 0.9900(64,720+\{41,548-29,750$$
$$-48,851*0.0100[1+0.0500]\}$$
$$[1-0.3000]-7,899*0.3000)$$
$$-34,196$$
$$= \underline{35,351} \text{ millions USD}$$

Corporate IFRS-GAAP (B/S-I/S),ISBN-13: **978-1721066308**, ISBN-10: **1721066306**

Law-8446:

If both (**M**= Margin of Contribution= 41,548 millions USD), (**S'**= Sales of Past Year= 48,851 millions USD), (**s**= [S/S']-1= Sales Growth= 5.00%), (**F**= Fixed Cost= 29,740 millions USD), (**Q**= Quoted Longterm Liabilities= 35,351 millions USD), (**t**= T/B= Tax Rate= 30.00%), (**d**= D/A= Dividend Portion or Payout= 30.00%), (**A**= After Tax Income= 7,899 millions USD), (**i**= I/S= Interest Portion= 1.00%), (**X**= Xpress or Current Debt= 34,196 millions USD), and (**U**= Utilized or Starting Capital= 64,720 millions USD), then it's (**I** = Leverage or Gearing Ratio planned), is:

$$I = [Q+X]/(U+\{M\text{-}F\text{-}S'i[1+s]\}[1\text{-}t]\text{-}Ad)$$
$$= [35,351+34,196]/(64,720+\{41,548$$
$$-29,750\text{-}48,851*0.0100$$
$$[1+0.0500]\}[1\text{-}0.3000]$$
$$-7,899*0.3000)$$
$$= 99.00\%$$

Corporate IFRS-GAAP (B/S-I/S),ISBN-13: **978-1721066308**, ISBN-10: **1721066306**

Law-8447:

If both (**M**= Margin of Contribution= 41,548 millions USD), (**S'**= Sales of Past Year= 48,851 millions USD), (**s**= [S/S']-1= Sales Growth= 5.00%), (**F**= Fixed Cost= 29,740 millions USD), (**Q**= Quoted Longterm Liabilities= 35,351 millions USD), (**t**= T/B= Tax Rate= 30.00%), (**I**= L/E= Leverage or Gearing Ratio= 99.00%), (**d**= D/A= Dividend Portion or Payout= 30.00%), (**A**= After Tax Income= 7,899 millions USD), (**i**= I/S= Interest Portion= 1.00%), and (**X**= Xpress or Current Debt= 34,196 millions USD), and then it's (**U** = Utilized or Starting Capital planned), is:

$$U= Ad+[Q+X]/I - \{M-F-S'i[1+s]\}[1-t]$$
$$= 7,899*0.3000+[35,351+34,196]$$
$$/0.9900-\{41,548-29,750$$
$$-48,851*0.0100[1+0.0500]\}$$
$$[1-0.3000]$$
$$= 64,720 \text{ millions USD}$$

Corporate IFRS-GAAP (B/S-I/S),ISBN-13: **978-1721066308**, ISBN-10: **1721066306**

Law-8448:

If both (**U**= Utilized or Starting Capital= 64,720 millions USD), (**S'**= Sales of Past Year= 48,851 millions USD), (**s**= [S/S']-1= Sales Growth= 5.00%), (**F**= Fixed Cost= 29,740 millions USD), (**Q**= Quoted Longterm Liabilities= 35,351 millions USD), (**t**= T/B= Tax Rate= 30.00%), (**l**= L/E= Leverage or Gearing Ratio= 99.00%), (**d**= D/A= Dividend Portion or Payout= 30.00%), (**A**= After Tax Income= 7,899 millions USD), (**i**= I/S= Interest Portion= 1.00%), and (**X**= Xpress or Current Debt= 34,196 millions USD), then it's (**M**= Margin of Contribution planned), is:

$$\mathbf{M}= \mathbf{F}+\mathbf{S'i}[1+\mathbf{s}]+\{\mathbf{Ad}+[\mathbf{Q}+\mathbf{X}]/\mathbf{l}-\mathbf{U}\}/[1-\mathbf{t}]$$

$$= 29,750+48,851*0.0100[1+0.0500]$$
$$+\{7,899*0.3000+[35,351$$
$$+34,196]/0.9900-64,720\}$$
$$/[1-0.3000]$$

$$= \underline{41,548} \text{ millions USD}$$

Corporate IFRS-GAAP (B/S-I/S),ISBN-13: **978-1721066308**, ISBN-10: **1721066306**

<u>Law-8449</u>:

If both (**U**= Utilized or Starting Capital= <u>64,720</u> millions USD), (**$'**= Sales of Past Year= <u>48,851</u> millions USD), (**s**= [S/S']-1= Sales Growth= <u>5.00%</u>), (**M**= Margin of Contribution= <u>41,548</u> millions USD), (**Q**= Quoted Longterm Liabilities= <u>35,351</u> millions USD), (**t**= T/B= Tax Rate= <u>30.00%</u>), (**l**= L/E= Leverage or Gearing Ratio= <u>99.00%</u>), (**d**= D/A= Dividend Portion or Payout= <u>30.00%</u>), (**A**= After Tax Income= <u>7,899</u> millions USD), (**i**= I/S= Interest Portion= <u>1.00%</u>), and (**X**= Xpress or Current Debt= <u>34,196</u> millions USD), then it's (**F**= Fixed Cost planned), is:

$$F= M-S'i[1+s]-\{Ad+[Q+X]/l-U\}/[1-t]$$
$$= 41,548-48,851*0.0100[1+0.0500]$$
$$-\{7,899*0.3000+[35,351$$
$$+34,196]/0.9900-64,720\}$$
$$/[1-0.3000]$$
$$= \underline{29,750} \text{ millions USD}$$

Corporate IFRS-GAAP (B/S-I/S),ISBN-13: **978-1721066308**, ISBN-10: **1721066306**

Law-8450:

If both (**U**= Utilized or Starting Capital= 64,720 millions USD), (**F**= Fixed Cost= 29,740 millions USD), (**s**= [S/S']-1= Sales Growth= 5.00%), (**M**= Margin of Contribution= 41,548 millions USD), (**Q**= Quoted Longterm Liabilities= 35,351 millions USD), (**t**= T/B= Tax Rate= 30.00%), (**ʄ**= L/E= Leverage or Gearing Ratio= 99.00%), (**d**= D/A= Dividend Portion or Payout= 30.00%), (**A**= After Tax Income= 7,899 millions USD), (**i**= I/S= Interest Portion= 1.00%), and (**X**= Xpress or Current Debt= 34,196 millions USD), then it's (**S'**= Sales Past), must be:

$$S' = (M\text{-}F\text{-}\{Ad+[Q+X]/\!\!f\text{-}U\}/[1\text{-}t])/\{i[1+s]\}$$
$$= (41,548\text{-}29,750\text{-}\{7,899*0.3000$$
$$+[35,351+34,196]/0.9900$$
$$-64,720\}/[1\text{-}0.3000])$$
$$/\{0.0100[1+0.0500]\}$$
$$= 48,851 \text{ millions USD}$$

Corporate IFRS-GAAP (B/S-I/S),ISBN-13: **978-1721066308**, ISBN-10: **1721066306**

Law-8451:

If both (**U**= Utilized or Starting Capital= 64,720 millions USD), (**F**= Fixed Cost= 29,740 millions USD), (**s**= [S/S']-1= Sales Growth= 5.00%), (**M**= Margin of Contribution= 41,548 millions USD), (**Q**= Quoted Longterm Liabilities= 35,351 millions USD), (**t**= T/B= Tax Rate= 30.00%), (**l**= L/E= Leverage or Gearing Ratio= 99.00%), (**d**= D/A= Dividend Portion or Payout= 30.00%), (**A**= After Tax Income= 7,899 millions USD), (**S'**= Sales of Past Year= 48,851 millions USD), and (**X**= Xpress or Current Debt= 34,196 millions USD), then it's (**i**= Interest Portion planned), is:

$$i= (M\text{-}F\text{-}\{Ad+[Q+X]/l\text{-}U\}/[1\text{-}t])/\{S'[1+s]\}$$
$$= (41,548\text{-}29,750\text{-}\{7,899*0.3000$$
$$+[35,351+34,196]/0.9900$$
$$-64,720\}/[1\text{-}0.3000])/\{48,851$$
$$[1+0.0500]\}$$
$$= \underline{1.00\%}$$

<u>Law-8452</u>:

If both (**U**= Utilized or Starting Capital= <u>64,720</u> millions USD), (**F**= Fixed Cost= <u>29,740</u> millions USD), (**i**= I/S= Interest Portion= <u>1.00%</u>), (**M**= Margin of Contribution= <u>41,548</u> millions USD), (**Q**= Quoted Longterm Liabilities= <u>35,351</u> millions USD), (**t**= T/B= Tax Rate= <u>30.00%</u>), (**I**= L/E= Leverage or Gearing Ratio= <u>99.00%</u>), (**d**= D/A= Dividend Portion or Payout= <u>30.00%</u>), (**A**= After Tax Income= <u>7,899</u> millions USD), (**S'**= Sales of Past Year= <u>48,851</u> millions USD), and (**X**= Xpress or Current Debt= <u>34,196</u> millions USD), then it's (**s**= Sales Growth planned), is:

$$s= (M-F-\{Ad+[Q+X]/I -U\}/[1-t])/[S'i]-1$$
$$= (41,548-29,750-\{7,899*0.3000$$
$$+[35,351+34,196]/0.9900$$
$$-64,720\}/[1-0.3000])/[48,851$$
$$*0.0100]-1$$
$$= \underline{5.00\%}$$

Corporate IFRS-GAAP (B/S-I/S),ISBN-13: **978-1721066308**, ISBN-10: **1721066306**

Law-8453:

If both (**U**= Utilized or Starting Capital= 64,720 millions USD), (**F**= Fixed Cost= 29,740 millions USD), (**i**= I/S= Interest Portion= 1.00%), (**M**= Margin of Contribution= 41,548 millions USD), (**Q**= Quoted Longterm Liabilities= 35,351 millions USD), (**s**= [S/S']-1= Sales Growth= 5.00%), (**l**= L/E= Leverage or Gearing Ratio= 99.00%), (**d**= D/A= Dividend Portion or Payout= 30.00%), (**A**= After Tax Income= 7,899 millions USD), (**S'**= Sales of Past Year= 48,851 millions USD), and (**X**= Xpress or Current Debt= 34,196 millions USD), then it's (**t**= Tax Rate planned), is:

$$t= 1-\{Ad+[Q+X]/l-U\}/\{M-F-S'i[1+s]\}$$
$$= 1-\{7,899*0.3000+[35,351+34,196]$$
$$/0.9900-64,720\}/\{41,548$$
$$-29,750-48,851*0.0100$$
$$[1+0.0500]\}$$
$$= 30.00\%$$

Law-8454:

If both (**U**= Utilized or Starting Capital= 64,720 millions USD), (**F**= Fixed Cost= 29,740 millions USD), (**i**= I/S= Interest Portion= 1.00%), (**M**= Margin of Contribution= 41,548 millions USD), (**Q**= Quoted Longterm Liabilities= 35,351 millions USD), (**s**= [S/S']-1= Sales Growth= 5.00%), (**I**= L/E= Leverage or Gearing Ratio= 99.00%), (**d**= D/A= Dividend Portion or Payout= 30.00%), (**t**= T/B= Tax Rate= 30.00%), (**S'**= Sales of Past Year= 48,851 millions USD), and (**X**= Xpress or Current Debt= 34,196 millions USD), then it's (**A**= After Tax Income planned), is:

$$A= (U+\{M\text{-}F\text{-}S'i[1+s]\}[1\text{-}t]\text{-}[Q+X]/I)/d$$
$$= (64,729+\{41,548\text{-}29,750\text{-}48,851$$
$$*0.0100[1+0.0500]$$
$$[1\text{-}0.3000]\text{-}\{[35,351$$
$$+34,196]/0.9900)$$
$$/0.3000$$
$$= \underline{7,899} \text{ millions USD}$$

Corporate IFRS-GAAP (B/S-I/S),ISBN-13: **978-1721066308**, ISBN-10: **1721066306**

<u>Law-8455</u>:

If both (U= Utilized or Starting Capital= <u>64,720</u> millions USD), (F= Fixed Cost= <u>29,740</u> millions USD), (i= I/S= Interest Portion= <u>1.00%</u>), (M= Margin of Contribution= <u>41,548</u> millions USD), (Q= Quoted Longterm Liabilities= <u>35,351</u> millions USD), (s= [S/S']-1= Sales Growth= <u>5.00%</u>), (I= L/E= Leverage or Gearing Ratio= <u>99.00%</u>), (A= After Tax Income= <u>7,899</u> millions USD), (t= T/B= Tax Rate= <u>30.00%</u>), (S'= Sales of Past Year= <u>48,851</u> millions USD), and (X= Xpress or Current Debt= <u>34,196</u> millions USD), then it's (d= Dividend Portion or Payout planned), is:

$$d= (U+\{M-F-S'i[1+s]\}[1-t]-[Q+X]/I)/A$$
$$= (64,729+\{41,548-29,750-48,851$$
$$*0.0100[1+0.0500]\}$$
$$[1-0.3000]-\{[35,351$$
$$+34,196]/0.9900)/7,899$$
$$= \underline{30.00\%}$$

Corporate IFRS-GAAP (B/S-I/S),ISBN-13: **978-1721066308**, ISBN-10: **1721066306**

Law-8456:

If both (**U**= Utilized or Starting Capital= 64,720 millions USD), (**F**= Fixed Cost= 29,740 millions USD), (**i**= I/S= Interest Portion= 1.00%), (**M**= Margin of Contribution= 41,548 millions USD), (**Q**= Quoted Longterm Liabilities= 35,351 millions USD), (**s**= [S/S']-1= Sales Growth= 5.00%), (**f**= L/E= Leverage or Gearing Ratio= 99.00%), (**A**= After Tax Income= 7,899 millions USD), (**t**= T/B= Tax Rate= 30.00%), (**S'**= Sales of Past Year= 48,851 millions USD), and (**d**= D/A= Dividend Portion or Payout= 30.00%), then it's (**X**= Xpress or Current Debt planned), is:

$$X= f(U+\{M-F-S'i[1+s]\}[1-t]-Ad)-Q$$
$$= 0.9900(64,729+\{41,548-29,750$$
$$-48,851*0.0100[1+0.0500]\}$$
$$[1-0.3000]-7,899*0.3000)$$
$$-35,351$$
$$= 34,196 \text{ millions USD}$$

Corporate IFRS-GAAP (B/S-I/S),ISBN-13: **978-1721066308**, ISBN-10: **1721066306**

Law-8457:

If both (**U**= Utilized or Starting Capital= 64,720 millions USD), (**F**= Fixed Cost= 29,740 millions USD), (**i**= I/S= Interest Portion= 1.00%), (**M**= Margin of Contribution= 41,548 millions USD), (**P**= Procured Inventories= 6,497 millions USD), (**q**= Quick or Acid Test Ratio= [C-P]/X= 1.01 times), (**c**= C/X= Current Ratio= 1.20 times), (**s**= [S/S']-1= Sales Growth= 5.00%), (**l**= L/E= Leverage or Gearing Ratio= 99.00%), (**A**= After Tax Income= 7,899 millions USD), (**t**= T/B= Tax Rate= 30.00%), (**S'**= Sales of Past Year= 48,851 millions USD), and (**d**= D/A= Dividend Portion or Payout= 30.00%), then it's (**Q**= Quoted Longterm Liabilities planned), is:

$$Q= l\,(U+\{M-F-S'i[1+s]\}[1-t]-Ad)-P/[c-q]$$
$$= 0.9900(64,729+\{41,548-29,750$$
$$-48,851*0.0100[1+0.0500]\}$$
$$[1-0.3000]-7,899*0.3000)$$
$$-6,497/[1.2000-1.0100]$$
$$= 35,351 \text{ millions USD}$$

Corporate IFRS-GAAP (B/S-I/S),ISBN-13: **978-1721066308**, ISBN-10: **1721066306**

Law-8458:

If both (**U**= Utilized or Starting Capital= 64,720 millions USD), (**F**= Fixed Cost= 29,740 millions USD), (**i**= I/S= Interest Portion= 1.00%), (**M**= Margin of Contribution= 41,548 millions USD), (**P**= Procured Inventories= 6,497 millions USD), (**q**= Quick or Acid Test Ratio= [C-P]/X= 1.01 times), (**c**= C/X= Current Ratio= 1.20 times), (**s**= [S/S']-1= Sales Growth= 5.00%), (**Q**= Quoted Longterm Liabilities= 35,351 millions USD), (**A**= After Tax Income= 7,899 millions USD), (**t**= T/B= Tax Rate= 30.00%), (**S'**= Sales of Past Year= 48,851 millions USD), and (**d**= D/A= Dividend Portion or Payout= 30.00%), then it's

(**I** = Leverage or Gearing Ratio planned), is:

$$I = \{Q+P/[c\text{-}q]\}/(U+\{M\text{-}F\text{-}S'i[1+s]\}[1\text{-}t]\text{-}Ad)$$

$$= \{35,351+6,497/[1.2000\text{-}1.0100]\}$$
$$/(64,729+\{41,548\text{-}29,750$$
$$-48,851*0.0100[1+0.0500]\}$$
$$[1\text{-}0.3000]\text{-} 7,899*0.3000)$$
$$= 99.00\%$$

Corporate IFRS-GAAP (B/S-I/S),ISBN-13: **978-1721066308**, ISBN-10: **1721066306**

<u>Law-8459</u>:

If both (**I**= L/E= Leverage or Gearing Ratio= 99.00%), (**F**= Fixed Cost= 29,740 millions USD), (**i**= I/S= Interest Portion= 1.00%), (**M**= Margin of Contribution= 41,548 millions USD), (**P**= Procured Inventories= 6,497 millions USD), (**q**= [C-P]/X= Quick or Acid Test Ratio= 1.01 times), (**c**= C/X= Current Ratio= 1.20 times), (**s**= [S/S']-1= Sales Growth= 5.00%), (**Q**= Quoted Longterm Liabilities= 35,351 millions USD), (**A**= After Tax Income= 7,899 millions USD), (**t**= T/B= Tax Rate= 30.00%), (**S'**= Sales of Past Year= 48,851 millions USD), and (**d**= D/A= Dividend Portion or Payout= 30.00%), then it's (**U**= Utilized or Starting Capital planned), is:

$$U= Ad+\{Q+P/[c-q]\}/I-\{M-F-S'i[1+s]\}[1-t]$$
$$= 7,899*0.3000+\{35,351+6,497$$
$$/[1.2000-1.0100]\}/0.9900$$
$$-\{41,548-29,750-48,851$$
$$*0.0100[1+0.0500]\}$$
$$[1-0.3000]-7,899*0.3000)$$
$$= \underline{64,720} \text{ millions USD}$$

Corporate IFRS-GAAP (B/S-I/S),ISBN-13: **978-1721066308**, ISBN-10: **1721066306**

Law-8460:

If both (I= L/E= Leverage or Gearing Ratio= 99.00%), (F= Fixed Cost= 29,740 millions USD), (i= I/S= Interest Portion= 1.00%), (U= Utilized or Starting Capital= 64,720 millions USD), (P= Procured Inventories= 6,497 millions USD), (q= Quick or Acid Test Ratio= [C-P]/X= 1.01 times), (c= C/X= Current Ratio= 1.20 times), (s= [S/S']-1= Sales Growth= 5.00%), (Q= Quoted Longterm Liabilities= 35,351 millions USD), (A= After Tax Income= 7,899 millions USD), (t= T/B= Tax Rate= 30.00%), (S'= Sales of Past Year= 48,851 millions USD), and (d= D/A= Dividend Portion or Payout= 30.00%), then it's (M= Margin of Contribution planned), is:

$$M= F+S'i[1+s]+(Ad+\{Q+P/[c-q]\}/I-U)$$
$$/[1-t]$$

$$= 29,750+48,851*0.0100[1+0.0500]$$
$$+(7,899*0.3000+\{35,351$$
$$+6,497/[1.2000-1.0100]\}$$
$$/0.9900-64,720)/[1-0.3000]$$

$$= 41,548 \text{ millions USD}$$

Corporate IFRS-GAAP (B/S-I/S),ISBN-13: **978-1721066308**, ISBN-10: **1721066306**

Law-8461:

If both (I= L/E= Leverage or Gearing Ratio= 99.00%), (M= Margin of Contribution= 41,548 millions USD), (i= I/S= Interest Portion= 1.00%), (U= Utilized or Starting Capital= 64,720 millions USD), (P= Procured Inventories= 6,497 millions USD), (q= [C-P]/X= Quick or Acid Test Ratio= 1.01 times), (c= C/X= Current Ratio= 1.20 times), (s= [S/S']-1= Sales Growth= 5.00%), (Q= Quoted Longterm Liabilities= 35,351 millions USD), (A= After Tax Income= 7,899 millions USD), (t= T/B= Tax Rate= 30.00%), (S'= Sales of Past Year= 48,851 millions USD), and (d= D/A= Dividend Portion or Payout= 30.00%), then it's (F= Fixed Cost planned), is:

$$F= M\text{-}S'i[1+s]\text{-}(Ad+\{Q+P/[c\text{-}q]\}/I\text{-}U)/[1\text{-}t]$$
$$= 41,548\text{-}48,851*0.0100[1+0.0500]$$
$$\text{-}(7,899*0.3000+\{35,351$$
$$+6,497/[1.2000\text{-}1.0100]\}$$
$$/0.9900\text{-}64,720)/[1\text{-}0.3000]$$
$$= 29,750 \text{ millions USD}$$

Corporate IFRS-GAAP (B/S-I/S),ISBN-13: **978-1721066308**, ISBN-10: **1721066306**

Law-8462:

If both (**I**= L/E= Leverage or Gearing Ratio= 99.00%), (**M**= Margin of Contribution= 41,548 millions USD), (**i**= I/S= Interest Portion= 1.00%), (**U**= Utilized or Starting Capital= 64,720 millions USD), (**P**= Procured Inventories= 6,497 millions USD), (**q**= [C-P]/X= Quick or Acid Test Ratio= 1.01 times), (**c**= C/X= Current Ratio= 1.20 times), (**s**= [S/S']-1= Sales Growth= 5.00%), (**Q**= Quoted Longterm Liabilities= 35,351 millions USD), (**A**= After Tax Income= 7,899 millions USD), (**t**= T/B= Tax Rate= 30.00%), (**F**= Fixed Cost= 29,740 millions USD), and (**d**= D/A= Dividend Portion or Payout= 30.00%), then it's (**S'**= Sales Past), must be:

$$S'= [M\text{-}F\text{-}(Ad+\{Q+P/[c\text{-}q]\}/I\text{-}U)/[1\text{-}t]]$$
$$/\{i[1+s]\}$$
$$= [41,548\text{-}29,750\text{-}(7,899*0.3000$$
$$+\{35,351+6,497/[1.2000$$
$$-1.0100]\}/0.9900\text{-}64,720)$$
$$/[1\text{-}0.3000]]$$
$$/\{0.0100[1+0.0500]\}$$
$$= \underline{48,851} \text{ millions USD}$$

Corporate IFRS-GAAP (B/S-I/S),ISBN-13: **978-1721066308**, ISBN-10: **1721066306**

Law-8463:

If both (I= L/E= Leverage or Gearing Ratio= 99.00%), (M= Margin of Contribution= 41,548 millions USD), (S'= Sales of Past Year= 48,851 millions USD), (U= Utilized or Starting Capital= 64,720 millions USD), (P= Procured Inventories= 6,497 millions USD), (q= [C-P]/X= Quick or Acid Test Ratio= 1.01 times), (c= C/X= Current Ratio= 1.20 times), (s= [S/S']-1= Sales Growth= 5.00%), (Q= Quoted Longterm Liabilities= 35,351 millions USD), (A= After Tax Income= 7,899 millions USD), (t= T/B= Tax Rate= 30.00%), (F= Fixed Cost= 29,740 millions USD), and (d= D/A= Dividend Portion or Payout= 30.00%), then it's (i= Interest Portion planned), is:

$$i= [M-F-(Ad+\{Q+P/[c-q]\}/I-U)/[1-t]]/\{S'[1+s]\}$$

$$= [41,548-29,750-(7,899*0.3000$$
$$+\{35,351+6,497/[1.2000$$
$$-1.0100]\}/0.9900-64,720)$$
$$/[1-0.3000]]/\{48,851$$
$$[1+0.0500]\}$$

$$= 1.00\%$$

Corporate IFRS-GAAP (B/S-I/S),ISBN-13: **978-1721066308**, ISBN-10: **1721066306**

<u>Law-8464</u>:

If both (**I**= L/E= Leverage or Gearing Ratio= <u>99.00%</u>), (**M**= Margin of Contribution= <u>41,548</u> millions USD), (**S'**= Sales of Past Year= <u>48,851</u> millions USD), (**U**= Utilized or Starting Capital= <u>64,720</u> millions USD), (**P**= Procured Inventories= <u>6,497</u> millions USD), (**q**= [C-P]/X= Quick or Acid Test Ratio= <u>1.01</u> times), (**c**= C/X= Current Ratio= <u>1.20</u> times), (**i**= I/S= Interest Portion= <u>1.00%</u>), (**Q**= Quoted Longterm Liabilities= <u>35,351</u> millions USD), (**A**= After Tax Income= <u>7,899</u> millions USD), (**t**= T/B= Tax Rate= <u>30.00%</u>), (**F**= Fixed Cost= <u>29,740</u> millions USD), and (**d**= D/A= Dividend Portion or Payout= <u>30.00%</u>), then it's (**s**= Sales Growth planned), is:

$$s = [M\text{-}F\text{-}(Ad+\{Q+P/[c\text{-}q]\}/I\text{-}U)/[1\text{-}t]]$$
$$/[S'i]\text{-}1$$
$$= [41,548\text{-}29,750\text{-}(7,899*0.3000$$
$$+\{35,351+6,497/[1.2000$$
$$-1.0100]\}/0.9900\text{-}64,720)$$
$$/[1\text{-}0.3000]]/[48,851)$$
$$0.0100]\text{-}1$$

$$= \underline{5.00\%}$$

Corporate IFRS-GAAP (B/S-I/S),ISBN-13: **978-1721066308**, ISBN-10: **1721066306**

Law-8465:

If both (I= L/E= Leverage or Gearing Ratio= 99.00%), (M= Margin of Contribution= 41,548 millions USD), (S'= Sales of Past Year= 48,851 millions USD), (U= Utilized or Starting Capital= 64,720 millions USD), (P= Procured Inventories= 6,497 millions USD), (q= [C-P]/X= Quick or Acid Test Ratio= 1.01 times), (c= C/X= Current Ratio= 1.20 times), (i= I/S= Interest Portion= 1.00%), (Q= Quoted Longterm Liabilities= 35,351 millions USD), (A= After Tax Income= 7,899 millions USD), (s= [S/S']-1= Sales Growth= 5.00%), (F= Fixed Cost= 29,740 millions USD), and (d= D/A= Dividend Portion or Payout= 30.00%), then it's (t= Tax Rate planned), is:

$$t= 1-(Ad+\{Q+P/[c-q]\}/I-U)$$
$$/\{M-F-S'i[1+s]\}$$
$$= 1-(7,899*0.3000+\{35,351+6,497$$
$$/[1.2000-1.0100]\}/0.9900$$
$$-64,720)/\{41,548-29,750$$
$$-48,851*0.0100[1+0.0500]\}$$
$$= 30.00\%$$

Corporate IFRS-GAAP (B/S-I/S),ISBN-13: **978-1721066308**, ISBN-10: **1721066306**

<u>Law-8466</u>:

If both (**I**= L/E= Leverage or Gearing Ratio= 99.00%), (**M**= Margin of Contribution= 41,548 millions USD), (**$'**= Sales of Past Year= 48,851 millions USD), (**U**= Utilized or Starting Capital= 64,720 millions USD), (**t**= T/B= Tax Rate= 30.00%), (**q**= [C-P]/X= Quick or Acid Test Ratio= 1.01 times), (**c**= C/X= Current Ratio= 1.20 times), (**i**= I/S= Interest Portion= 1.00%), (**Q**= Quoted Longterm Liabilities= 35,351 millions USD), (**A**= After Tax Income= 7,899 millions USD), (**s**= [S/S']-1= Sales Growth= 5.00%), (**F**= Fixed Cost= 29,740 millions USD), and (**d**= D/A= Dividend Portion or Payout= 30.00%), then it's (**P**= Procured Inventories planned), is:

$$\mathbf{P} = [\mathbf{c}\text{-}\mathbf{q}][\mathbf{I}\,(\mathbf{U} + \{\mathbf{M}\text{-}\mathbf{F}\text{-}\mathbf{\$'i}[1 + \mathbf{s}]\}[1 - \mathbf{t}] - \mathbf{Ad}) - \mathbf{Q}]$$

$$= [1.2000 - 1.0100][0.9900(64,720$$
$$+ \{41,548 - 29,750 - 48,851$$
$$*0.0100[1 + 0.0500]\}$$
$$[1 - 0.3000] - 7,899$$
$$*0.3000) - 35,351]$$

$$= \underline{6,497} \text{ millions USD}$$

Corporate IFRS-GAAP (B/S-I/S),ISBN-13: **978-1721066308**, ISBN-10: **1721066306**

Law-8467:

If both (**I**= L/E= Leverage or Gearing Ratio= 99.00%), (**M**= Margin of Contribution= 41,548 millions USD), (**S'**= Sales of Past Year= 48,851 millions USD), (**U**= Utilized or Starting Capital= 64,720 millions USD), (**t**= T/B= Tax Rate= 30.00%), (**q**= [C-P]/X= Quick or Acid Test Ratio= 1.01 times), (**P**= Procured Inventories= 6,497 millions USD), (**i**= I/S= Interest Portion= 1.00%), (**Q**= Quoted Longterm Liabilities= 35,351 millions USD), (**A**= After Tax Income= 7,899 millions USD), (**s**= [S/S']-1= Sales Growth= 5.00%), (**F**= Fixed Cost= 29,740 millions USD), and (**d**= D/A= Dividend Portion or Payout= 30.00%), then it's (**c**= Current Ratio planned), is:

$$c= q+P/[I(U+\{M-F-S'i[1+s]\}[1-t]-Ad)-Q]$$

$$= 1.0100+6,497/[0.9900(64,720$$
$$+\{41,548-29,750-48,851$$
$$*0.0100[1+0.0500]\}$$
$$[1-0.3000]-7,899*0.3000)$$
$$-35,351]$$

$$= \underline{1.20} \text{ times}$$

Corporate IFRS-GAAP (B/S-I/S),ISBN-13: **978-1721066308**, ISBN-10: **1721066306**

Law-8468:

If both (I= L/E= Leverage or Gearing Ratio= 99.00%), (M= Margin of Contribution= 41,548 millions USD), (S'= Sales of Past Year= 48,851 millions USD), (U= Utilized or Starting Capital= 64,720 millions USD), (t= T/B= Tax Rate= 30.00%), (c= C/X= Current Ratio= 1.20 times), (P= Procured Inventories= 6,497 millions USD), (i= I/S= Interest Portion= 1.00%), (Q= Quoted Longterm Liabilities= 35,351 millions USD), (A= After Tax Income= 7,899 millions USD), (s= [S/S']-1= Sales Growth= 5.00%), (F= Fixed Cost= 29,740 millions USD), and (d= D/A= Dividend Portion or Payout= 30.00%), then it's (q= Quick or Acid Test Ratio planned), is:

$$q= c-P/[I(U+\{M-F-S'i[1+s]\}[1-t]-Ad)-Q]$$
$$= 1.2000-6,497/[0.9900(64,720$$
$$+\{41,548-29,750-48,851$$
$$*0.0100[1+0.0500]\}$$
$$[1-0.3000]-7,899$$
$$*0.3000)-35,351]$$

$$= 1.01 \text{ times}$$

Corporate IFRS-GAAP (B/S-I/S),ISBN-13: **978-1721066308**, ISBN-10: **1721066306**

Law-8469:

If both (**I**= L/E= Leverage or Gearing Ratio= 99.00%), (**M**= Margin of Contribution= 41,548 millions USD), (**S'**= Sales of Past Year= 48,851 millions USD), (**U**= Utilized or Starting Capital= 64,720 millions USD), (**t**= T/B= Tax Rate= 30.00%), (**q**= [C-P]/X= Quick or Acid Test Ratio= 1.01 times), (**p**= 360 P/V= Procured Inventory Days= 240 Days), (**V**= Variable Cost= 9,746 millions USD), (**i**= I/S= Interest Portion= 1.00%), (**c**= C/X= Current Ratio= 1.20 times), (**A**= After Tax Income= 7,899 millions USD), (**s**= [S/S']-1= Sales Growth= 5.00%), (**F**= Fixed Cost= 29,740 millions USD), and (**d**= D/A= Dividend Portion or Payout= 30.00%), then it's (**Q**= Quoted Longterm Liabilities planned), is:

$$Q= \textbf{\textit{I}}\,(\textbf{U}+\{\textbf{M-F-S'i}[1+\textbf{s}]\}[1\text{-}\textbf{t}]\text{-}\textbf{Ad})$$
$$-\textbf{Vp}/\{360[\textbf{c-q}]\}$$
$$= 0.9900(64,720+\{41,548\text{-}29,750$$
$$-48,851*0.0100[1+0.0500]\}$$
$$[1\text{-}0.3000]\text{-}7,899*0.3000)$$
$$-9,746*240/\{360[1.2000$$
$$-1.0100]\}$$
$$= \underline{35,351} \text{ millions USD}$$

Corporate IFRS-GAAP (B/S-I/S),ISBN-13: **978-1721066308**, ISBN-10: **1721066306**

Law-8470:

If both (**Q**= Quoted Longterm Liabilities= 35,351 millions USD), (**M**= Margin of Contribution= 41,548 millions USD), (**S'**= Sales of Past Year= 48,851 millions USD), (**U**= Utilized or Starting Capital= 64,720 millions USD), (**t**= T/B= Tax Rate= 30.00%), (**q**= [C-P]/X= Quick or Acid Test Ratio= 1.01 times), (**p**= 360 P/V= Procured Inventory Days= 240 Days), (**V**= Variable Cost= 9,746 millions USD), (**i**= I/S= Interest Portion= 1.00%), (**c**= C/X= Current Ratio= 1.20 times), (**A**= After Tax Income= 7,899 millions USD), (**s**= [S/S']-1= Sales Growth= 5.00%), (**F**= Fixed Cost= 29,740 millions USD), and (**d**= D/A= Dividend Portion or Payout= 30.00%), then it's (**l** = Leverage or Gearing Ratio planned), is:

$$l = (Q+Vp/\{360[c-q]\})/(U+\{M-F-S'i[1+s]\}$$
$$[1\ t]\ Ad)$$
$$= (35,351+9,746*240/\{360[1.2000$$
$$-1.0100]\}\ )/(64,720+\{41,548$$
$$-29,750-48,851*0.0100$$
$$[1+0.0500]\}[1-0.3000]$$
$$-7,899*0.3000)$$
$$= 99.00\%$$

Corporate IFRS-GAAP (B/S-I/S),ISBN-13: **978-1721066308**, ISBN-10: **1721066306**

Law-8471:

If both (I= L/E= Leverage or Gearing Ratio= 99.00%), (M= Margin of Contribution= 41,548 millions USD), (S'= Sales of Past Year= 48,851 millions USD), (Q= Quoted Longterm Liabilities= 35,351 millions USD), (t= T/B= Tax Rate= 30.00%), (q= [C-P]/X= Quick or Acid Test Ratio= 1.01 times), (p= 360 P/V= Procured Inventory Days= 240 Days), (V= Variable Cost= 9,746 millions USD), (i= I/S= Interest Portion= 1.00%), (c= C/X= Current Ratio= 1.20 times), (A= After Tax Income= 7,899 millions USD), (s= [S/S']-1= Sales Growth= 5.00%), (F= Fixed Cost= 29,740 millions USD), and (d= D/A= Dividend Portion or Payout= 30.00%), then it's (U= Utilized or Starting Capital planned), is:

$$U = Ad+(Q+Vp/\{360[c-q]\})/I$$
$$-\{M-F-S'i[1+s]\}[1-t]$$
$$= 7,899*0.3000 +(35,351+9,746$$
$$*240/\{360[1.2000-1.0100]\})$$
$$/0.9900-\{41,548-29,750$$
$$-48,851*0.0100[1+0.0500]\}$$
$$[1-0.3000]$$
$$= 64,720 \text{ millions USD}$$

Corporate IFRS-GAAP (B/S-I/S),ISBN-13: **978-1721066308**, ISBN-10: **1721066306**

<u>Law-8472</u>:

If both (I= L/E= Leverage or Gearing Ratio= <u>99.00%</u>), (U= Utilized or Starting Capital= <u>64,720</u> millions USD), (S'= Sales of Past Year= <u>48,851</u> millions USD), (Q= Quoted Longterm Liabilities= <u>35,351</u> millions USD), (t= T/B= Tax Rate= <u>30.00%</u>), (q= [C-P]/X= Quick or Acid Test Ratio= <u>1.01</u> times), (p= 360 P/V= Procured Inventory Days= <u>240</u> Days), (V= Variable Cost= <u>9,746</u> millions USD), (i= I/S= Interest Portion= <u>1.00%</u>), (c= C/X= Current Ratio= <u>1.20</u> times), (A= After Tax Income= <u>7,899</u> millions USD), (s= [S/S']-1= Sales Growth= <u>5.00%</u>), (F= Fixed Cost= <u>29,740</u> millions USD), and (d= D/A= Dividend Portion or Payout= <u>30.00%</u>), then it's (M= Margin of Contribution planned), is:

$$M= F+S'i[1+s]+[Ad+(Q+Vp$$
$$/\{360[c-q])/I-U]/[1-t]$$
$$= 29,750+48,851*0.0100[1+0.0500]$$
$$+[7,899*0.3000+(35,351$$
$$+9,746*240/\{360[1.2000$$
$$-1.0100]\})/0.9900-64,720]$$
$$/[1-0.3000]$$
$$= \underline{41,548} \text{ millions USD}$$

Corporate IFRS-GAAP (B/S-I/S),ISBN-13: **978-1721066308**, ISBN-10: **1721066306**

<u>Law-8473</u>:

If both (I= L/E= Leverage or Gearing Ratio= 99.00%), (U= Utilized or Starting Capital= 64,720 millions USD), (S'= Sales of Past Year= 48,851 millions USD), (Q= Quoted Longterm Liabilities= 35,351 millions USD), (t= T/B= Tax Rate= 30.00%), (q= [C-P]/X= Quick or Acid Test Ratio= 1.01 times), (p= 360 P/V= Procured Inventory Days= 240 Days), (V= Variable Cost= 9,746 millions USD), (i= I/S= Interest Portion= 1.00%), (c= C/X= Current Ratio= 1.20 times), (A= After Tax Income= 7,899 millions USD), (s= [S/S']-1= Sales Growth= 5.00%), (M= Margin of Contribution= 41,548 millions USD), and (d= D/A= Dividend Portion or Payout= 30.00%), then it's (F= Fixed Cost planned), is:

$$F = M-S'i[1+s]-[Ad+(Q+Vp$$
$$/\{360[c-q])/I -U]/[1-t]$$
$$= 41,548-48,851*0.0100[1+0.0500]$$
$$+[7,899*0.3000+(35,351$$
$$+9,746*240/\{360[1.2000$$
$$-1.0100]\})/0.9900-64,720]$$
$$/[1-0.3000]$$
$$= 29,750 \text{ millions USD}$$

Corporate IFRS-GAAP (B/S-I/S),ISBN-13: **978-1721066308**, ISBN-10: **1721066306**

Law-8474:

If both (**I**= L/E= Leverage or Gearing Ratio= 99.00%), (**U**= Utilized or Starting Capital= 64,720 millions USD), (**F**= Fixed Cost= 29,740 millions USD), (**Q**= Quoted Longterm Liabilities= 35,351 millions USD), (**t**= T/B= Tax Rate= 30.00%), (**q**= [C-P]/X= Quick or Acid Test Ratio= 1.01 times), (**p**= 360 P/V= Procured Inventory Days= 240 Days), (**V**= Variable Cost= 9,746 millions USD), (**i**= I/S= Interest Portion= 1.00%), (**c**= C/X= Current Ratio= 1.20 times), (**A**= After Tax Income= 7,899 millions USD), (**s**= [S/S']-1= Sales Growth= 5.00%), (**M**= Margin of Contribution= 41,548 millions USD), and (**d**= D/A= Dividend Portion or Payout= 30.00%), then it's (**S'**= Sales Past), must be:

$$S'= \{M\text{-}F\text{-}[Ad+(Q+Vp/\{360[c\text{-}q]\})/I\text{-}U]$$
$$/[1\text{-}t]\}/\{i[1+s]\}$$
$$= \{41{,}548\text{-}29{,}750\text{-}[7{,}899*0.3000$$
$$+(35{,}351+9{,}746*240/\{360$$
$$[1.2000\text{-}1.0100]\})/0.9900$$
$$-64{,}720]/[1\text{-}0.3000]\}$$
$$/\{0.0100[1+0.0500]\}$$
$$= \underline{48{,}851} \text{ millions USD}$$

Corporate IFRS-GAAP (B/S-I/S),ISBN-13: **978-1721066308**, ISBN-10: **1721066306**

<u>Law-8475</u>:

If both (**I**= L/E= Leverage or Gearing Ratio= 99.00%), (**U**= Utilized or Starting Capital= 64,720 millions USD), (**F**= Fixed Cost= 29,740 millions USD), (**Q**= Quoted Longterm Liabilities= 35,351 millions USD), (**t**= T/B= Tax Rate= 30.00%), (**q**= [C-P]/X= Quick or Acid Test Ratio= 1.01 times), (**p**= 360 P/V= Procured Inventory Days= 240 Days), (**V**= Variable Cost= 9,746 millions USD), (**S'**= Sales of Past Year= 48,851 millions USD), (**c**= C/X= Current Ratio= 1.20 times), (**A**= After Tax Income= 7,899 millions USD), (**s**= [S/S']-1= Sales Growth= 5.00%), (**M**= Margin of Contribution= 41,548 millions USD), and (**d**= D/A= Dividend Portion or Payout= 30.00%), then it's (**i**= Interest Portion planned), is:

$$i= \{M\text{-}F\text{-}[Ad+(Q+Vp/\{360[c\text{-}q]\})/I\text{-}U]$$
$$/[1\text{-}t]\}/\{S'[1+s]\}$$
$$= \{41,548\text{-}29,750\text{-}[7,899*0.3000$$
$$+(35,351+9,746*240/\{360$$
$$[1.2000\text{-}1.0100]\})/0.9900$$
$$-64,720]/[1\text{-}0.3000]\}/\{48,851$$
$$[1+0.0500]\}$$
$$= \underline{1.00\%}$$

Corporate IFRS-GAAP (B/S-I/S),ISBN-13: **978-1721066308**, ISBN-10: **1721066306**

Law-8476:

If both (I= L/E= Leverage or Gearing Ratio= 99.00%), (U= Utilized or Starting Capital= 64,720 millions USD), (F= Fixed Cost= 29,740 millions USD), (Q= Quoted Longterm Liabilities= 35,351 millions USD), (t= T/B= Tax Rate= 30.00%), (q= [C-P]/X= Quick or Acid Test Ratio= 1.01 times), (p= 360 P/V= Procured Inventory Days= 240 Days), (V= Variable Cost= 9,746 millions USD), (S'= Sales of Past Year= 48,851 millions USD), (c= C/X= Current Ratio= 1.20 times), (A= After Tax Income= 7,899 millions USD), (i= I/S= Interest Portion= 1.00%), (M= Margin of Contribution= 41,548 millions USD), and (d= D/A= Dividend Portion or Payout= 30.00%), then it's (s= Sales Growth planned), is:

$$s= \{M\text{-}F\text{-}[Ad+(Q+Vp/\{360[c\text{-}q]\})/I\text{-}U]$$
$$/[1\text{-}t]\}/[S'i]\text{-}1$$
$$= \{41,548\text{-}29,750\text{-}[7,899*0.3000$$
$$+(35,351+9,746*240/\{360$$
$$[1.2000\text{-}1.0100]\})/0.9900$$
$$-64,720]/[1\text{-}0.3000]\}/[48,851$$
$$*0.0100]\text{-}1$$
$$= 5.00\%$$

Corporate IFRS-GAAP (B/S-I/S),ISBN-13: **978-1721066308**, ISBN-10: **1721066306**

<u>Law-8477</u>:

If both (I= L/E= Leverage or Gearing Ratio= 99.00%), (U= Utilized or Starting Capital= 64,720 millions USD), (F= Fixed Cost= 29,740 millions USD), (Q= Quoted Longterm Liabilities= 35,351 millions USD), (s= [S/S']-1= Sales Growth= 5.00%), (q= [C-P]/X= Quick or Acid Test Ratio= 1.01 times), (p= 360 P/V= Procured Inventory Days= 240 Days), (V= Variable Cost= 9,746 millions USD), (S'= Sales of Past Year= 48,851 millions USD), (c= C/X= Current Ratio= 1.20 times), (A= After Tax Income= 7,899 millions USD), (i= I/S= Interest Portion= 1.00%), (M= Margin of Contribution= 41,548 millions USD), and (d= D/A= Dividend Portion or Payout= 30.00%), then it's (t= Tax Rate planned), is:

$$t= 1-[Ad+(Q+Vp/\{360[c-q])/I-U]$$
$$/\{M-F-S'i[1+s]\}$$
$$= 1-[7,899*0.3000+(35,351+9,746$$
$$*240/\{360[1.2000-1.0100]\})$$
$$/0.9900-64,720]/\{41,548$$
$$-29,750-48,851*0.0100$$
$$[1+0.0500]\}$$
$$= \underline{30.00\%}$$

Corporate IFRS-GAAP (B/S-I/S),ISBN-13: **978-1721066308**, ISBN-10: **1721066306**

Law-8478:

If both (I= L/E= Leverage or Gearing Ratio= 99.00%), (U= Utilized or Starting Capital= 64,720 millions USD), (F= Fixed Cost= 29,740 millions USD), (Q= Quoted Longterm Liabilities= 35,351 millions USD), (s= [S/S']-1= Sales Growth= 5.00%), (t= T/B= Tax Rate= 30.00%), (q= [C-P]/X= Quick or Acid Test Ratio= 1.01 times), (p= 360 P/V= Procured Inventory Days= 240 Days), (V= Variable Cost= 9,746 millions USD), (S'= Sales of Past Year= 48,851 millions USD), (c= C/X= Current Ratio= 1.20 times), (i= I/S= Interest Portion= 1.00%), (M= Margin of Contribution= 41,548 millions USD), and (d= D/A= Dividend Portion or Payout= 30.00%), then it's (A= After Tax Income planned), is:

$$A= [U+\{M-F-S'i[1+s][1-t]$$
$$-(Q+Vp/\{360[c-q]\})/I]/d$$
$$= [64,720+\{41,548-29,750-48,851$$
$$*0.0100[1+0.0500]$$
$$[1-0.3000]-(35,351+9,746$$
$$*240/\{360[1.2000$$
$$-1.0100]\})/0.9900]/0.3000$$
$$= 7,899 \text{ millions USD}$$

Corporate IFRS-GAAP (B/S-I/S),ISBN-13: **978-1721066308**, ISBN-10: **1721066306**

Law-8479:

If both (I= L/E= Leverage or Gearing Ratio= 99.00%), (U= Utilized or Starting Capital= 64,720 millions USD), (F= Fixed Cost= 29,740 millions USD), (Q= Quoted Longterm Liabilities= 35,351 millions USD), (s= [S/S']-1= Sales Growth= 5.00%), (t= T/B= Tax Rate= 30.00%), (q= [C-P]/X= Quick or Acid Test Ratio= 1.01 times), (p= 360 P/V= Procured Inventory Days= 240 Days), (V= Variable Cost= 9,746 millions USD), (S'= Sales of Past Year= 48,851 millions USD), (c= C/X= Current Ratio= 1.20 times), (i= I/S= Interest Portion= 1.00%), (M= Margin of Contribution= 41,548 millions USD), and (A= After Tax Income= 7,899 millions USD), then it's (d= Dividend Portion or Payout planned), is:

$$d = [U+\{M-F-S'i[1+s][1-t]$$
$$-(Q+Vp/\{360[c-q])/I]/A$$
$$= [64,720+\{41,548-29,750-48,851$$
$$*0.0100[1+0.0500]$$
$$[1-0.3000]-(35,351+9,746$$
$$*240/\{360[1.2000$$
$$-1.0100]\})/0.9900]/7,899$$
$$= 30.00\%$$

Corporate IFRS-GAAP (B/S-I/S),ISBN-13: **978-1721066308**, ISBN-10: **1721066306**

Law-8480:

If both (**I**= L/E= Leverage or Gearing Ratio= 99.00%), (**U**= Utilized or Starting Capital= 64,720 millions USD), (**F**= Fixed Cost= 29,740 millions USD), (**Q**= Quoted Longterm Liabilities= 35,351 millions USD), (**s**= [S/S']-1= Sales Growth= 5.00%), (**t**= T/B= Tax Rate= 30.00%), (**q**= [C-P]/X= Quick or Acid Test Ratio= 1.01 times), (**p**= 360 P/V= Procured Inventory Days= 240 Days), (**d**= D/A= Dividend Portion or Payout= 30.00%), (**S'**= Sales of Past Year= 48,851 millions USD), (**c**= C/X= Current Ratio= 1.20 times), (**i**= I/S= Interest Portion= 1.00%), (**M**= Margin of Contribution= 41,548 millions USD), and (**A**= After Tax Income= 7,899 millions USD), then it's (**V**= Variable Cost planned), is:

$$V= 360[c-q][I(U+\{M-F-S'i[1+s]\}$$
$$[1 t]-Ad)-Q]/p$$
$$= 360[1.2000-1.0100][0.9900$$
$$(64,720+\{41,548-29,750$$
$$-48,851*0.0100[1+0.0500]\}$$
$$[1-0.3000]-7.899*0.3000)$$
$$-35,351]/240$$
$$= 9,746 \text{ millions USD}$$

Corporate IFRS-GAAP (B/S-I/S),ISBN-13: **978-1721066308**, ISBN-10: **1721066306**

Law-8481:

If both (**I**= L/E= Leverage or Gearing Ratio= 99.00%), (**U**= Utilized or Starting Capital= 64,720 millions USD), (**F**= Fixed Cost= 29,740 millions USD), (**Q**= Quoted Longterm Liabilities= 35,351 millions USD), (**s**= [S/S']-1= Sales Growth= 5.00%), (**t**= T/B= Tax Rate= 30.00%), (**q**= [C-P]/X= Quick or Acid Test Ratio= 1.01 times), (**V**= Variable Cost= 9,746 millions USD), (**d**= D/A= Dividend Portion or Payout= 30.00%), (**S'**= Sales of Past Year= 48,851 millions USD), (**c**= C/X= Current Ratio= 1.20 times), (**i**= I/S= Interest Portion= 1.00%), (**M**= Margin of Contribution= 41,548 millions USD), and (**A**= After Tax Income= 7,899 millions USD), then it's (**p**= Procured Inventory Days planned), is:

$$\mathbf{p}= 360[\mathbf{c}\text{-}\mathbf{q}][\mathbf{I}/(\mathbf{U}+\{\mathbf{M}\text{-}\mathbf{F}\text{-}\mathbf{S'i}[1+\mathbf{s}]\}$$
$$[1\text{-}\mathbf{t}]\text{-}\mathbf{Ad})\text{-}\mathbf{Q}]/\mathbf{V}$$
$$= 360[1.2000\text{-}1.0100][0.9900$$
$$(64,720+\{41,548\text{-}29,750$$
$$-48,851*0.0100[1+0.0500]\}$$
$$[1\text{-}0.3000]\text{-}7,899*0.3000)$$
$$-35,351]/9,746$$

$$= \underline{240} \text{ days}$$

Corporate IFRS-GAAP (B/S-I/S),ISBN-13: **978-1721066308**, ISBN-10: **1721066306**

<u>Law-8482</u>:

If both ($\textit{\textbf{F}}$= L/E= Leverage or Gearing Ratio= 99.00%), ($\textbf{U}$= Utilized or Starting Capital= <u>64,720</u> millions USD), ($\textbf{F}$= Fixed Cost= <u>29,740</u> millions USD), ($\textbf{Q}$= Quoted Longterm Liabilities= <u>35,351</u> millions USD), ($\textbf{s}$= [S/S']-1= Sales Growth= <u>5.00%</u>), ($\textbf{t}$= T/B= Tax Rate= <u>30.00%</u>), ($\textbf{q}$= [C-P]/X= Quick or Acid Test Ratio= <u>1.01</u> times), ($\textbf{V}$= Variable Cost= <u>9,746</u> millions USD), ($\textbf{d}$= D/A= Dividend Portion or Payout= <u>30.00%</u>), ($\textbf{S'}$= Sales of Past Year= <u>48,851</u> millions USD), ($\textbf{p}$= 360 P/V= Procured Inventory Days= <u>240</u> Days), ($\textbf{i}$= I/S= Interest Portion= <u>1.00%</u>), ($\textbf{M}$= Margin of Contribution= <u>41,548</u> millions USD), and ($\textbf{A}$= After Tax Income= <u>7,899</u> millions USD), then it's ($\textbf{c}$= Current Ratio planned), is:

$$c = q + Vp / \{360[\textit{\textbf{F}}/(U + \{M\text{-}F\text{-}S'i[1+s]\}$$
$$[1\text{-}t]\text{-}Ad) - Q]\}$$
$$= 1.0100 + 9,746*240 / \{360[0.9900$$
$$(64,720 + \{41,548\text{-}29,750$$
$$-48,851*0.0100[1+0.0500]\}$$
$$[1\text{-}0.3000]\text{-}7,899*0.3000)$$
$$-35,351]\}$$
$$= \underline{1.20} \text{ times}$$

Corporate IFRS-GAAP (B/S-I/S),ISBN-13: **978-1721066308**, ISBN-10: **1721066306**

Law-8483:

If both (I= L/E= Leverage or Gearing Ratio= 99.00%), (U= Utilized or Starting Capital= 64,720 millions USD), (F= Fixed Cost= 29,740 millions USD), (Q= Quoted Longterm Liabilities= 35,351 millions USD), (s= [S/S']-1= Sales Growth= 5.00%), (t= T/B= Tax Rate= 30.00%), (c= C/X= Current Ratio= 1.20 times), (V= Variable Cost= 9,746 millions USD), (d= D/A= Dividend Portion or Payout= 30.00%), (S'= Sales of Past Year= 48,851 millions USD), (p= 360 P/V= Procured Inventory Days= 240 Days), (i= I/S= Interest Portion= 1.00%), (M= Margin of Contribution= 41,548 millions USD), and (A= After Tax Income= 7,899 millions USD), then it's (q= Quick or Acid Test Ratio planned), is:

$$q= c\text{-}Vp/\{360[I\,(U+\{M\text{-}F\text{-}S'i[1+s]\}\\[1\text{-}t]\text{-}Ad)\text{-}Q]\}$$

$$= 1.2000\text{-}9,746*240/\{360[0.9900\\(64,720+\{41,548\text{-}29,750\\-48,851*0.0100[1+0.0500]\}\\[1\text{-}0.3000]\text{-}7,899*0.3000)\\-35,351]\}$$

$$= 1.01 \text{ times}$$

Corporate IFRS-GAAP (B/S-I/S),ISBN-13: **978-1721066308**, ISBN-10: **1721066306**

<u>Law-8484</u>:

If both (**I**= L/E= Leverage or Gearing Ratio= <u>99.00%</u>), (**U**= Utilized or Starting Capital= <u>64,720</u> millions USD), (**F**= Fixed Cost= <u>29,740</u> millions USD), (**q**= [C-P]/X= Quick or Acid Test Ratio= <u>1.01</u> times), (**s**= [S/S']-1= Sales Growth= <u>5.00%</u>), (**t**= T/B= Tax Rate= <u>30.00%</u>), (**c**= C/X= Current Ratio= <u>1.20</u> times), (**$**= Sales or Revenues= <u>51,294</u> millions USD), (**v**= V/S= Variable Portion= <u>19.00%</u>), (**d**= D/A= Dividend Portion or Payout= <u>30.00%</u>), (**$'**= Sales of Past Year= <u>48,851</u> millions USD), (**p**= 360 P/V= Procured Inventory Days= <u>240</u> Days), (**i**= I/S= Interest Portion= <u>1.00%</u>), (**M**= Margin of Contribution= <u>41,548</u> millions USD), and (**A**= After Tax Income= <u>7,899</u> millions USD), then it's (**Q**= Quoted Longterm Liabilities planned), is:

$$\mathbf{Q}= \mathbf{I}\,(\mathbf{U}+\{\mathbf{M}\text{-}\mathbf{F}\text{-}\mathbf{\$'i}[1+\mathbf{s}]\}\,[1\text{-}\mathbf{t}]\text{-}\mathbf{Ad})$$
$$-\mathbf{\$vp}/\{360[\mathbf{c}\text{-}\mathbf{q}]\}$$
$$= 0.9900(64,720+\{41,548\text{-}29,750$$
$$-48,851*0.0100[1+0.0500]\}$$
$$[1\text{-}0.3000]\text{-}7,899*0.3000)$$
$$- 51,294*0.1900*240/\{360$$
$$[1.2000\text{-}1.0100]\}$$
$$= \underline{35,351}\ \text{millions USD}$$

Corporate IFRS-GAAP (B/S-I/S),ISBN-13: **978-1721066308**, ISBN-10: **1721066306**

Law-8485:

If both ($\mathbf{Q}$= Quoted Longterm Liabilities= 35,351 millions USD), ($\mathbf{U}$= Utilized or Starting Capital= 64,720 millions USD), ($\mathbf{F}$= Fixed Cost= 29,740 millions USD), ($\mathbf{q}$= [C-P]/X= Quick or Acid Test Ratio= 1.01 times), ($\mathbf{s}$= [S/S']-1= Sales Growth= 5.00%), ($\mathbf{t}$= T/B= Tax Rate= 30.00%), ($\mathbf{c}$= C/X= Current Ratio= 1.20 times), ($\mathbf{S}$= Sales or Revenues= 51,294 millions USD), ($\mathbf{v}$= V/S= Variable Portion= 19.00%), ($\mathbf{d}$= D/A= Dividend Portion or Payout= 30.00%), ($\mathbf{S'}$= Sales of Past Year= 48,851 millions USD), ($\mathbf{p}$= 360 P/V= Procured Inventory Days= 240 Days), ($\mathbf{i}$= I/S= Interest Portion= 1.00%), ($\mathbf{M}$= Margin of Contribution= 41,548 millions USD), and ($\mathbf{A}$= After Tax Income= 7,899 millions USD), then it's ($\mathbf{I}$ = Leverage or Gearing Ratio planned), is:

$$\mathbf{I} = (\mathbf{Q}+\mathbf{Svp}/\{360[\mathbf{c}\text{-}\mathbf{q}]\})/(\mathbf{U}+\{\mathbf{M}\text{-}\mathbf{F}\text{-}\mathbf{S'i}[1+\mathbf{s}]\}[1\text{-}\mathbf{t}]\text{-}\mathbf{Ad})$$

$$= (35,351+51,294*0.1900*240/\{360[1.2000-1.0100]\})/(64,720+\{41,548-29,750-48,851*0.0100[1+0.0500]\}[1-0.3000]-7,899*0.3000)$$

$$= 99.00\%$$

Corporate IFRS-GAAP (B/S-I/S),ISBN-13: **978-1721066308**, ISBN-10: **1721066306**

Law-8486:

If both (**Q**= Quoted Longterm Liabilities= 35,351 millions USD), (**l**= L/E= Leverage or Gearing Ratio= 99.00%), (**F**= Fixed Cost= 29,740 millions USD), (**q**= [C-P]/X= Quick or Acid Test Ratio= 1.01 times), (**s**= [S/S']-1= Sales Growth= 5.00%), (**t**= T/B= Tax Rate= 30.00%), (**c**= C/X= Current Ratio= 1.20 times), (**$**= Sales or Revenues= 51,294 millions USD), (**v**= V/S= Variable Portion= 19.00%), (**d**= D/A= Dividend Portion or Payout= 30.00%), (**$'**= Sales of Past Year= 48,851 millions USD), (**p**= 360 P/V= Procured Inventory Days= 240 Days), (**i**= I/S= Interest Portion= 1.00%), (**M**= Margin of Contribution= 41,548 millions USD), and (**A**= After Tax Income= 7,899 millions USD), then it's (**U**= Utilized or Starting Capital planned), is:

$$U= Ad+(Q+$vp/\{360[c\ q]\})/l$$
$$-\{M-F-$'i[1+s]\}[1-t]$$
$$= 7,899*0.3000 +(35,351+51,294$$
$$*0.1900*240/\{360[1.2000$$
$$-1.0100]\})/0.9900-\{41,548$$
$$-29,750-48,851*0.0100$$
$$[1+0.0500]\}[1-0.3000]$$
$$= 64,720 \text{ millions USD}$$

Corporate IFRS-GAAP (B/S-I/S),ISBN-13: **978-1721066308**, ISBN-10: **1721066306**

Law-8487:

If both (**Q**= Quoted Longterm Liabilities= 35,351 millions USD), (**l**= L/E= Leverage or Gearing Ratio= 99.00%), (**F**= Fixed Cost= 29,740 millions USD), (**q**= [C-P]/X= Quick or Acid Test Ratio= 1.01 times), (**s**= [S/S']-1= Sales Growth= 5.00%), (**t**= T/B= Tax Rate= 30.00%), (**c**= C/X= Current Ratio= 1.20 times), (**$**= Sales or Revenues= 51,294 millions USD), (**v**= V/S= Variable Portion= 19.00%), (**d**= D/A= Dividend Portion or Payout= 30.00%), (**$'**= Sales of Past Year= 48,851 millions USD), (**p**= 360 P/V= Procured Inventory Days= 240 Days), (**i**= I/S= Interest Portion= 1.00%), (**U**= Utilized or Starting Capital= 64,720 millions USD), and (**A**= After Tax Income= 7,899 millions USD), then it's (**M**= Margin of Contribution planned), is:

$$M= F+\$'i[1+s]+[Ad+(Q+\$vp$$
$$/\{360[c-q]\})/l -U]/[1-t]$$
$$= 29,750+48,851*0.0100[1+0.0500]$$
$$+[7,899*0.3000+(35,351$$
$$+51,294*0.1900*240/\{360$$
$$[1.2000-1.0100]\})/0.9900$$
$$-64,720]/[1-0.3000]$$
$$= 41,548 \text{ millions USD}$$

Corporate IFRS-GAAP (B/S-I/S),ISBN-13: **978-1721066308**, ISBN-10: **1721066306**

<u>Law-8488</u>:

If both ($\mathbf{Q}$= Quoted Longterm Liabilities= <u>35,351</u> millions USD), ($\mathbf{l}$= L/E= Leverage or Gearing Ratio= <u>99.00%</u>), ($\mathbf{M}$= Margin of Contribution= <u>41,548</u> millions USD), ($\mathbf{q}$= [C-P]/X= Quick or Acid Test Ratio= <u>1.01</u> times), ($\mathbf{s}$= [S/S']-1= Sales Growth= <u>5.00%</u>), ($\mathbf{t}$= T/B= Tax Rate= <u>30.00%</u>), ($\mathbf{c}$= C/X= Current Ratio= <u>1.20</u> times), ($\mathbf{S}$= Sales or Revenues= <u>51,294</u> millions USD), ($\mathbf{v}$= V/S= Variable Portion= <u>19.00%</u>), ($\mathbf{d}$= D/A= Dividend Portion or Payout= <u>30.00%</u>), ($\mathbf{S'}$= Sales of Past Year= <u>48,851</u> millions USD), ($\mathbf{p}$= 360 P/V= Procured Inventory Days= <u>240</u> Days), ($\mathbf{i}$= I/S= Interest Portion= <u>1.00%</u>), ($\mathbf{U}$= Utilized or Starting Capital= <u>64,720</u> millions USD), and ($\mathbf{A}$= After Tax Income= <u>7,899</u> millions USD), then it's ($\mathbf{F}$= Fixed Cost planned), is:

$$F= M-S'i[1+s]-[Ad+(Q+Sup$$
$$/\{360[c-q]\})/l-U]/[1-t]$$
$$= 41,548-48,851*0.0100[1+0.0500]$$
$$-[7,899*0.3000 +(35,351$$
$$+51,294*0.1900*240/\{360$$
$$[1.2000-1.0100]\})/0.9900$$
$$-64,720]/[1-0.3000]$$
$$= \underline{29,750} \text{ millions USD}$$

Corporate IFRS-GAAP (B/S-I/S),ISBN-13: **978-1721066308**, ISBN-10: **1721066306**

Law-8489:

If both (**Q**= Quoted Longterm Liabilities= 35,351 millions USD), (**l**= L/E= Leverage or Gearing Ratio= 99.00%), (**M**= Margin of Contribution= 41,548 millions USD), (**q**= [C-P]/X= Quick or Acid Test Ratio= 1.01 times), (**s**= [S/S']-1= Sales Growth= 5.00%), (**t**= T/B= Tax Rate= 30.00%), (**c**= C/X= Current Ratio= 1.20 times), (**$**= Sales or Revenues= 51,294 millions USD), (**v**= V/S= Variable Portion= 19.00%), (**d**= D/A= Dividend Portion or Payout= 30.00%), (**F**= Fixed Cost= 29,740 millions USD), (**p**= 360 P/V= Procured Inventory Days= 240 Days), (**i**= I/S= Interest Portion= 1.00%), (**U**= Utilized or Starting Capital= 64,720 millions USD), and (**A**= After Tax Income= 7,899 millions USD), then it's (**$'**= Sales Past), must be:

$$\$'= \{M-F-[Ad+(Q+\$vp/\{360[c-q]\})/l-U]$$
$$/[1-t]\}/\{i[1+s]\}$$
$$= \{41,548-29,750-[7,899*0.3000$$
$$+(35,351+51,294*0.1900$$
$$*240/\{360[1.2000$$
$$-1.0100]\})/0.9900-64,720]$$
$$/[1-0.3000]\}/\{0.0100$$
$$[1+0.0500]\}$$
$$= 48,851 \text{ millions USD}$$

Corporate IFRS-GAAP (B/S-I/S),ISBN-13: **978-1721066308**, ISBN-10: **1721066306**

Law-8490:

If both (**Q**= Quoted Longterm Liabilities= 35,351 millions USD), (**I**= L/E= Leverage or Gearing Ratio= 99.00%), (**M**= Margin of Contribution= 41,548 millions USD), (**q**= [C-P]/X= Quick or Acid Test Ratio= 1.01 times), (**s**= [S/S']-1= Sales Growth= 5.00%), (**t**= T/B= Tax Rate= 30.00%), (**c**= C/X= Current Ratio= 1.20 times), (**S**= Sales or Revenues= 51,294 millions USD), (**v**= V/S= Variable Portion= 19.00%), (**d**= D/A= Dividend Portion or Payout= 30.00%), (**F**= Fixed Cost= 29,740 millions USD), (**p**= 360 P/V= Procured Inventory Days= 240 Days), (**S'**= Sales of Past Year= 48,851 millions USD), (**U**= Utilized or Starting Capital= 64,720 millions USD), and (**A**= After Tax Income= 7,899 millions USD), then it's (**i**= Interest Portion planned), is:

$$i= \{M-F-[Ad+(Q+Svp/\{360[c-q]\})/I-U]$$
$$/[1-t]\}/\{S'[1+s]\}$$
$$= \{41,548-29,750-[7,899*0.3000$$
$$+(35,351+51,294*0.1900$$
$$*240/\{360[1.2000-1.0100]\})$$
$$/0.9900-64,720]/[1-0.3000]\}$$
$$/\{48,851[1+0.0500]\}$$
$$= 1.00\%$$

Corporate IFRS-GAAP (B/S-I/S),ISBN-13: **978-1721066308**, ISBN-10: **1721066306**

Law-8491:

If both (**Q**= Quoted Longterm Liabilities= 35,351 millions USD), (**I**= L/E= Leverage or Gearing Ratio= 99.00%), (**M**= Margin of Contribution= 41,548 millions USD), (**q**= [C-P]/X= Quick or Acid Test Ratio= 1.01 times), (**i**= I/S= Interest Portion= 1.00%), (**t**= T/B= Tax Rate= 30.00%), (**c**= C/X= Current Ratio= 1.20 times), (**S**= Sales or Revenues= 51,294 millions USD), (**v**= V/S= Variable Portion= 19.00%), (**d**= D/A= Dividend Portion or Payout= 30.00%), (**F**= Fixed Cost= 29,740 millions USD), (**p**= 360 P/V= Procured Inventory Days= 240 Days), (**S'**= Sales of Past Year= 48,851 millions USD), (**U**= Utilized or Starting Capital= 64,720 millions USD), and (**A**= After Tax Income= 7,899 millions USD), then it's (**s**= Sales Growth planned), is:

$$s= \{M\text{-}F\text{-}[Ad+(Q+Svp/\{360[c\text{-}q]\})/I\text{-}U]$$
$$/[1\text{-}t]\}/[S'i]\text{-}1$$
$$= \{41,548\text{-}29,750\text{-}[7,899*0.3000$$
$$+(35,351+51,294*0.1900$$
$$*240/\{360[1.2000\text{-}1.0100]\})$$
$$/0.9900\text{-}64,720]/[1\text{-}0.3000]\}$$
$$/[48,851*0.0100]\text{-}1$$
$$= 5.00\%$$

Corporate IFRS-GAAP (B/S-I/S),ISBN-13: **978-1721066308**, ISBN-10: **1721066306**

<u>Law-8492</u>:

If both (**Q**= Quoted Longterm Liabilities= <u>35,351</u> millions USD), (**I**= L/E= Leverage or Gearing Ratio= <u>99.00%</u>), (**M**= Margin of Contribution= <u>41,548</u> millions USD), (**q**= [C-P]/X= Quick or Acid Test Ratio= <u>1.01</u> times), (**i**= I/S= Interest Portion= <u>1.00%</u>), (**s**= [S/S']-1= Sales Growth= <u>5.00%</u>), (**c**= C/X= Current Ratio= <u>1.20</u> times), (**S**= Sales or Revenues= <u>51,294</u> millions USD), (**v**= V/S= Variable Portion= <u>19.00%</u>), (**d**= D/A= Dividend Portion or Payout= <u>30.00%</u>), (**F**= Fixed Cost= <u>29,740</u> millions USD), (**p**= 360 P/V= Procured Inventory Days= <u>240</u> Days), (**S'**= Sales of Past Year= <u>48,851</u> millions USD), (**U**= Utilized or Starting Capital= <u>64,720</u> millions USD), and (**A**= After Tax Income= <u>7,899</u> millions USD), then it's (**t**= Tax Rate planned), is:

$$t= 1-[Ad+(Q+Svp/\{360[c-q]\})/I-U]$$
$$/\{M-F-S'i[1+s]\}$$
$$= 1-[7,899*0.3000+(35,351+51,294$$
$$*0.1900*240/\{360[1.2000$$
$$-1.0100]\})/0.9900-64,720]$$
$$/\{41,548-29,750-48,851$$
$$*0.0100[1+0.0500]\}$$
$$= \underline{30.00\%}$$

Corporate IFRS-GAAP (B/S-I/S),ISBN-13: **978-1721066308**, ISBN-10: **1721066306**

Law-8493:

If both (**Q**= Quoted Longterm Liabilities= 35,351
millions USD), (**l**= L/E= Leverage or Gearing Ratio=
99.00%), (**M**= Margin of Contribution= 41,548
millions USD), (**q**= [C-P]/X= Quick or Acid Test
Ratio= 1.01 times), (**i**= I/S= Interest Portion= 1.00%),
(**s**= [S/S']-1= Sales Growth= 5.00%), (**c**= C/X=
Current Ratio= 1.20 times), (**S**= Sales or Revenues=
51,294 millions USD), (**v**= V/S= Variable Portion=
19.00%), (**d**= D/A= Dividend Portion or Payout=
30.00%), (**F**= Fixed Cost= 29,740 millions USD),
(**p**= 360 P/V= Procured Inventory Days= 240 Days),
(**S'**= Sales of Past Year= 48,851 millions USD), and
(**t**= T/B= Tax Rate= 30.00%), then it's (**A**= After Tax
Income planned), is:

$$A= [U+\{M\text{-}F\text{-}S'i[1+s]\}[1\text{-}t]$$
$$-(Q+Svp/\{360[c\text{-}q]\})/l]/d$$
$$= [64,720+\{41,548\text{-}29,750\text{-}48,851$$
$$*0.0100[1+0.0500]\}$$
$$[1\text{-}0.3000]\text{-}(35,351+51,294$$
$$*0.1900*240/\{360[1.2000$$
$$-1.0100]\})/0.9900]/0.3000$$
$$= 7,899 \text{ millions USD}$$

Corporate IFRS-GAAP (B/S-I/S),ISBN-13: **978-1721066308**, ISBN-10: **1721066306**

Law-8494:

If both (**Q**= Quoted Longterm Liabilities= 35,351 millions USD), (**I**= L/E= Leverage or Gearing Ratio= 99.00%), (**M**= Margin of Contribution= 41,548 millions USD), (**q**= [C-P]/X= Quick or Acid Test Ratio= 1.01 times), (**i**= I/S= Interest Portion= 1.00%), (**s**= [S/S']-1= Sales Growth= 5.00%), (**c**= C/X= Current Ratio= 1.20 times), (**S**= Sales or Revenues= 51,294 millions USD), (**v**= V/S= Variable Portion= 19.00%), (**A**= After Tax Income= 7,899 millions USD), (**F**= Fixed Cost= 29,740 millions USD), (**p**= 360 P/V= Procured Inventory Days= 240 Days), (**S'**= Sales of Past Year= 48,851 millions USD), and (**t**= T/B= Tax Rate= 30.00%), then it's (**d**= Dividend Portion or Payout planned), is:

$$d= [U+\{M\text{-}F\text{-}S'i[1+s]\}[1\text{-}t]$$
$$-(Q+Svp/\{360[c\text{-}q]\})/I]/A$$
$$= [64,720+\{41,548\text{-}29,750\text{-}48,851$$
$$*0.0100[1+0.0500]\}$$
$$[1\text{-}0.3000]\text{-}(35,351+51,294$$
$$*0.1900*240/\{360[1.2000$$
$$-1.0100]\})/0.9900]/7,899$$
$$= 30.00\%$$

Corporate IFRS-GAAP (B/S-I/S),ISBN-13: **978-1721066308**, ISBN-10: **1721066306**

Law-8495:

If both (**Q**= Quoted Longterm Liabilities= 35,351 millions USD), (**f**= L/E= Leverage or Gearing Ratio= 99.00%), (**M**= Margin of Contribution= 41,548 millions USD), (**q**= [C-P]/X= Quick or Acid Test Ratio= 1.01 times), (**i**= I/S= Interest Portion= 1.00%), (**s**= [S/S']-1= Sales Growth= 5.00%), (**c**= C/X= Current Ratio= 1.20 times), (**d**= D/A= Dividend Portion or Payout= 30.00%), (**v**= V/S= Variable Portion= 19.00%), (**A**= After Tax Income= 7,899 millions USD), (**F**= Fixed Cost= 29,740 millions USD), (**p**= 360 P/V= Procured Inventory Days= 240 Days), (**$'**= Sales of Past Year= 48,851 millions USD), and (**t**= T/B= Tax Rate= 30.00%), then it's (**$**= Sales or Revenues planned), is:

$$\$= 360[\textbf{c-q}][\textbf{\textit{I}}\,(\textbf{U}+\{\textbf{M-F-$'i}[1+\textbf{s}]\}$$
$$[1\textbf{-t}]\textbf{-Ad})\textbf{-Q}]/[\textbf{vp}]$$
$$= 360[1.2000-1.0100][0.9900(64,720$$
$$+\{41,548-29,750-48,851$$
$$*0.0100[1+0.0500]\}$$
$$[1-0.3000]- 7,899*0.3000)$$
$$-35,351]/[0.1900*240]$$
$$= 51,294 \text{ millions USD}$$

Corporate IFRS-GAAP (B/S-I/S),ISBN-13: **978-1721066308**, ISBN-10: **1721066306**

Law-8496:

If both (**Q**= Quoted Longterm Liabilities= 35,351 millions USD), (**I**= L/E= Leverage or Gearing Ratio= 99.00%), (**M**= Margin of Contribution= 41,548 millions USD), (**q**= [C-P]/X= Quick or Acid Test Ratio= 1.01 times), (**i**= I/S= Interest Portion= 1.00%), (**s**= [S/S']-1= Sales Growth= 5.00%), (**c**= C/X= Current Ratio= 1.20 times), (**d**= D/A= Dividend Portion or Payout= 30.00%), (**S**= Sales or Revenues= 51,294 millions USD), (**A**= After Tax Income= 7,899 millions USD), (**F**= Fixed Cost= 29,740 millions USD), (**p**= 360 P/V= Procured Inventory Days= 240 Days), (**S'**= Sales of Past Year= 48,851 millions USD), and (**t**= T/B= Tax Rate= 30.00%), then it's (**v**= Variable Portion planned), is:

$$v= 360[\textbf{c-q}][\textbf{\textit{I}}\,(\textbf{U}+\{\textbf{M-F-S'i}[1+\textbf{s}]\}$$
$$[1\textbf{-t}]\textbf{-Ad})\ \textbf{Q}]/[\textbf{Sp}]$$
$$= 360[1.2000\text{-}1.0100][0.9900(64,720$$
$$+\{41,548\text{-}29,750\text{-}48,851$$
$$*0.0100[1+0.0500]\}$$
$$[1\text{-}0.3000]\text{-}7,899*0.3000)$$
$$-35,351]/[51,294*240]$$
$$= 19.00\%$$

Corporate IFRS-GAAP (B/S-I/S),ISBN-13: **978-1721066308**, ISBN-10: **1721066306**

Law-8497:

If both (**Q**= Quoted Longterm Liabilities= 35,351 millions USD), (**l**= L/E= Leverage or Gearing Ratio= 99.00%), (**M**= Margin of Contribution= 41,548 millions USD), (**q**= [C-P]/X= Quick or Acid Test Ratio= 1.01 times), (**i**= I/S= Interest Portion= 1.00%), (**s**= [S/S']-1= Sales Growth= 5.00%), (**c**= C/X= Current Ratio= 1.20 times), (**d**= D/A= Dividend Portion or Payout= 30.00%), (**S**= Sales or Revenues= 51,294 millions USD), (**A**= After Tax Income= 7,899 millions USD), (**F**= Fixed Cost= 29,740 millions USD), (**v**= V/S= Variable Portion= 19.00%), (**S'**= Sales of Past Year= 48,851 millions USD), and (**t**= T/B= Tax Rate= 30.00%), then it's (**p**= Procured Inventory Days planned), is:

$$p= 360[\textbf{c-q}][\textbf{\textit{l}} (\textbf{U}+\{\textbf{M-F-S'i}[1+\textbf{s}]\}$$
$$[1-\textbf{t}]-\textbf{Ad})-\textbf{Q}]/[\textbf{Sv}]$$
$$= 360[1.2000-1.0100][0.9900(64,720$$
$$+\{41,548-29,750-48,851$$
$$*0.0100[1+0.0500]\}$$
$$[1-0.3000]- 7,899*0.3000)$$
$$-35,351]/[51,294*0.1900]$$
$$= \underline{240} \text{ days}$$

Corporate IFRS-GAAP (B/S-I/S),ISBN-13: **978-1721066308**, ISBN-10: **1721066306**

Law-8498:

If both (**Q**= Quoted Longterm Liabilities= 35,351 millions USD), (**l**= L/E= Leverage or Gearing Ratio= 99.00%), (**M**= Margin of Contribution= 41,548 millions USD), (**q**= [C-P]/X= Quick or Acid Test Ratio= 1.01 times), (**i**= I/S= Interest Portion= 1.00%), (**s**= [S/S']-1= Sales Growth= 5.00%), (**p**= 360 P/V= Procured Inventory Days= 240 Days), (**d**= D/A= Dividend Portion or Payout= 30.00%), (**S**= Sales or Revenues= 51,294 millions USD), (**A**= After Tax Income= 7,899 millions USD), (**F**= Fixed Cost= 29,740 millions USD), (**v**= V/S= Variable Portion= 19.00%), (**S'**= Sales of Past Year= 48,851 millions USD), and (**t**= T/B= Tax Rate= 30.00%), then it's (**c**= Current Ratio planned), is:

$$c= q+Svp/\{360[l\,(U+\{M\text{-}F\text{-}S'i[1+s]\}$$
$$[1\text{-}t]\text{-}Ad)\text{-}Q]\}$$
$$= 1.0100+51,294*0.1900*240/\{360$$
$$[0.9900(64,720+\{41,548$$
$$-29,750\,-48,851*0.0100$$
$$[1+0.0500]\}[1\text{-}0.3000]$$
$$-7,899*0.3000)\text{-}35,351]\}$$
$$= 1.20\ \text{days}$$

Corporate IFRS-GAAP (B/S-I/S),ISBN-13: **978-1721066308**, ISBN-10: **1721066306**

Law-8499:

If both (**Q**= Quoted Longterm Liabilities= 35,351 millions USD), (**I**= L/E= Leverage or Gearing Ratio= 99.00%), (**M**= Margin of Contribution= 41,548 millions USD), (**c**= C/X= Current Ratio= 1.20 times), (**i**= I/S= Interest Portion= 1.00%), (**s**= [S/S']-1= Sales Growth= 5.00%), (**p**= 360 P/V= Procured Inventory Days= 240 Days), (**d**= D/A= Dividend Portion or Payout= 30.00%), (**S**= Sales or Revenues= 51,294 millions USD), (**A**= After Tax Income= 7,899 millions USD), (**F**= Fixed Cost= 29,740 millions USD), (**v**= V/S= Variable Portion= 19.00%), (**S'**= Sales of Past Year= 48,851 millions USD), and (**t**= T/B= Tax Rate= 30.00%), then it's (**q**= Quick or Acid Test Ratio planned), is:

$$q= c\text{-}Svp/\{360[I\,(U+\{M\text{-}F\text{-}S'i[1+s]\}$$
$$[1\text{-}t]\text{-}Ad)\text{-}Q]\}$$
$$= 1.2000\text{-}51,294*0.1900*240/\{360$$
$$[0.9900(64,720+\{41,548$$
$$-29,750\text{-}48,851*0.0100$$
$$[1+0.0500]\}[1\text{-}0.3000]$$
$$- 7,899*0.3000)\text{-}35,351]\}$$
$$= 1.01 \text{ days}$$

Corporate IFRS-GAAP (B/S-I/S),ISBN-13: **978-1721066308**, ISBN-10: **1721066306**

Law-8500:

If both (**q**= [C-P]/X= Quick or Acid Test Ratio= 1.01 times), (**U**= Utilized or Starting Capital= 64,720 millions USD), (**t**= T/B= Tax Rate= 30.00%), (**f**= L/E= Leverage or Gearing Ratio= 99.00%), (**M**= Margin of Contribution= 41,548 millions USD), (**c**= C/X= Current Ratio= 1.20 times), (**i**= I/S= Interest Portion= 1.00%), (**s**= [S/S']-1= Sales Growth= 5.00%), (**p**= 360 P/V= Procured Inventory Days= 240 Days), (**d**= D/A= Dividend Portion or Payout= 30.00%), (**A**= After Tax Income= 7,899 millions USD), (**F**= Fixed Cost= 29,740 millions USD), (**v**= V/S= Variable Portion= 19.00%), (**$'**= Sales of Past Year= 48,851 millions USD), and (**t**= T/B= Tax Rate= 30.00%), then it's (**Q**= Quoted Longterm Liabilities planned), is:

$$Q = f(U+\{M-F-\$'i[1+s]\}[1-t]-Ad)$$
$$-\$'vp[1+s]/\{360[c-q]\}$$
$$= 0.9900(64,720+\{41,548-29,750$$
$$-48,851*0.0100[1+0.0500]\}$$
$$[1-0.3000]- 7,899*0.3000)$$
$$-48,851*0.1900*240/\{360$$
$$[1.2000-1.0100]\}$$
$$= 35,351 \text{ millions USD}$$

Corporate IFRS-GAAP (B/S-I/S),ISBN-13: **978-1721066308**, ISBN-10: **1721066306**

The Rest will be Continued in
The Series of Book

Finance Construction 13

Corporate IFRS-GAAP (B/S-I/S) Engineering
Technologies No. 8,501-9,000 of 111,111 Laws

At
http: //www.Amazon.Com

Corporate IFRS-GAAP (B/S-I/S),ISBN-13: **978-1721066308**, ISBN-10: **1721066306**

REFERENCES:

A. Books:

1) **Amin, A.R. & Asikin, S.**(2011a- 110 pages), *CEO Time Matrix: Productivity Management Skill that Increase CEO Effectiveness by 12400%*, Seattle (US): Amazon Createspace. http://www.amazon.com/CEO-Time-Matrix-Productivity-Effectiveness/dp/1461066069/ref=sr_1_1?s=books&ie=UTF8&qid=1387091880&sr=1-1&keywords=ceo+time+matrix. ISBN-13: 978-1461066064, ISBN-10: 1461066069

2) **Amin, A.R. & Asikin, S.** (2014- 76 pages), *No Urgency CEO: A High-Tech Book in Modern Management*, Seattle (US): Amazon Createspace. http://www.amazon.com/No-Urgency-CEO-Hi-Tech-Management/dp/1489509356/ref=sr_1_1?s=books&ie=UTF8&qid=1396440688&sr=1-1&keywords=no+urgency. ISBN-13: 978-1489509352, ISBN-10: 1489509356

3} **Amin, A.R. & Asikin, S.** (2011b- 292 pages), *The Celestial Management: Spiritual Wisdom for All Human Beings*, Seattle (US): Amazon Createspace. http://www.amazon.com/Celestial-Management-Islamic-Wisdom-Beings/dp/1456450700/ref=sr_1_1?s=books&ie=UTF8&qid=1387092036&sr=1-1&keywords=celestial+management. ISBN-13: 978-1456450700, ISBN-10: 1456450700

4) **Asikin, S.** (2009- 548 pages), *Arigato, Obrigado... (Thanks): Portuguese-Japan 1550 1647 Historic Novel*, Seattle (US): Amazon Createspace. http://www.amazon.com/Arigato-Obrigado-Thanks-Portuguese-Japan-1550-1647/dp/1453786368/ref=sr_1_1?s=books&ie=UTF8&qid=1396442162&sr=1-1&keywords=arigato. ISBN-13: 978-1453786369, ISBN-10: 1453786368

5) **Asikin, S.** (2013a- 824 pages), *Econometric Rex: 888 Marketing Promo 6666 Six Sigma Parametric Theories (of 10 Data)*, Seattle (US): Amazon Createspace. http://www.amazon.co.uk/Econometric-Rex-Marketing-Parametric-Theories/dp/1482745259. ISBN-13: 978-1482745252, ISBN-10: 1482745259

6) **Asikin, S.** (2010b- 224 pages), *FINANCE Architecture: VISUAL art, for Corporate Finance's Beauty, Safety and Simplicity*, Seattle (US): Amazon Createspace. http://www.amazon.co.uk/FINANCE-Architecture-Corporate-Finances-Simplicity/dp/1453829547/ref=sr_1_14?s=books&ie=UTF8&qid=1385371842&sr=1-14&keywords=finance+architecture. ISBN-13: 978-1453829547, ISBN-10: 1453829547

7) **Asikin, S.** (2010c- 150 pages), *Geometric FINANCE: Useful concepts that work better than Economic Noble Laureates'* Seattle (US): Amazon Createspace. http://www.amazon.co.uk/Magic-Geometric-FINANCE-concepts-Laureates/dp/1453790284/ref=tmm_pap_title_0. ISBN-13: 978-1453790281, ISBN-10: 1453790284

Corporate IFRS-GAAP (B/S-I/S),ISBN-13: **978-1721066308**, ISBN-10: **1721066306**

8) **Asikin, S.** (2010a- 290 pages), *Investment ALGEBRA: Strategic Profitable Comprehensive Business Finance Engineering Technology*, Seattle (US): Amazon Createspace. http://www.amazon.co.uk/Investment-ALGEBRA-Profitable-Comprehensive-Engineering/dp/1453856811/ref=tmm_pap_title_0?ie=UTF8&qid=1385371309&sr=1-1, ISBN-13: 978-1453856819, ISBN-10: 1453856811

9) **Asikin, S.** (2011a- 352 pages), *State Economics: Comprehensive Macro-Micro Economics' Simple Fiscal-Monetary Export-Import Accouting, Integrated Supply-Demand Managerial ... Mathematical Engineering Visual* . Seattle (US): Amazon Createspace. http://www.amazon.co.uk/STATE-ECONOMICS-Steve-Asikin-ebook/dp/B00G89Q5SS, http://www.amazon.co.uk State-Economics-Steve-Asikin-ebook/dp/B00BBPLQDI, ISBN-13: 978-1461066095, ISBN-10: 1461066093

10) **Asikin, S.** (2013b- 260 pages), *Technoeconomistat: Economic &Statistical Technology for Financial Planning*, Seattle (US): Amazon Createspace. http://www.amazon.co.uk/Technoeconomistat-Economic-Statistical-Technology-Financial/dp/1482304244, ISBN-13: 978-1482304244, ISBN-10: 1482304244

11) **Asikin, S.** (2011b- 540 pages), *Terima Kasih: Novel Internasional Sejarah Jepang-Portugis 1550-1647*, Seattle (US): Amazon Createspace. http://www.amazon.com/Terima-Kasih-Internasional-Jepang-Portugis-1550-1647/dp/146351106X/ref=sr_1_1?s=books&ie=UTF8&qid=1396442361&sr=1-1&keywords=terima, ISBN-13: 978-1463511067, ISBN-10: 146351106X

12) **Asikin, S.** (2010d- 294 pages), *TREASURY Planology: Strategic Profitable Comprehensive Business Finance Engineering Technology*, Seattle (US): Amazon Createspace. http://www.amazon.co.uk/TREASURY-PLANOLOGY-Steve-Asikin-ebook/dp/B00G88MTFW/ref=sr_1_1?s=books&ie=UTF8&qid=1385403690&sr=1-1&keywords=Treasury+Planology. ISBN-13: 978-1453891476, ISBN-10: 1453891471

13) **Asikin, S.** (2014a- 190 pages), *Calculus Accountancy: Leibnitz Newton Pacioli's Polynomial Quantitative Finance Risk Modelling Optimization*, Seattle (US): Amazon Createspace. http://www.amazon.com/Calculus-Accountancy-Polynomial-Quantitative-Optimization/dp/1495463028/ref=sr_1_1?s=books&ie=UTF8&qid=1396441338&sr=1-1&keywords=calculus+accountancy, ISBN-13: 978-1495463020, ISBN-10: 1495463028

14) **Asikin, S.**(2014b- 318 pages), *No Parametric: No Parametric: Hyperbolic Octahedron Central 3-Dimensional Orthogonal Risk Distribution Topology*, Seattle (US): Amazon Createspace. http://www.amazon.com/Parametric-Hyperbolic-Octahedron-3-Dimensional-Distribution/dp/1496089731/ref=sr_1_1?s=books&ie=UTF8&qid=1396441543&sr=1-1&keywords=No+Parametric, ISBN-13: 978-1496089731, ISBN-10: 1496089731

Corporate IFRS-GAAP (B/S-I/S),ISBN-13: **978-1721066308**, ISBN-10: **1721066306**

15) **Asikin, S.** (2014c- 138 pages), *Quick Help: Religion vs Real-Legion Philosophy*, Seattle (US): Amazon Createspace.
http://www.amazon.com/Quick-Help-Religion-Real-Legion-Philosophy/dp/1497466520/ref=sr_1_2?s=books&ie=UTF8&qid=1396441740&sr=1-2&keywords=quick+help, ISBN-13: 978-1497466524, ISBN-10: 1497466520

16) **Asikin, S.** (2014d- 824 pages), *Anne of Denmark: King James I the Bible, Historic Plays Drama*, Seattle (US): Amazon Createspace.
http://www.amazon.com/Anne-Denmark-James-Bible-Drama/dp/149299734X/ref=sr_1_1?s=books&ie=UTF8&qid=1396441857&sr=1-1&keywords=anne+asikin, ISBN-13: 978-1492997344, ISBN-10: 149299734X

17) **Asikin, S.** (2014e- 184 pages), *Constructive Spells: Positive Spiritual Mentality Personal Development*: Seattle (US): Amazon Createspace.
http://www.amazon.com/Constructive-Spells-Spiritual-Mentality-Development/dp/1497498899/ref=sr_1_1?s=books&ie=UTF8&qid=1397664512&sr=1-1&keywords=constructive+spells; ISBN-13: 978-1497498891, ISBN-10: 1497498899

18) **Asikin, S.** (2014f- 466 pages), *Elizabeth of Russia: Another Virgin Queen Gloariana: A Romanov Plays Drama*, Seattle (US): Amazon Createspace.
http://www.amazon.com/Elizabeth-RUSSIA-Another-Gloriana-Romanov/dp/1505542480/ref=sr_1_4?s=books&ie=UTF8&qid=1423888482&sr=1-4&keywords=elizabeth+russia; ISBN-13: 978-1505542486, ISBN-10: 1505542480

19) **Asikin, S.** (2014f- 708 pages), *Shinmen Munisai:Miyamoto Musashi's Father, A Sword Samurai Plays Drama*, Seattle (US): Amazon Createspace. http://www.amazon.com/Shinmen-Munisai-Miyamoto-Musashis-Samurai/dp/1507857799/ref=sr_1_1?s=books&ie=UTF8&qid=1428094280&sr=1-1&keywords=munisai; ISBN-13: 978-1507857793, ISBN-10: 1507857799

20) **Asikin, S.** (2015a-767 pages), *Math Fin Law-1: Mathematical Financial Law for Public Listed Firms Rule 1-4441*, Seattle (US): Amazon Createspace.
http://www.amazon.com/Math-Fin-Law-Mathematical-Financial-ebook/dp/B00WLNX3Y4/ref=asap_bc?ie=UTF8; ISBN-13: 978-1511792219, ISBN-10: 1511792213

21) **Asikin, S.** (2015b-813 pages), *Math Fin Law-2: Mathematical Financial Law for Public Listed Firms Rule 4442-8712*, Seattle (US): Amazon Createspace. http://www.amazon.com/Math-Fin-Law-Mathematical-No-4442-8712/dp/1512285080/ref=sr_1_1?s=books&ie=UTF8&qid=1432732199&sr=1-1&keywords=math+fin+law+2; ISBN-13: 978-1512285086, ISBN-10: 1512285080

22) **Asikin, S.** (2015c-813 pages), *Math Fin Law-3: Mathematical Financial Law for Public Listed Firms Rule 8713-12575*, Seattle (US): Amazon Createspace. http://www.amazon.com/Math-Fin-Law-Mathematical-8712-12575/dp/1514276763/ref=sr_1_1?s=books&ie=UTF8&qid=1433982149&sr=1-1&keywords=math+fin+law+3, ISBN-13: 978-1514276761. ISBN-10: 1514276763

Corporate IFRS-GAAP (B/S-I/S),ISBN-13: **978-1721066308**, ISBN-10: **1721066306**

23) Asikin, S. (2015d-813 pages), *Math Fin Law-4: Mathematical Financial Law for Public Listed Firms Rule 12576-16333,* Seattle (US): Amazon Createspace. http://www.amazon.com/Math-Fin-Law-Mathematical-12576-16333/dp/1514685132/ref=sr_1_1?s=books&ie=UTF8&qid=1435393628&sr=1-1&keywords=math+fin+law+4, ISBN-13: 978-1514685136, ISBN-10: 1514685132

24) Asikin, S. (2015d-816 pages), *Math Fin Law-5: Mathematical Financial Law for Public Listed Firms Rule 16334-19904,* Seattle (US): Amazon Createspace. http://www.amazon.com/Math-Fin-Law-Mathematical-No-16334-19904/dp/1515073009/ref=sr_1_1?s=books&ie=UTF8&qid=1437224693&sr=1-1&keywords=math+fin+law+5. ISBN-13: 978-1515073000, ISBN-10: 1515073009

25) Asikin, S. (2015d-728 pages), *Math Fin Law-6: Mathematical Financial Law for Public Listed Firms Rule 19905-23237,* Seattle (US): Amazon Createspace. https://www.amazon.com/Math-Fin-Law-Mathematical-No-19905-23238-ebook/dp/B0129ILVNK/ref=sr_1_2?ie=UTF8&qid=1478017624&sr=8-2&keywords=Math+Fin+Law+Asikin. ISBN-13: 978-1515158691, ISBN-10: 1515158691

26) Asikin, S. (2016a-728 pages), *Math Fin Law-7: Mathematical Financial Law for Public Listed Firms Rule 23238-27115,* Seattle (US): Amazon Createspace. https://www.amazon.com/Math-Fin-Law-Mathematical-No-23238-27115/dp/1508518521/ref=la_B00428V6JU_1_27?s=books. ISBN-13: 978-1508518525, ISBN-10: 1508518521

27) Asikin, S. (2016b-728 pages), *Math Fin Law-8: Mathematical Financial Law for Public Listed Firms Rule 27116-30330,* Seattle (US): Amazon Createspace. https://www.amazon.com/dp/1541091884. https://www.amazon.com/Math-Fin-Law-Mathematical-No-27116-30330/dp/1541091884/ref=la_B00428V6JU_1_27?s=books&ie=UTF8&qid=1479398538&sr=1-27&refinements=p_82%3AB00428V6JU. ISBN-13: 978-1508518525, ISBN-10: 1508518521

28) Asikin, S. (2017a-828 pages), *Math Fin Law-9: Mathematical Financial Law for Public Listed Firms Rule 30330-33373,* Seattle (US): Amazon Createspace. https://www.amazon.com/Math-Fin-Law-Mathematical-30331-33373-ebook/dp/B01MRA8HZQ/ref=la_B00428V6JU_1_4?s=books&ie=UTF8&qid=1485921040&sr=1-4. ISBN-13: 978-1541092433, ISBN-10: 1541092430

29) Asikin, S. (2017b-823 pages), *Math Fin Law-10: Mathematical Financial Law for Public Listed Firms Rule 33374-36242,* Seattle (US): Amazon Createspace. https://www.amazon.com/s/ref=nb_sb_noss?url=search-alias%3Dstripbooks&field-keywords=math+fin+law+10. https://www.amazon.com/s/ref=nb_sb_noss?url=search-alias%3Dstripbooks&field-keywords=math+fin+law+10, ISBN-13: 978-1541092761, ISBN-10: 1541092767

Corporate IFRS-GAAP (B/S-I/S),ISBN-13: **978-1721066308**, ISBN-10: **1721066306**

30) **Asikin, S**. (2017c-823 pages), *Math Finance Law-11: Mathematical Financial Law for Public Listed Firms Rule 36243-39158,* Seattle (US): Amazon Createspace. https://www.amazon.com/s/ref=nb_sb_noss?url=search-alias%3Dstripbooks&field-keywords=math+finance+law+11, https://www.amazon.com/dp/1541215265/ref=sr_1_1?s=books&ie=UTF8&qid=1480029978& sr=1-1&keywords=math+finance+law, ISBN-13: 978-1541215269. ISBN-10: 1541215265

31) **Asikin, S**. (2017d-823 pages), *Math Finance Law-12: Mathematical Financial Law for Public Listed Firms Rule 30150-42152,* Seattle (US): Amazon Createspace. https://www.amazon.com/s/ref=nb_sb_noss?url=search-alias%3Dstripbooks&field-keywords=math+finance+law+12, https://www.amazon.com/Math-Finance-Law-Mathematical-39159-42152/dp/1541215516/ref=sr_1_2?s=books&ie=UTF8&qid=1490990342&sr=1-2&keywords=math+finance+law+12 ISBN-13: 978-1541215511, ISBN-10: 1541215516

32) **Asikin, S**. (2017e-824 pages), *Math Finance Law-13: Mathematical Financial Law for Public Listed Firms Rule 42153-44938,* Seattle (US): Amazon Createspace. https://www.amazon.com/dp/1542675219/ref=sr_1_1?s=books&ie=UTF8&qid=1491146227&s r=1-1&keywords=math+finance+law+13 . ISBN-13: 978-1542675215, ISBN-10: 1542675219

33) **Asikin, S**. (2017f-824 pages), *Math Finance Law-14: Mathematical Financial Law for Public Listed Firms Rule 44939-47745,* Seattle (US): Amazon Createspace. httpœ://www.amazon.com/dp/1542675219/ref=sr_1_1?s=books&ie=UTF8&qid=1491146227&s r=1-1&keywords=math+finance+law+14 . ISBN-13: 978-1544628967, ISBN-10: 154462896X

34) **Asikin, S**. (2017g-810 pages), *Math Finance Law-15: Mathematical Financial Law for Public Listed Firms Rule 47746-50465,* Seattle (US): Amazon Createspace. https://www.amazon.com/dp/1545118647/ref=sr_1_1?s=books&ie=UTF8&qid=1493626425&s r=1-1&keywords=math+finance+law+15, . ISBN-13: 978-1545118641, ISBN-10: 1545118647

35) **Asikin, S**. (2017h-816 pages), *Math Finance Law 16: Mathematical Financial Law Public Listed Firm Rule No.50,466-53,320,* Seattle (US), Amazon CreateSpace, https://www.amazon.com/dp/1545320179/ref=sr_1_1?s=books&ie=UTF8&qid=1495692695& sr=1-1&keywords=math+finance+law+16. ISBN-13: 978-1545320174, ISBN-10: 1545320179

Corporate IFRS-GAAP (B/S-I/S),ISBN-13: **978-1721066308**, ISBN-10: **1721066306**

36) **Asikin, S**. (2017i-809 pages), *Math Finance Law 17*: *Mathematical Financial Law Public Listed Firm Rule No.53,231-56,124*, Seattle (US), Amazon CreateSpace, https://www.amazon.com/dp/1545320179/ref=sr_1_1?s=books&ie=UTF8&qid=14956926956 sr=1-1&keywords=math+finance+law+17. ISBN-13: 978-1545486474, ISBN-10: 1545486476

37) **Asikin, S**. (2017j-815 pages), *Math Finance Law 18*: *Mathematical Financial Law Public Listed Firm Rule No.56,125-58,942*, Seattle (US), Amazon CreateSpace, https://www.amazon.com/dp/1545320179/ref=sr_1_1?s=books&ie=UTF8&qid=14956926956 sr=1-1&keywords=math+finance+law+18. ISBN-13: 978-1546912361, ISBN-10: 1546912363

38) **Asikin, S**. (2014-47 pages), *Doing Business in Comprehensive Interactive Interdisciplinary Micro-Macro Eco nomic International Accounting Architecture of Serial Well Known General Great Theories*, Proceedings in Finance and Risk Perspective 13, ACRN Oxford Publishing House, UK, p.197-243, https://books.google.com/books?id=4e4KAwAAQBAJ&pg=PA219&lpg=PA219&dq=steve+asikin+oxford&source=bl&ots=qVq9KyAmem&sig=ABdBliwuzUwRigIsuLimuJ5ZdW4&hl=en&sa=X&s qi=2&ved=0ahUKEwjZkMrF7KjTAhUDSiYKHREDB-qQ6AEILjAC#v=onepage&q=steve%20asikin%20oxford&f=false. ISBN 978-3-9503518-1-1.

39) **Asikin, T.& Asikin, S**. (2013- 828 pages), *448 Poems: KARAOKE Song Book: 24 Hours Story of 30 Languages*, Seattle (US): Amazon Createspace. http:/www.amazon.com/448-Poems-KARAOKE-Song-Book/dp/1480034363/ref=sr_1_1?s=books&ie=UTF8&qid=1396442034&sr=1-1&keywords=448+poems, ISBN-13: 978-1480034365, ISBN-10: 1480034363

40) **Asikin, T, Asikin, S, Senihardja, I**. (2017a- 508 pages), *Finance Construction 1*: *Corporate IFRS-GAAP (B/S-I/S) Engineering Technologies No. 1-1,000 of 111,111 Laws*, Seattle (US): Amazon Createspace. https://www.amazon.com/s/ref=nb_sb_noss?url=search-alias%3Dstripbooks&field-keywords=finance+construction+1+asikin. ISBN-13: 978-1544059525, ISBN-10: 1544059523

41) **Asikin, T, Asikin, S, Senihardja, I**. (2017b- 526 pages), *Finance Construction 2*: *Corporate IFRS-GAAP (B/S-I/S) Engineering Technologies No. 1,001-2,000 of 111,111 Laws*, Seattle (US): Amazon Createspace. https://www.amazon.com/s/ref=nb_sb_noss?url=search-alias%3Dstripbooks&field-keywords=finance+construction+2+asikin. ISBN-13: 978-1542402415, ISBN-10: 1542402417

Corporate IFRS-GAAP (B/S-I/S),ISBN-13: **978-1721066308**, ISBN-10: **1721066306**

42) **Asikin, T, Asikin, S, Senihardja, I.** (2017c- 738 pages), *Finance Construction 3: Corporate IFRS-GAAP (B/S-I/S) Engineering Technologies No. 2,001-3,000 of 111,111 Laws*, Seattle (US): Amazon Createspace. https://www.amazon.com/s/ref=nb_sb_noss?url=search-alias%3Dstripbooks&field-keywords=finance+construction+3+asikin. ISBN-13: 978-1975999292, ISBN-10: 1975999290

43) **Asikin, T, Asikin, S, Senihardja, I.** (2018a- 656 pages), *Finance Construction 4: Corporate IFRS-GAAP (B/S-I/S) Engineering Technologies No. 3,001-4,000 of 111,111 Laws*, Seattle (US): Amazon Createspace. https://www.amazon.com/s/ref=nb_sb_noss?url=search-alias%3Dstripbooks&field-keywords=finance+construction+4+asikin. ISBN-13: 978-1977659187, ISBN-10: 1977659187

44) **Asikin, T, Asikin, S, Senihardja, I.** (2018b- 656 pages), *Finance Construction 5: Corporate IFRS-GAAP (B/S-I/S) Engineering Technologies No. 4,001-5,000 of 111,111 Laws*, Seattle (US): Amazon Createspace. https://www.amazon.com/s/ref=nb_sb_noss?url=search-alias%3Dstripbooks&field-keywords=finance+construction+5+asikin. ISBN-978-1983566141, ISBN-10: 1983566144

45) **Asikin, T, Asikin, S, Senihardja, I.** (2018c- 524 pages), *Finance Construction 6: Corporate IFRS-GAAP (B/S-I/S) Engineering Technologies No. 5,001-5,500 of 111,111 Laws*, Seattle (US): Amazon Createspace. https://www.amazon.com/s/ref=nb_sb_noss?url=search-alias%3Dstripbooks&field-keywords=finance+construction+6+asikin. ISBN-978-1983566332, ISBN-10: 1983566330

46) **Asikin, T, Asikin, S, Senihardja, I.** (2018d- 532 pages), *Finance Construction 7: Corporate IFRS-GAAP (B/S-I/S) Engineering Technologies No. 5,501-6,000 of 111,111 Laws*, Seattle (US): Amazon Createspace. https://www.amazon.com/s/ref=nb_sb_noss?url=search-alias%3Dstripbooks&field-keywords=finance+construction+7+asikin. ISBN-978-1987571509, ISBN-10: 1987571509

47) **Asikin, T, Asikin, S, Senihardja, I.** (2018e- 534 pages), *Finance Construction 8: Corporate IFRS-GAAP (B/S-I/S) Engineering Technologies No. 6,001-6,500 of 111,111 Laws*, Seattle (US): Amazon Createspace. https://www.amazon.com/s/ref=nb_sb_noss?url=search-alias%3Dstripbooks&field-keywords=finance+construction+8+asikin. ISBN-13: 978-1718712300, ISBN-10: 1718712308

48) **Asikin, T, Asikin, S, Senihardja, I.** (2018f- 536 pages), *Finance Construction 9: Corporate IFRS-GAAP (B/S-I/S) Engineering Technologies No. 6,501-7,000 of 111,111 Laws*, Seattle (US): Amazon Createspace. https://www.amazon.com/s/ref=nb_sb_noss?url=search-alias%3Dstripbooks&field-keywords=finance+construction+9+asikin. ISBN-13: 978-1718784055, ISBN-10: 1718784058

Corporate IFRS-GAAP (B/S-I/S),ISBN-13: **978-1721066308**, ISBN-10: **1721066306**

49) **Asikin, T, Asikin, S, Senihardja, I.** (2018g- 536 pages), *Finance Construction 10*: *Corporate IFRS-GAAP (B/S-I/S) Engineering Technologies No. 7,001-7,500 of 111,111 Laws*, Seattle (US): Amazon Createspace. https://www.amazon.com/s/ref=nb_sb_noss?url=search-alias%3Dstripbooks&field-keywords=finance+construction+10+asikin. ISBN-13: 978-1719046626, ISBN-10: 171904662X

50) **Asikin, T, Asikin, S, Senihardja, I.** (2018h- 536 pages), *Finance Construction 11*: *Corporate IFRS-GAAP (B/S-I/S) Engineering Technologies No. 7,501-8,000 of 111,111 Laws*, Seattle (US): Amazon Createspace. https://www.amazon.com/s/ref=nb_sb_noss?url=search-alias%3Dstripbooks&field-keywords=finance+construction+11+asikin. ISBN-13: 978-1720792789, ISBN-10: 172079278X

51) **Brigham, EF & Gapenski, LC, .** (1988- Textbook), *Financial Management: Theory and Practice (Study Guide)*, NY (US): Dryden Press. http://www.amazon.com/Financial-Management-Theory-Practice-Study/dp/0030125391/ref=sr_1_1?s=books&ie=UTF8&qid=1429380835&sr=1-1 ;

52) **Chandonet, RE, .** (2011-32 pages), *Balancesheet Management*, Trends & Strategies, NY (US): Sandler O'Neil. https://www.ncbankers.org/uploads/File/Meetings/Presentations/2011CONV-Chandonnet.pdf;

53) **Cornuejols G & Tutuncu, R, .** (2006-358 pages), *Optimization Methods in Finance*, Cambridge (UK): Cambridge University Press. http://www.math.ku.dk/~rolf/CT_FinOpt.pdf;

54) **Crook, M.** (2015-6 pages), *Optimization Methods in Finance*, UBS FS (Swiss): CIO WM Research. https://www.ubs.com/content/dam/WealthManagementAmericas/documents/investment-strategy-insights-2015-01-09-balance-sheet-optimization.pdf;

55) **Fee, Greg.** (2012-127 pages), *Catalan's Constant: (Ramanujan's Formula}: Catalan Constant to 300,000 digits*, Seattle (US): Amazon Kindle. http://www.Amazon.com/gp/aw/d/B0082751EW/ref=mp_sa_1_4?ie=UTF8%Qqid=14843678186sr=8-46pi=AC_SX236_SY340_QL656keywords=ramanujan&dpPl=1&dpID=51FqIwDZpRL&ref=plSrch

56) **Gardenfors, Peter.** (2014) *The Geometry of Meaning: Semantics Based on Conceptual Spaces*. Cambridge, Mass.: MIT Press. ISBN 0-262-02678-3.

57) **Gardenfors, Peter.** (2000), *Conceptual Spaces: The Geometry of Thought*. Cambridge, Mass.: MIT Press. ISBN 0-262-07199-1.

Corporate IFRS-GAAP (B/S-I/S),ISBN-13: **978-1721066308**, ISBN-10: **1721066306**

58) **Kanigel, Robert.** (2013-465 pages), *The Man Who Knew Infinity:A Life of the Genius Ramanujan*, Washington (US): Washington Square Press&Amazon Kindle.
http://www.Amazon.com/gp/aw/d/B0008BW4VEGM/ref=pd_aw_sim351_1?ie=UTF8&psc=1&refRID=VYN345GVDBJ6ATDQ35W2&dpID=91Zgpc4ETdL

59) **Limlingan, Victor S.** (1993-1 page), *The Limlingan Financial Model*, Manila: Asian Manager. http://limlingan.com/images/article.gif

60) **Newton, Isaac.** (2015-631 pages), *Principia:The Mathematical Principles of Natural Philosophy (Active Content)*, London (UK): RSM&Amazon Kindle.
http://www.Amazon.com/gp/aw/sitb/B01ISFJSJO?ref=sib_dp_aw_kb_udp

61) **Nikisa, Evets.** (2014g- 828 pages), *Wonderful Live Spells: 144-Steps Recreative Nice Children Story, Succesful Development:* Seattle (US): Amazon Createspace. http://www.amazon.com/WONDERFUL-Live-Spells-Recreative-Development/dp/1499248423/ref=sr_1_1?s=books&ie=UTF8&qid=1418698379&sr=1-1&keywords=wonderful+asikin, ISBN-13: 978-1499248425, ISBN-10: 1499248423

62) **Nielson, SS.** (2007- 430 pages), *Practical Financial Optimization: Decision Making for Financial Engineers:* NY (US): Wiley.
https://books.google.com/books/about/Practical_Financial_Optimization.html?id=_XkwLAAACAAJ.

63) **Niswonger CR, Fess PE & Warren CE.** (1993-Textbook), *Accounting Principles:* Seattle (US): South Western Publishing Co.
http://www.amazon.com/Accounting-Principles-Chapters-C-Rollin-Niswonger/dp/0538818603

64) **Omerod, P.** (1993-Textbook), (1994), *The Death of Ecoomics*, St Martin's Press, London, . http://www.amazon.com/Accounting-Principles-Chapters-C-Rollin-Niswonger/dp/0538818603, ISBN-13: 978-0312134648, ISBN-10: 0312134649

65) **Paccioli, L. (1989-Textbook),** *Summa de Arithmetica Geometria Proportioni et Proportionalita*, Venice 1494.
http://www.amazon.com/Arithmetica-Geometria-Proportioni-Proportionalita-Venice/dp/B00T4HXTIW/ref=sr_1_fkmr0_2?s=books&ie=UTF8&qid=1429597966&sr=1-2-fkmr0&keywords=summa+mhematica+arithmetica

66) **Persson, Torsten & Tabellini, Guido** (2000), *Political Economics – Explaining Economic Policy*, MIT Press.
https://mitpress.mit.edu/authors/torsten-persson

67) **Persson, Torsten & Tabellini, Guido** (2003), *The Economic Effects of Constitutions*, MIT Press. https://mitpress.mit.edu/authors/torsten-persson

68) **Persson, Torsten & Tabellini, Guido** (2000), *Monetary and Fiscal Policy –1 Credibility*, MIT Press. https://mitpress.mit.edu/authors/torsten-persson

69) **Persson, Torsten & Tabellini, Guido** (2003), *Monetary and Fiscal Policy –2 Politics*, MIT Press. https://mitpress.mit.edu/authors/torsten-persson

`

Corporate IFRS-GAAP (B/S-I/S),ISBN-13: **978-1721066308**, ISBN-10: **1721066306**

70) **Puts, J** (2012-102 pages), *Balancesheet Optimization Under Basel III*, Amsterdam (Netherlads): Vrije Universiteit.
http://www.math.vu.nl/~sbhulai/papers/thesis-puts.pdf

71) **Timmermans, K** (2012-30 pages), *Balancesheet Optimization Under Basel III*, Amsterdam (Netherlads): ING Investor Day.
https://www.ing.com/web/file?uuid=09c803d9-79a1-47ec-bfe5...owner...

72) **Weston JF, Copeland TE & Shastri, K.** (2004-Textbook), *Financial Theory and Corporate Policy:* NY (US): Addison Wesley.
http:/www.amazon.com/Financial-Corporate-Copeland-Kuldeep-Paperback/dp/B008YSUE4M/ref=sr_1_1?s=books&ie=UTF8&qid=1429381700&sr=1-1&keywords=weston+copeland

73) **Zenios SA.** (2002-350 pages), *Financial Optimization:* Cambridge (UK): Cambridge University Press.
https://books.google.com/books/about/Financial_Optimization.html?id=-FJdgjZOqSAC,.

Corporate IFRS-GAAP (B/S-I/S),ISBN-13: **978-1721066308**, ISBN-10: **1721066306**

B. Intellectual Properties:

1)<u>Indonesian Banker Certification 1993</u>, *Sertifikasi Bankir Indonesia 1993*;
2)<u>MATHFINPLAN</u> (Mathematical Financial Planning);
    Perencanaan Keuangan Matematis
3)<u>BANK MATHPLAN</u> (Bank's Mathematical Planning);
    Perencanaan Matematis Bank
4)<u>TECHNO-ECONOMISTAT</u> (Technology for Economical Statistics);
    Teknologi Statistik untuk Ekonomi
5)<u>FINANCIAL GEOMETRY</u> (IT Program); *Ilmu Geometri Keuangan*
6)<u>BAK DOBEL</u> (Car Gasoline Tank: Left and Right Side Fuel Fill-in).
    Tangki Mobil yang bisa diisi dari Kiri atau Kanan
7)<u>Mister GO-CHENG</u> (Cheap Food Franchising). *Waralaba Pangan Murah*

Corporate IFRS-GAAP (B/S-I/S),ISBN-13: **978-1721066308**, ISBN-10: **1721066306**

C. SCIENTIFIC PAPERS:

1)**Eichhornia Crassipes**, as Anti Polutant and Fertile Sediment, *Pemanfaatan Eceng Gondok untuk Menetralkan Polusi dan Pengurukan Subur pada Genangan Air,* LKIR, Lomba Karya Ilmiah Remaja, LIPI (1979)

2)**Archimedes Catesian Buoyancy Law**, *Penyimpangan Hukum Archimedes pada Model Pelampung Cartesian,* Lomba Karya Ilmiah Remaja, Depdikbud-RI (1980)

3)**SURAMADU**, Surabaya-Madura Inter Island Bridge, Design Supervisor, *Sumarsono Sumali Engineering Contractor,* Tambak-V, Tandes, Surabaya (1982),

4)**GAZOLINE Taxi-Pool Depot**, Jemur Handayani, Surabaya, *Sumarsono Sumali Engineering Contractor,* Tambak-V, Tandes, Surabaya (1982),

5)**SOEKARNO-HATTA**, Jakarta International Airport, Garuda Maintenance Facilities, *Junior Architect,* As Bulit Drawings, Coinstruction Management, Aerospatiale-Encona Engineering (1983-1985),

6)**ADISUCIPTO**, Yogya International Airport, *Junior Architect,* Preliminary Technical Drawings, Budiono Surasno-Encona Engineering (1986),

7)**BLOK-M PLAZA**, "New Garden Hall", Kebayoran Baru Supermall, *Preliminary Design Architect,* Danisworo-James Ferry- Jasa Ferry-Encona (1986),

8)**NORTH JAKARTA PUBLIC LIBRARY**, Semper, Tanjung Priok, *Chief Architect,* Agung Darma Sarana-Asep Hermawan (1987),

9)**The Banker's Trainer**: A Manual for The Training Head in Banking. *Profesi Bankir Pelatih, Tuntunan bagi Kepala Diklat Perbankan,* (1200 Pages, 1993),

10)**Indonesian Bankers Certification**, a Strategic Move for the National Banking Improvement. *Sertifikasi Bankir Indonesia, Langkah Strategis Pembenahan Perbankan Nasional,* (KOMPAS, 20 Juli 1993).

11)**Indonesian Rupiah's Exchange Rate**, on a Quiet but Steady Movement. *Nilai Tukar Rupiah, Diam-diam Meningkat Pesat,* (MANAJEMEN, Jan-Feb 1994).

12)**The Bank's Credit**, for the Best Practices. *Kredit Praktis* (200 pages, 1994),

13)**Down to Earth Credit Meeting Credo**, *Kiat Rapat Pemberian Kredit yang Realistik.* (MANAJEMEN, Sep-Oct 1994).

14)**SCHIPHOL**, Amsterdam International Airport, 1st Anthony Fokker Weg, Sloten, *Airport Engineer,* Samsung- Fokker-Beurlin, BV (1996),

15)**Ultrapeneurship: The Secret of Starting Business Without Capital**. Ultrapreneur: *Kiat Usaha Tanpa Modal,* (MANAJEMEN, Nov-Dec 1996),

16)**Recipe from the Corporate Raider's Kitchen**, *Menguak Dapur Corporate Raiders,* (MANAJEMEN, Jan-Feb 1997).

17)**The 10-Tips and Tricks of Engineered Conglomeration**. *Sepuluh Rekayasa Keuangan Konglomerat,* (MANAJEMEN, Mar-Apr 1997),

18)**The Management of Succession in Five Parts of the World**. *Manajemen Suksesi di Lima bagian Dunia,* (MANAJEMEN, Mei-Jun 1997),

19)**Auditing**: The Global Language for the 2003's Human Resources. *Audit : Bahasa Global untuk SDM 2003,* (MANAJEMEN, Jul-Aug 1997),

Corporate IFRS-GAAP (B/S-I/S),ISBN-13: **978-1721066308**, ISBN-10: **1721066306**

20)**The Elected Parliament 1997-2002**, and the Hope of Professionals. *DPR 1997–2002 dan Harapan Profesional*, (MANAJEMEN, Sep-Okt 1997),
21)**Disappearance of World's Soft Currencies**, and Proposition for Global Economic Thinking. *Lenyapnya Mata Uang Lunak Dunia dan Usulan Pemikiran Ekonomi Global*, (MANAJEMEN, Nov-Dec 1997),
22)**The Human Resources and Its Accountability**, in Critical Situation. *Akuntabilitas Sumberdaya Manusia di Masa Krisis*, (MANAJEMEN, Jan-Feb 1998), '
23)**The Investment Economic Model**, and the Failure of Growth Economic Theories. *Model Ekonomi Investasi dan Kegagalan Teori Ekonomi Pertumbuhan*, (MANAJEMEN, Nov 1998),
24)**The Exact Equilibrium of Capital Budgeting**, *Persamaan-persamaan akurat dalam Penganggaran Modal Perusahaan.* (MANAGEMENT EXPOSE, November 1998),
25)**The Corruptive Economic Model**, and Its Solution for Long Term Debt's Problem. *Model Ekonomi Korupsi dan Penyelesaian Utang Jangka Panjang*, (MANAJEMEN, Dec 1998),
26)**The Self Equity Budgeting**, and Operations Research in Investments. *Penganggaran Modal Sendiri dan Riset Operasi dalam Investasi*, (MANAJEMEN, Jan 1999),
27)**Actual Impact of Inflation**, and the Need of Geometric Approach in Statistic. *Pengaruh Riil Inflasi dan Perlunya Statistika Geometri*, (MANAJEMEN, Feb 1999),
28)**Multiplication of Money**, to Overcome the Crisis. *Melipat Gandakan Uang untuk Mengusir Krisis*, (MANAJEMEN, Apr 1999),
29)**The Interactive Mathematical Model**, for Multi Constraints Financial Budgeting, *Model Matematika Ekonomi Interaktif bagi Penganggaran Keuangan Multi Kendala.* (SCRIPTA ECONOMICA, Vol. 2, April 1999)
30)**Application of Ethics**, for the Management Accounting Practices. *Aplikasi Etika dalam Praktek Akuntansi Manajemen*, (MANAJEMEN, May 1999),
31)**The Curriculum of Magistrate of Management in Finance**: Its Needs and Practices. *Kurikulum Magister Manajemen Keuangan 1988-1998: Antara Kebutuhan dan Praktek*, (MANAJEMEN, Jun 1999),
32)**The Attractiveness**, Personal Power and Management. *Pesona Pribadi: "Personal Power" dan Manajemen*, (MANAJEMEN, Jul 1999),
33)**The Market Research**, for Magistrate Degree in Management. *Soal Jawab Riset Pasar untuk Pascasarjana Manajemen* (126 Pages, 1999),
34)**The Thesis Manual**, Question and Answer for Magistrate Degree in Management. *Soal Jawab Tugas Akhir untuk Pascasarjana Manajemen* (126 Pages, 1999),
35)**The Risk Management**, Question and Answers for Magistrate Degree in Management. *Soal Jawab Manajemen Resiko untuk Pascasarjana Manajemen* (126 Pages, 1999),
36)**The Credit Manual**, Question and Answers for Magistrate Degree in Management. *Soal Jawab Manajemen Kredit untuk Pascasarjana Manajemen* (126 Pages, 1999),

Corporate IFRS-GAAP (B/S-I/S),ISBN-13: **978-1721066308**, ISBN-10: **1721066306**

37)**Enhancing the Morality Structure**. *Solidkan Struktur Moralitas,* (MANAJEMEN, Oct 1999)

38)**Performance of PT. Astra Otoparts**, Plc, in its Press Releases and Public Financial Reports. *Kinerja PT. Astra Otoparts dalam Siaran Pers dan Laporan Keuangan Publik,* (MANAJEMEN, Oct 1999)

39)**The Management of Conflicts**, for the New Indonesian's "Rainbow Colored" Cabinet. *Manajemen Konflik dan Kabinet Pelangi,* (MANAJEMEN, Dec 1999),

40)**Technoeconometric:** An Engineering Elaboration to Matrix Multi Variate Econometrics, *Tekno Ekonometrik sebagai sumbangan teknologi untuk Ekonometrika Matrik Multi Variat.* (MANAGEMENT EXPOSE, November 1999),

41)**Technoeconometric** : A Mathematical Shortcut for Non Linear Bivariate Econometrics, *Tekno Ekonometrik sebagai Usulan Jalan Pintas Matematik untuk Ekonometrika non Linear Bivariat.* (SCRIPTA ECONOMICA, Vol. 2, No. 3, Desember 1999),

42)**Developing "New Image"**, for the New Indonesia. *Membangun Citra Indonesia Baru,* (MANAJEMEN, Jan 2000),

43)**Segregations of Duties**, the Needs of Both Indonesian Governments and Private Sectors. *Pemerintah dan Swasta Perlu Berbagi Tugas,* MANAJEMEN, Mar 2000),

44)**The 812.500 Financial Models**, for the Autonomous Country Governments. *Tentang 812.500 Model Keuangan Otonomi Daerah,* (MANAJEMEN, Mar 2000),

45)**The Forecast of 10 Most Attractive Share Values**, in the Jakarta Stock Exchange, April 2000. *Prakiraan Harga Sepuluh Saham Unggulan di Bursa Efek Jakarta, April 2000,* (MANAJEMEN, Mar 2000),

46)**Let Us Merchandise the Intelectual Properties!**. *Jual "Intelectual Property" Saja!* (MANAJEMEN, Apr 2000),

47)**The Forecast of ten Most Attractive Share Values**, in the Jakarta Stock Exchange May 2000. *Prakiraan Harga Sepuluh Saham Unggulan di Bursa Efek Jakarta Mei 2000,* (MANAJEMEN, Apr 2000),

48)**HEATHROW**, London International Airport, Second Runway, *Q-Basic Algorithmic Engineer,* Beasley-Krishnamoorthy, Imperial College, Cambridge, Post Doctoral Project (2000),

49)**Why the Banking Sector Should be Healthy**, Before Others?. *Mengapa Perbankan Harus Sehat?* (MANAJEMEN, Mei 2000),

50)**Recycling of Wastes as a Negative Costing**. *Memanfaatkan Sampah atau Negative Costing,* (MANAJEMEN, Agt 2000),

51)**The Guidance for Indonesia Corporate Restructuring**. *Kiat Restrukturisasi Perusahaan Indonesia,* (MANAJEMEN, Agt 2000),

52)**Redefinition of the Ethical Comprehension**. *Pemaknaan Kembali atas Pemahaman Etika,* (MANAJEMEN, Sep 2000),

53)**The Critical View the 1999's Indonesians Central Bank Financial Statement**. *Analisa Laporan Keuangan Bank Indonesia 1999,* (MANAJEMEN, Oct 2000),

Corporate IFRS-GAAP (B/S-I/S),ISBN-13: **978-1721066308**, ISBN-10: **1721066306**

54)**Becoming an "Achiever"**, not Just Common "Worker". *Bukan Sekadar "Kerja", Tapi Menghasilkan Karya!* (MANAJEMEN, Feb 2001),

55)**Ten Steps to Real Wealth!** *Sepuluh Langkah Menjadi Kaya,* (MANAJEMEN, Jun 2001),

56)**Creative Financial Breakthrough**, *Seminar Sehari tentang Terobosan Keuangan yang Kreatif,* (Gramedia-PPM dan MANAJEMEN, Jun 2001),

57)**The Marketing Practices of Financial Softwares**, for Publics Listed Corporation. *Praktek Pemasaran Program Perangkat Lunak untuk Keuangan Perusahaan Publik* (Univ Satyagama, Jakarta Juni 24, 2001).

58)**The Mathematical Financial Planning**, *Perancangan Keuangan Secara Eksak dengan Matematik.* (MANAJEMEN, Jul 2001),

59)**The Real Research and Development, not Rework and Destruction**. *Litbang yang Bukan "Sulit Berkembang",* (MANAJEMEN, Jul 2001),

60)**The Treatise on the Advantages and Disadvantages**, of Revolutionary Learning Methods. *Baik Buruknya: Revolusi Cara Belajar,* (MANAJEMEN, Jul 2001),

61)**Better Make an Indonesian "Corruption Prevention" Body**, not Just a "Corruption Watch". *Buat Indonesian Corruption Prevention (ICP), Bukan Indonesian Corruption Watch, ICW!* (MANAJEMEN, Okt 2001),

62)**Preparation of the Worlds Business Leaders**, for our Century and the Century After. *Mempersiapkan Manusia-manusia yang akan menjadi Khafilah Bisnis Dunia Abad ini dan Abad Mendatang.* (Univ Satyagama – Gregorio Araneta Univ, Jakarta, Oktober 27, 2001).

63)**The Good Corporate Governance**, is not just a Morale Condition. *Kepemerintahan Perusahaan yang Baik, Bukan Sekedar Moralitas,* (MANAJEMEN, Nov 2001),

64)**Corporate Healthiness and Cleanliness as a Challenge**. *Bersih, Sehat Perusahaan, sebuah Tantangan* (MANAJEMEN, Jan 2002),

65)**Developing the Cultural Base for Future Competitions**. *Budaya untuk Bersaing di Masa Depan* (MANAJEMEN, Jan 2002),

66)**"Catapult", "Bullet" or "Ballistic Missiles" Approac**h, for Corporate Healthiness and Cleanliness. *Pola "Ketapel", "Peluru Biasa" atau "Rudal" untuk Keuangan Perusahaan yang Bersih Sehat.* (MANAJEMEN, Feb 2002),

67)**The Mathematical Invesment Solution Mode**l, for Indonesian Foreign Debt and Its Corruption, *Solusi Matematika Investasi bagi Penyelesaian Masalah Utang Luar Negeri dan Korupsi di Indonesia.* (MANAGEMENT EXPOSE, Mar 2002),

68)**The Attitude, Skills and Knowledge (ASK)**, in Educational System Approach. *Sikap, Keterampilan dan Pengetahuan Menurut Pendekatan Sistemik Pendidikan,* (MANAJEMEN, Apr 2002),

Corporate IFRS-GAAP (B/S-I/S),ISBN-13: **978-1721066308**, ISBN-10: **1721066306**

69)<u>**Strategic Treatise on Regional Autonomy**</u>, in Futurization, Digitalization, Securitization and Globalization Era in Accordance to the Awakening of the Mercantile Nations. *Kajian Strategik Otonomi Daerah : Era Futurisasi, Digitalisasi, Strukturisasi, dan Globalisasi dalam Kebangkitan Negara-Negara Dagang.* (Training for Provincial Government Regional Management, Indonesian Association of Provincial Governments, *Pelatihan Pengembangan Manajemen Pemerintah Daerah bagi Perangkat Pemerintah Propinsi, Angkatan II,* (APPSI), June 27, 2002.

70)<u>**The Suggestion for the Doctoral Course**</u>, for Islamic Financial Management System. *Usulan-usulan Perbaikan dalam Kurikulum Doktor Manajemen Keuangan Islam.* (December, 17-19, 2002, Canadian International Development Agency - CIDA, ADSGM, Universiti Utara Malaysia, Kedah, Da'arul Sintok).

71)<u>**The Defensive Strategy for the Market Leader**</u>, a Nokia Cellular Telephone's Case. *Strategi Bertahan Sang Pemimpin Pasar dengan Menggunakan Multi Merk, Kasus Telepon Seluler Nokia.* (March, 12, 2004, Univ Tarumanegara, Jakarta).

Corporate IFRS-GAAP (B/S-I/S),ISBN-13: **978-1721066308,** ISBN-10: **1721066306**

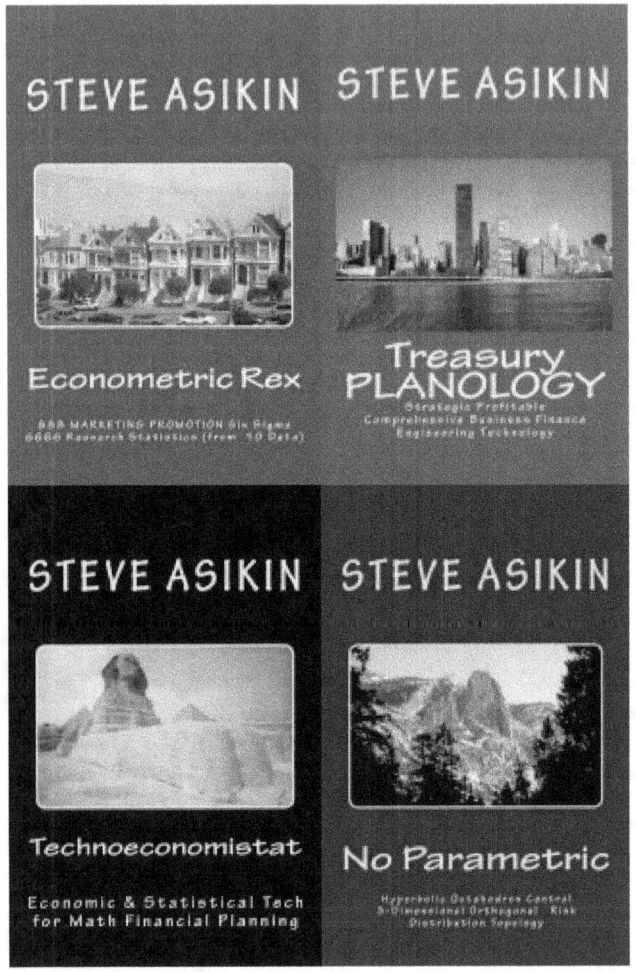

Corporate IFRS-GAAP (B/S-I/S),ISBN-13: **978-1721066308**, ISBN-10: **1721066306**

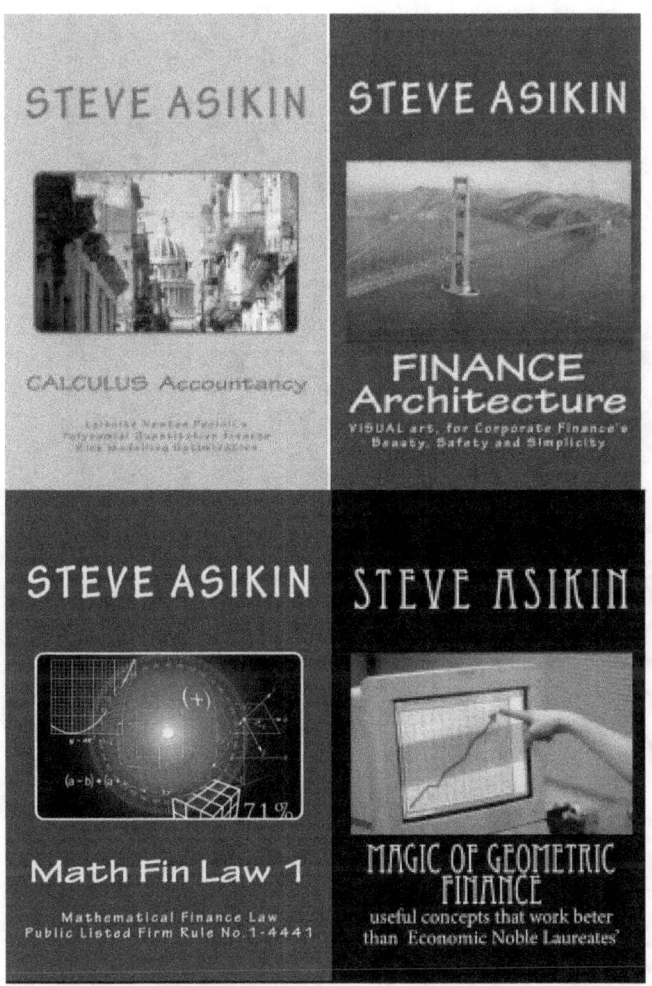

Corporate IFRS-GAAP (B/S-I/S),ISBN-13: **978-1721066308**, ISBN-10: **1721066306**

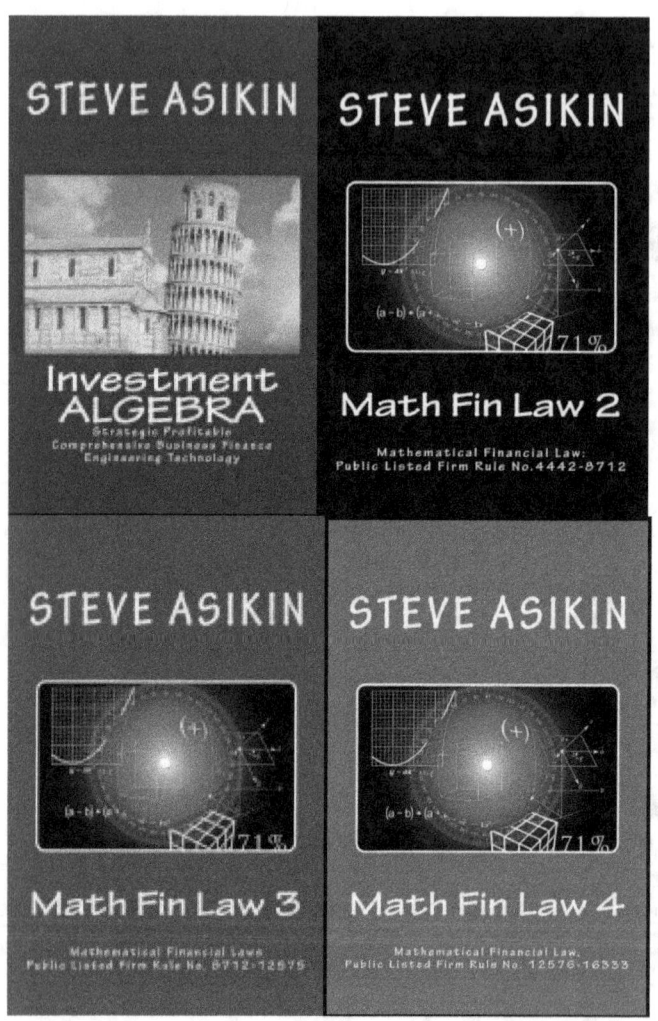

Corporate IFRS-GAAP (B/S-I/S),ISBN-13: **978-1721066308**, ISBN-10: **1721066306**

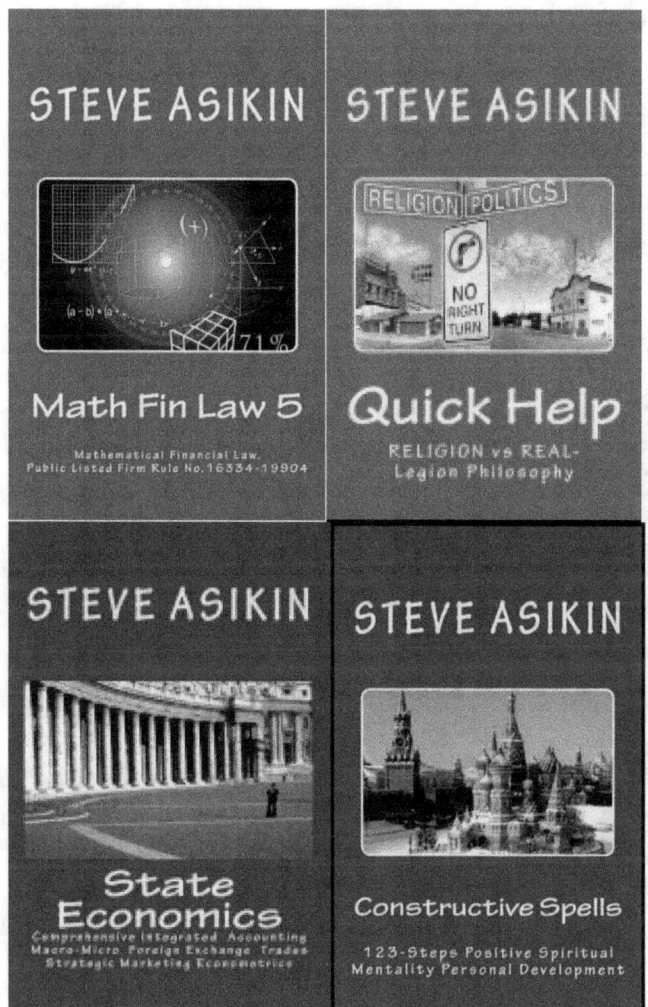

Corporate IFRS-GAAP (B/S-I/S),ISBN-13: **978-1721066308**, ISBN-10: **1721066306**

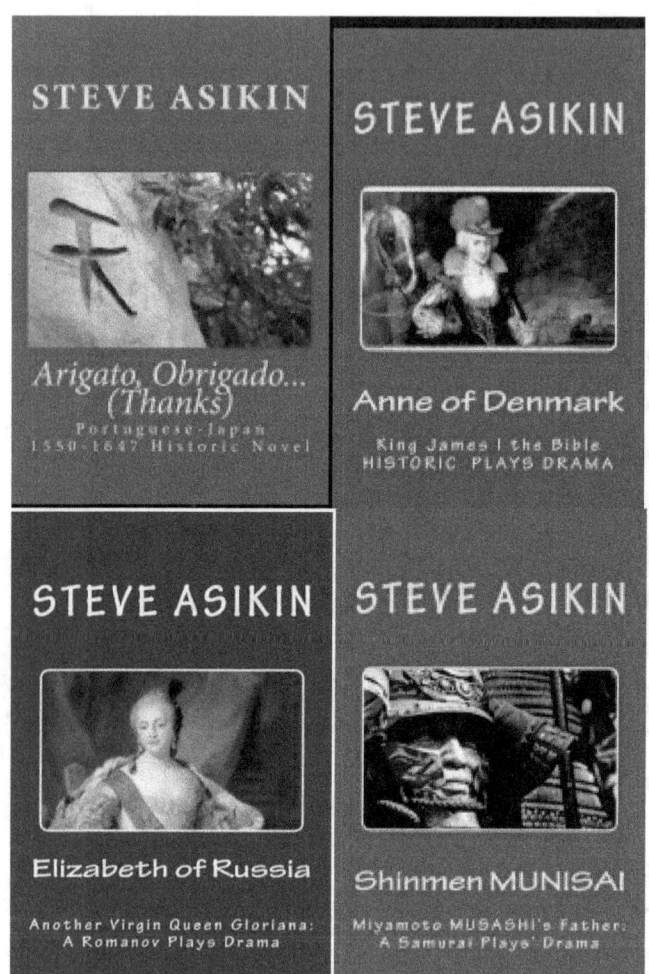

Corporate IFRS-GAAP (B/S-I/S),ISBN-13: **978-1721066308**, ISBN-10: **1721066306**

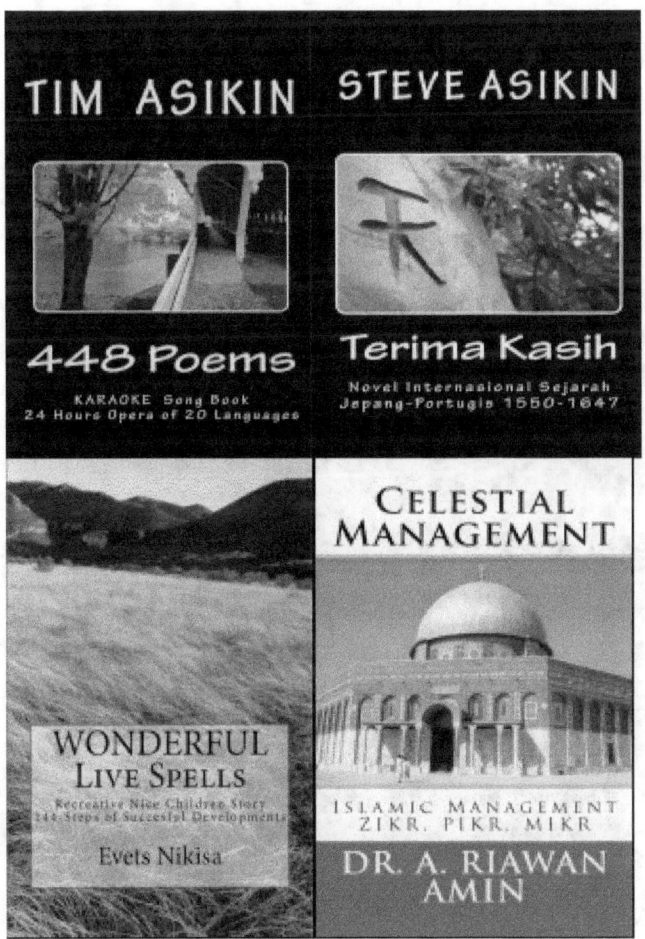

Corporate IFRS-GAAP (B/S-I/S),ISBN-13: **978-1721066308**, ISBN-10: **1721066306**

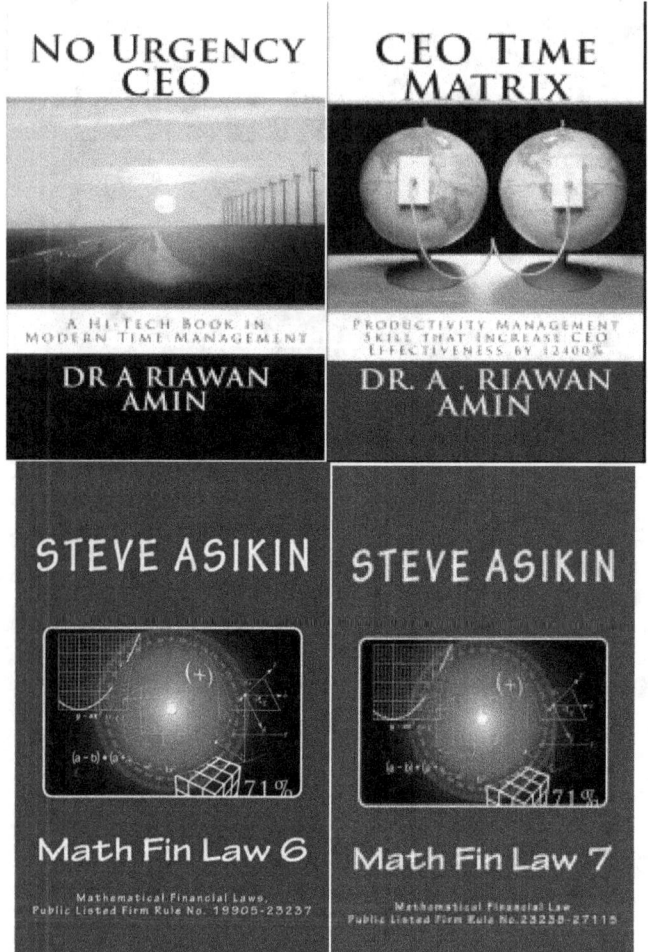

`

Corporate IFRS-GAAP (B/S-I/S),ISBN-13: **978-1721066308**, ISBN-10: **1721066306**

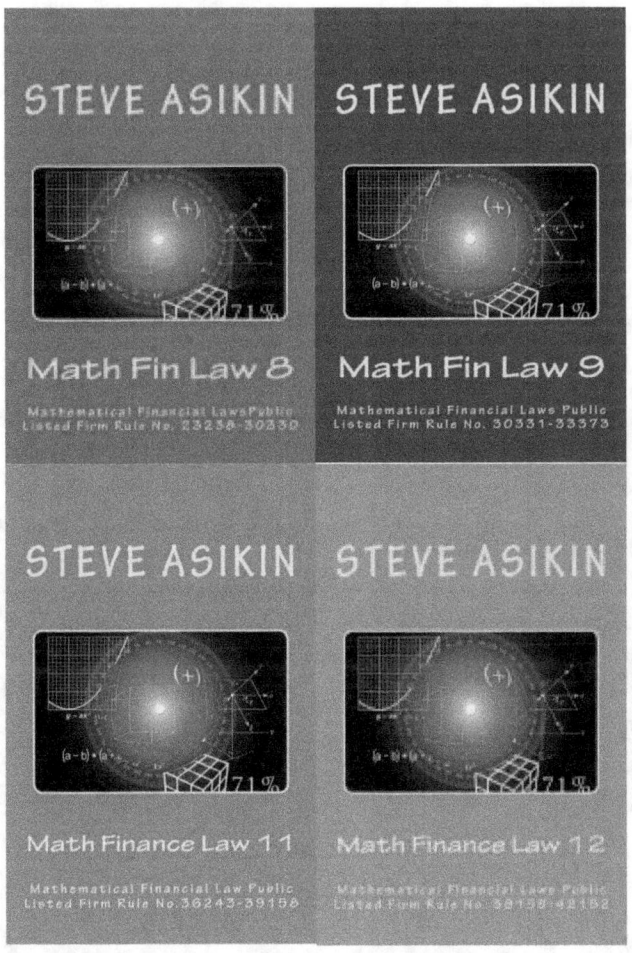

Corporate IFRS-GAAP (B/S-I/S),ISBN-13: **978-1721066308**, ISBN-10: **1721066306**

Corporate IFRS-GAAP (B/S-I/S),ISBN-13: **978-1721066308**, ISBN-10: **1721066306**

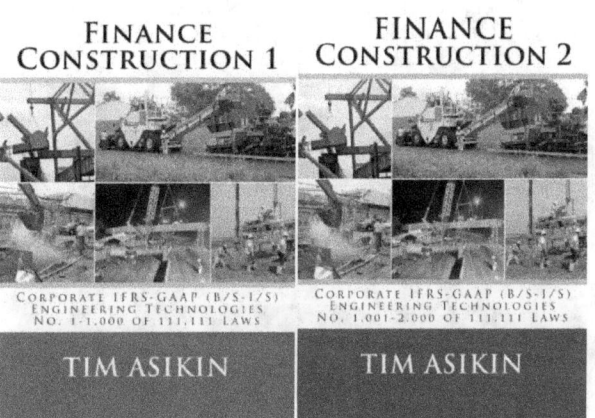

Corporate IFRS-GAAP (B/S-I/S),ISBN-13: **978-1721066308**, ISBN-10: **1721066306**

Finance Construction-12, *Tim Asikin, Steve Asikin, Indra Senihardja*

Corporáte IFRS-GAAP (B/S-I/S),ISBN-13: **978-1721066308**, ISBN-10: **1721066306**

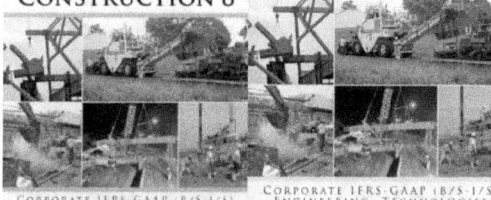

Corporate IFRS-GAAP (B/S-I/S),ISBN-13: **978-1721066308**, ISBN-10: **1721066306**

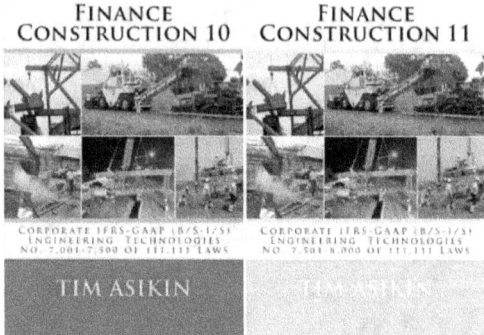

Corporate IFRS-GAAP (B/S-I/S),ISBN-13: **978-1721066308**, ISBN-10: **1721066306**